VLSI AND HARDWARE IMPLEMENTATIONS USING MODERN MACHINE LEARNING METHODS

VLSI AND HARDWARE IMPLEMENTATIONS USING MODERN MACHINE LEARNING METHODS

Edited by
Sandeep Saini, Kusum Lata, and
G.R. Sinha

CRC Press
Taylor & Francis Group
Boca Raton London New York

CRC Press is an imprint of the
Taylor & Francis Group, an **informa** business

First edition published 2022
by CRC Press
6000 Broken Sound Parkway NW, Suite 300, Boca Raton, FL 33487-2742

and by CRC Press
2 Park Square, Milton Park, Abingdon, Oxon, OX14 4RN

Library of Congress Cataloging-in-Publication Data
Names: Saini, Sandeep, editor. | Lata, Kusum (Electronics engineer), editor. | Sinha, G. R., 1975- editor.
Title: VLSI and hardware implementations using modern machine learning methods / edited by Sandeep Saini, Kusum Lata and G.R. Sinha.
Description: First edition. | Boca Raton, FL : CRC Press, 2022. | Includes bibliographical references and index. |
Summary: "Machine learning is a potential solution to resolve bottleneck issues in VLSI via optimizing tasks in the design process. This book aims to provide the latest machine learning based methods, algorithms, architectures, and frameworks designed for VLSI design. Focus is on digital, analog, and mixed-signal design techniques, device modeling, physical design, hardware implementation, testability, reconfigurable design, synthesis and verification, and related areas. It contains chapters on case studies as well as novel research ideas in the given field. Overall, the book provides practical implementations of VLSI design, IC design and hardware realization using machine learning techniques"-- Provided by publisher.
Identifiers: LCCN 2021037483 (print) | LCCN 2021037484 (ebook) | ISBN 9781032061719 (hbk) | ISBN 9781032061726 (pbk) | ISBN 9781003201038 (ebk)
Subjects: LCSH: Integrated circuits--Very large scale integration--Design and construction--Data processing. | Machine learning.
Classification: LCC TK7874.75 .V564 2022 (print) | LCC TK7874.75 (ebook) | DDC 006.3/1--dc23/eng/20211014
LC record available at https://lccn.loc.gov/2021037483
LC ebook record available at https://lccn.loc.gov/2021037484

ISBN: 978-1-032-06171-9 (hbk)
ISBN: 978-1-032-06172-6 (pbk)
ISBN: 978-1-003-20103-8 (ebk)

DOI: 10.1201/9781003201038

Typeset in Times
by MPS Limited, Dehradun

Contents

Preface

VLSI is a well-established field of research that ignited the modern computing revolution. Moore's law provided the direction in this field for a number of decades, which allowed the shrinking size and increasing speed of the next generation of circuits. In every future generation, the computing hardware is becoming more compact and faster. Designing and manufacturing chips below 10 nm and 7 nm is very challenging; thus, alternate ways are being explored for higher performance IC designs. Machine learning is emerging as one of the potential solutions to resolve some of the bottleneck issues in VLSI as well. The machine-learning–based architecture and models help in optimizing or deciding a few tasks in the design process, at various stages of system design, to achieve the goals. In this book, we provide a compilation of the latest machine-learning–based methods for the VLSI design domain.

The computing era is making a transition from conventional computing to cognitive computing. Artificial intelligence and its subset, machine-learning–based approaches, are providing solutions for all future technologies and filling the gap to move towards the cognitive era. In verification and fabrication as well, the time taken to process the whole system at 10 nm and 7 nm is very large. Machine-learning–based methods help to reduce characterization time by weeks as well as reduce the number of resources. Thus, machine-learning–based approaches will be extensively used in the future in this field, and this book provides future paths for such developments.

This book aims to provide the latest machine-learning–based methods, algorithms, architectures, and frameworks designed for VLSI design and implementations of hardware. The scope of the book includes machine-learning–based methods and models for digital, analog, and mixed-signal design techniques; device modeling; physical design; hardware implementation; IC testing; the manufacturing process; reconfigurable design; FPGA-based systems; machine-learning–based IoT; VLSI implementation of ANN; machine-learning–based image processing systems; synthesis and verification; and related areas. The book contains chapters on case studies as well as novel research ideas in the given field.

About the Editors

Sandeep Saini received his B.Tech. degree in electronics and communication engineering from the International Institute of Information Technology, Hyderabad, India, in 2008. He completed his M.S. from the same institute in 2010. He earned his Ph.D. from Malaviya National Institute of Technology, Jaipur, in 2020.

He is working at LNM Institute of Information Technology, Jaipur, as an Assistant Professor from 2011 onward. He has worked as adjunct faculty at the International Institute of Information Technology (IIIT), Bangalore (Deputation at Myanmar Institute of Information Technology, Mandalay, Myanmar) for two years, and a Lecturer at Jaypee University of Engineering and Technology, Guna, for 3 semesters. His research interests are in deep learning, machine learning, natural language processing, cognitive modeling of language learning models, biomedical and agricultural applications of deep learning. Sandeep is a member of IEEE since 2009 and an active member of ACM as well.

Kusum Lata received her M.Tech. and Ph.D. degrees from Indian Institute of Technology (IIT), Roorkee, India, and Indian Institute of Science (IISc), Bangalore, India, in 2003 and 2010. She also worked as a Research Associate in the Centre of Electronics Design and Technology, IISc Bangalore, India, for six months after completing her PhD. Since June 2010, she has worked as Lecturer for three years at the Indian Institute of Information Technology, Allahabad (IIIT-A), India.

She has worked as Assistant Professor from December 2013 to February 2016, and since March 2016, she has been working as an Associate Professor in the Department of Electronics and Communication Engineering at The LNM Institute of Information Technology, Jaipur. She is the recipient of the *Outstanding Research Paper Award* in *1ˢᵗAsia Symposium on Quality Electronic Design (ASQED-2009),* July 15–16, 2009, Kula Lumpur, Malaysia. Her research interests include digital circuit design using FPGAs, design for testability, formal verification of analog and mixed signal designs, and hardware security. Kusum is a member of IEEE since 2003 and an active member of ACM since 2011. She is also a lifetime member Computer Society of India.

G. R. Sinha is an Adjunct Professor at the International Institute of Information Technology Bangalore (IIITB) and Professor at Myanmar Institute of Information Technology (MIIT) Mandalay Myanmar. He obtained his B.E. (Electronics Engineering) and M.Tech. (Computer Technology) with Gold Medal from National Institute of Technology Raipur, India. He received his Ph.D. in electronics & telecommunication engineering from Chhattisgarh Swami Vivekanand Technical University (CSVTU) Bhilai, India. He is Visiting Professor (Honorary) in Sri Lanka Technological Campus Colombo for one year 2019-2020.

He has published 258 research papers, book chapters, and books at the International level that includes *Biometrics,* published by Wiley India, a subsidiary of John Wiley; *Medical Image Processing,* published by Prentice Hall of India, and 5 edited books on *Cognitive Science* – two volumes (Elsevier), *Optimization Theory (IOP)* and *Biometrics* (Springer). He is currently editing 6 more books on biomedical signals; brain and behavior computing; modern Sensors, and data deduplication with Elsevier, IOP, and CRC Press. He is an active reviewer and editorial member of more than 12 reputed international journals in his research areas, such as IEEE Transactions, Elsevier Journals, Springer Journals, etc.

He has teaching and research experience of 21 years. He has been Dean of Faculty and Executive Council Member of CSVTU and currently is a member of the Senate of MIIT. Dr. Sinha has been delivering ACM lectures as ACM Distinguished Speaker in the field of DSP since 2017 across the world. His few more important assignments include Expert Member for Vocational Training Programme by Tata Institute of Social Sciences (TISS) for two years (2017–2019); Chhattisgarh Representative of IEEE MP Sub-Section Executive Council (2016–2019); Distinguished Speaker in the field of digital image processing by Computer Society of India (2015).

He is the recipient of many awards and recognitions like TCS Award 2014 for outstanding contributions in Campus Commune of TCS, Rajaram Bapu Patil ISTE National Award 2013 for Promising Teacher in Technical Education by ISTE New Delhi, Emerging Chhattisgarh Award 2013, Engineer of the Year Award 2011, Young Engineer Award 2008, Young Scientist Award 2005, IEI Expert Engineer Award 2007, ISCA Young Scientist Award 2006 Nomination and Deshbandhu Merit Scholarship for 5 years. He served as Distinguished IEEE Lecturer in IEEE India Council for the Bombay section. He is a Senior Member of IEEE, Fellow of Institute of Engineers India, and Fellow of IETE India.

He has delivered more than 50 keynote/invited talks and chaired many technical sessions in International Conferences across the world, such as Singapore, Myanmar, Sri Lanka, Bangalore, Mumbai, Trivandrum, Hyderabad, Mysore, Allahabad, Nagpur, Yangon, and Meikhtila. His special session on "Deep Learning in Biometrics" was included in IEEE International Conference on Image Processing 2017. He is also a member of many national professional bodies like ISTE, CSI, ISCA, and IEI. He is a member of the university's various committees and has been Vice President of Computer Society of India for Bhilai Chapter for two consecutive years. He is a Consultant of various skill development initiatives of NSDC, Govt. of India. He is a regular Referee of Project Grants under the DST-EMR scheme and several other schemes of the Govt. of India. He received a few important consultancy supports as grants and travel support.

Dr. Sinha has supervised 8 PhD Scholars, 15 M. Tech. Scholars, and has been supervising 1 more PhD Scholar. His research interest includes biometrics, cognitive science, medical image processing, computer vision, outcome-based education (OBE), and ICT tools for developing employability skills.

Contributors

Madhvi Agarwal
Indraprastha Institute of Information
  Technology
Delhi, India

Rangel Arthur
Faculty of Technology (FT)
State University of Campinas (UNICAMP)
São Paulo, Brazil

Dinesh Bhatia
Malaviya National Institute of Technology
Jaipur, India

Ana Carolina Borges Monteiro
School of Electrical Engineering and
  Computing (FEEC)
State University of Campinas (UNICAMP)
São Paulo, Brazil

Giulliano Paes Carnielli
Faculty of Technology (FT)
State University of Campinas (UNICAMP)
São Paulo, Brazil

Rajat Subhra Chakraborty
Indian Institute of Technology Kharagpur
Kharagpur, India

Abhishek Choubey
Department of Electronics and
  Communication
Sreenidhi Institute of Science and
  Technology
Hyderabad, India

Shruti Bhargava Choubey
Department of Electronics and
  Communication
Sreenidhi Institute of Science and
  Technology
Hyderabad, India

Kamal Kishor Choure
Malaviya National Institute of Technology
Jaipur, India

Philemon Daniel
National Institute of Technology
Hamirpur, India

S. Dillibabu Shanmugam
Society for Electronic Transactions and
  Security
Chennai, India

Reinaldo Padilha França
School of Electrical Engineering and
  Computing (FEEC)
State University of Campinas (UNICAMP)
São Paulo, Brazil

Bibhas Ghoshal
Department of Information Technology
Indian Institute of Information
  Technology Allahabad
Allahabad, India

Ankur Gogoi
Department of Information Technology
Indian Institute of Information
  Technology Allahabad
Allahabad, India

Yogendra Gupta
Swami Keshwanand Institute of
  Technology Jaipur
Jaipur, India

Yuzo Iano
School of Electrical Engineering and
  Computing (FEEC)
State University of Campinas (UNICAMP)
São Paulo, Brazil

Pranav Jain
Indraprastha Institute of Information
    Technology
Delhi, India

Rajalakshmi K
Dept of ECE
PSG College of Technology
Tamil Nadu, India

Sushant Kumar
Banasthali University
Banasthali, India

Varsha Satheesh Kumar
Carnegie Mellon University
Pennsylvania, USA

Lavanya T
Dept of ECE
PSG College of Technology
Tamil Nadu, India

Shubhankar Majumdar
ECE Dept
National Institute of Technology
    Meghalaya
Meghalaya, India

Bodhisatwa Mazumdar
Department of Computer Science and
    Engineering
Indian Institute of Technology
Indore, India

Shaik Mohammed Waseem
VLSI Systems Research Group
IIIT Bangalore
Banglore, India

Nitesh Mudgal
Malaviya National Institute of Technology
Jaipur, India

Saurabh Mukherjee
Banasthali University
Banasthali, India

Kattekola Naresh
ECE Department
Vallurupalli Nageswara Rao Vignana
    Jyothi Institute of Engineering &
    Technology
Hyderabad, India

Rahul Pandey
Swami Keshwanand Institute of
    Technology Jaipur
Jaipur, India

Sree Ranjani Rajendran
RISE Lab
Indian Institute of Technology Madras
Madras, India

OVS Shashank Ram
Indraprastha Institute of Information
    Technology
Delhi, India

Subir Kumar Roy
VLSI Systems Research Group
IIIT Bangalore
Bangalore, India

Soma Saha
Department of Computer Engineering
Shri Govindram Seksaria Institute of
    Technology and Science
Indore, India

Ankur Saharia
Manipal University Jaipur
Jaipur, India

N. Sarat Chandra Babu
Society for Electronic Transactions and
    Security
Chennai, India

Sneh Saurabh
Indraprastha Institute of Information
    Technology
Delhi, India

Ashish Sharma
Indian Institute of Information Technology
  Kota
Kota, India

Harish Sharma
Rajasthan Technical University Kota
Kota, India

Niketa Sharma
Swami Keshwanand Institute of
  Technology Jaipur
Jaipur, India

Ghanshyam Singh
Malaviya National Institute of Technology
Jaipur, India

Manish Tiwari
Manipal University Jaipur
Jaipur, India

1 VLSI and Hardware Implementation Using Machine Learning Methods: A Systematic Literature Review

Kusum Lata[1], Sandeep Saini[1], and G. R. Sinha[2]
[1]Department of Electronics and Communication Engineering, The LNM Institute of Information Technology, Jaipur, India
[2]Department of Electronics and Communication Engineering, Myanmar Institute of Information Technology, Mandalay, Myanmar

CONTENTS

DOI: 10.1201/9781003201038-1

1

1.1 INTRODUCTION

Machine learning (ML) technologies have gained a lot of traction as a result of advancement in design technology of computer systems in terms of design constraints like functioning, power consumption, and area used over the last decade. In order to get better outcomes than conventional methods, a wide range of applications began to use ML algorithms. There are many applications, including image processing in health systems such as cancer detection [1], image classifications [2], banking and risk management [3], healthcare and clinical note analysis [4], managing energy efficiency of the public sector towards smart cities [5], and automatic database management systems [6]. Integrated vision systems provide hassle-free integration for various industries, such as full/semi-autonomous vehicles [7,8], and many security [9] applications, including cyber security [10].

In the application domain, such as autonomous and semi-autonomous vehicles, ML is completely transforming the field [11]. ML is also used in location-based services (LBS), such as global positioning system (GPS)-based vehicle navigation. Individual users can also utilize LBS to get information on nearby live entertainment and intelligent path navigation [12]. There are numerous applications where accurate, precisely calculated values that improve efficiency also can lead to the drastic difference between life and death. ML is at the heart of many such industrial applications. This pattern demonstrates that ML technologies continue to pique people's interest and have a lot of promise. Every electrical or embedded system is becoming more and more reliant on these technologies. The showstopper to this new paradigm is that ML algorithms are power hungry when it comes to their hardware implementation. Because of this fact, lots of improvements can be seen in developing hardware accelerators to provide essential computing power to such applications.

By reducing the power consumption of boosted performance and lowering essential design resources, intelligently tailored and optimized hardware implementation can significantly lower overall system design costs [13]. Field-programmable gate arrays (FPGAs), graphics processing units (GPUs), and application-specific integrated circuits (ASICs) are being considered for hardware implementation, each with their own set of advantages and disadvantages. In comparison to regular central processing units (CPUs), these hardware accelerators take advantage of parallelism to boost throughput and deliver substantially greater performance [14].

We give a complete survey that includes a systematic literature review (SLR) to aid academics working on ML hardware acceleration. The survey includes hardware implementation research for ML algorithms from 2011 through 2020. We collected a total of 150 different research papers, out of which 113 papers addressed hardware implementation of ML algorithms. We adopted narrow exclusion criteria for this work to consider as many papers as possible in order to cover a wide perspective on the issue at hand.

1.2 MOTIVATION

ML algorithms have been implemented using a variety of design strategies for a variety of purposes. General-purpose processors (GPPs), ASICs, and scalable/reconfigurable hardware architectures such as FPGAs are commonly used in these techniques/methods. GPPs offer scalability and flexibility, albeit at the expense of performance. Similarly, ASICs provide better performance, but the scalability parameter is rarely taken into account. FPGA-based hardware solutions, on the other hand, offer a large design area for balancing performance and scalability. As a result, scalable hardware architectures for pattern-matching algorithm implementation are becoming increasingly common. As a result, a thorough investigation is required to classify and define the current scalable hardware architectures.

1.3   CONTRIBUTIONS

The SLR focuses mainly on different types of hardware implementations of widely used ML methods for various applications. Kitchenham and Charter's [15] methodology is followed for the SLR, which is illustrated in Figure 1.1.

The main objective of this review article is to provide answers to a number of research questions involving ML hardware accelerators, which include the following:

Q1. What are the most common applications for which ML methods are used in hardware implementation?
Q2. Identify the ML algorithms and techniques produced between 2011 and 2020 that are most frequently used for hardware implementation?
Q3. What are the advantages and disadvantages of employing multiple hardware platforms for ML acceleration?

1.4   LITERATURE REVIEW

In recent years, research into the hardware implementation of ML algorithms has gotten a lot of attention. Many survey publications have been reported in the last decade. The focus of this research is on ML algorithms that are computationally expensive and power hungry. This is because large-scale problems necessitate specialized hardware with sufficient computational power to run ML algorithms.

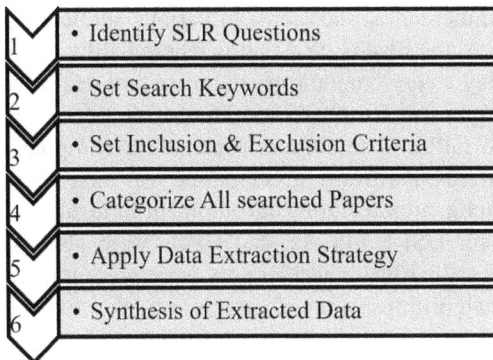

FIGURE 1.1   Systematic Literature Review (SLR) approach.

For these algorithms, hardware acceleration is considered to be the best option [16–18]. The goal is to make these ML algorithms process faster and more efficiently. In this regard, various hardware implementations for different applications have been reported in the last decade [19–22].

There were many survey articles published between 2011 and 2020 that discuss hardware implementation of ML algorithms. In this section, we report only those related survey articles. Some surveys cover ML accelerators and also discuss their benchmarking in general [21–23], whereas others discuss FPGA-based accelerators and implementation of ML algorithms [19,20,24,25]. In HajiRassouliha et al. [26], the authors deliberated upon the practical information related to state-of-the-art FPGAs, GPUs, and digital signal processors (DSPs) related to image processing algorithms and computer vision. A more comprehensive comparison to select the most suitable accelerator for the specific application is also given for the various hardware accelerators [26].

Zhen et al. in Li et al. [23] present the features of neural networks and introduce some recent efforts for accelerating ML techniques. Presenting neural network accelerators in economical ways, such as high performance, reduced area, and low energy consumption, is explored through numerous research papers in this review article. Reuther et al. [21] surveyed accelerators and processors for ML, specifically deep learning neural networks, along with benchmarking results that were conducted on commercial low-power processing systems. For that, they overviewed ML processors, and the authors found that the trends in many processors, which include area, power consumption, speed, and number of cores, were not increasing anymore. They also surveyed and analyzed ML processors and then finally presented benchmarking results for two low-power ML accelerator systems. Mittal [19] surveyed accelerators for convolutional neural networks (CNNs), which are based on FPGAs. The author also reported that with the limited hardware resources available on a single FPGA, usage of multiple FPGAs is essential for designing large-scale neural network model accelerators. With this, the author also concluded that, because of the manual partitioning approach, automated partitioning tools are required. Being an alternate and attractive option for the designer, more resource-rich FPGAs are required for designing accelerators of large-scale neural networks [19]. Image-processing algorithms and computer vision have much wider applications in many industry, healthcare, and commercial applications. In various applications, fast processing is required to analyze the results, and that is where hardware implementations of these algorithms play a very crucial role. In this regard, HajiRassouliha et al. [26] surveyed different types of hardware accelerators that could be based on CPUs, DSPs, FPGAs, and GPUs. The authors presented a comprehensive review and also compared various hardware accelerators for their specific applications. Practical details of chip architectures, development time, along with the available tools and utilities for DSPs, FPGAs, and GPUs were also discussed in detail. In the studied surveys, hardware accelerators were presented mainly for artificial intelligence (AI) algorithms. Our focus is on the existing literature that reports ML algorithms and their hardware implementations using different approaches towards many practical applications. The chapter provides a much-needed comprehensive review and comparative study that analyzes the existing work that has been done in the last decade from GPU, FPGA, and ASIC perspectives.

This work complements the existing SLR surveys reported in the literature and contributes towards providing the complete background of hardware implementations of ML algorithms.

1.5 METHODS

Systematic literature studies entail that the technique outlined below is strictly followed. In this section, main components of the review process are described. It starts with the considered data sources and search strategy, the inclusion/exclusion criteria, and then search methods. Later, data screening methodologies and data synthesis are described.

1.5.1 SEARCH STRATEGY

To perform the search for the review chapter, first we identified and finalized databases. Since our research focused on VLSI and hardware implementations of ML algorithms specifically, which broadly lies in the domain of electronics and computer system design, we chose the specific databases accordingly to get the right articles to review for this chapter. Databases used for performing the literature search were as follows:

- Google Scholar
- ACM Digital Library
- IEEE Explorer Digital Library
- Springer Digital Library
- Elsevier Digital Library

For performing the related search, we used various keywords that are mostly related to ML algorithms and hardware implementation approaches for them:

ML keywords: machine learning, global positioning system, deep learning, navigation system, clustering algorithms, path prediction, parallel machine learning, distributed machine learning, reinforcement, SVM, support vector machine, neural network, recurrent, convolution, LSTM, extreme learning, CNN, RNN, random forest

Hardware keywords: FPGA, ASIC, CPU, GPU, graphics processing unit, Internet of Things, field-programmable gate array, low power, reconfigurable devices, SOC-FPGA, device, raspberry-pi, RISC-V, microcontroller, FPU, wearable

All searches were carried out regarding title, abstract, and keywords in conjunction. If this combination was not possible, we used the title for the search. The searches were conducted between January 2011 and December 2020. Hence, articles published beyond January 1, 2021, were out of the scope. Any article with a posterior published date was manually included and used for reference purposes. Our search was limited to VLSI and hardware implementations of various ML algorithms. Further search was also limited to journal and conference papers only.

1.5.2 INCLUSION AND EXCLUSION RULES

Based on the previously given search terms, a total of 113 documents were initially gathered. The papers were then sorted into more specific categories. The following were the steps in the selection and filtration process:

Step 1. Remove the review/survey paper from the collected papers.

Step 2. To avoid any irrelevant documents, follow the inclusion and exclusion criteria.

Step 3. Remove articles that are duplicated or have several versions.

Step 4. Using the reference lists of the collected papers, look for any new similar publications and repeat the process.

The following principles were used for the inclusion and exclusion of research work available in the literature:

- **Relevant Work:** Include only reported literature works that are relevant to our designed queries (Q1–Q3). Exclude all studies that do not satisfy any of these predefined categories.
- **Most Recent (2011–2020):** To make sure only recent studies are in our chapter, choose studies reported from 2011 to 2020. Exclude all other studies that are published before 2011.
- **Publishing Organizations:** Select only reported works that are present in any of these five scientific databases, i.e. IEEE, SPRINGER, ELESVIER, ACM, and Google Scholar.
- **VLSI and Hardware Implementation-based Results:** Select only works that provide VLSI and hardware implementation details of ML algorithms for any application domain. Hardware implementation could be based on GPUs, FPGAs, and ASIC or complementary metal-oxide-semiconductor (CMOS)-based designs.
- **Publications Format:** Select only journals and conference papers that discuss the VLSI or hardware implementation of ML algorithms using FPGAs, GPUs, CPUs, or ASICs.

1.5.3 DATA EXTRACTION STRATEGY

In this step, we scrutinized the final list of publications and tried extracting the information needed to answer the research questions. To begin, we extracted general information, such as paper title, kind of publication, year of publication, etc., about each document. Then, more detailed information was needed, such as type of algorithm and associated hardware accelerator category, such as FPGA, ASIC-based implementation, GPU, CPU, or a mix of these. In the end, any relevant information that was directly connected to the study topics was sought. Due to the underlying unstructured nature of the required data, extracting such details is difficult. For example, some studies [27–29] discussed compiling software into a hardware description language that might be implemented on FPGAs, and some studies presented the automated design methodology to implement these algorithms on FPGAs. It is also obvious that all the papers are not responding to all three research questions that are set to complete this systematic review.

1.5.4 SYNTHESIS OF EXTRACTED DATA

We use a variety of methods to collect and create the necessary responses to the survey questionnaire in order to synthesize the information obtained from the selected publications. For Q1 and Q2, a qualitative analysis is utilized to get the necessary data. Quantitative analysis, on the other hand, is employed in Q3 to assess and evaluate the hardware implementations from various articles. The comparisons are given in a comparative tabulated format using typical figures of merit calculations. From each article, several challenges and trends were extracted for Q4.

1.5.5 RESULTS AND DISCUSSIONS

The results of the predefined questions are presented in this section. We present a study of hardware implementations of ML techniques between 2011 and 2020 based on the publications we selected. The sub-sections provide the summary of the collected articles, followed by the detailed study of each, and ten key observations and highlights of the articles are presented and discussed.

1.5.6 STUDY OVERVIEW

Out of a total of 150 research publications surveyed, this chapter examines 113 of them. These publications have been grouped based on the hardware that was used to implement them, for example (FPGA, ASIC, GPU, RISC-V, Raspberry Pi, etc.). Of the hardware platforms used in various collected research articles, the majority of the hardware implementations of ML/AI algorithms are based on FPGAs. Since FPGAs provide the rapid prototyping environment, the majority of the research papers used them to implement the many ML/AI algorithms for many applications. Apart from easy and fast prototyping, FPGAs also provide many additional features like powerful resources, and also many embedded devices are the integral part of SoC-FPGAs in recent times. Because of that, it becomes very useful for the researchers to design and verify their hardware implementations of these algorithms [30,31]. As a result of advancement in the hardware resources available in the latest FPGAs, implementation of complex algorithms such as ML and AI is evolving at the very fast pace [32,33]. FPGAs have flexibility to implement such types of algorithms, and moreover, the latest FPGAs have reached to the level where complete systems can be designed using their embedded features. Recent design trends show that FPGAs have gained significant attention for the implementation of ML/AI algorithms, specifically deep learning research [34,35].

GPU-based hardware enables the processing of large data sets in reasonable time scales. As a result of that, many applications that involve ML/AI algorithms and computer vision are evolving rapidly. Therefore, the development of various tools and methodologies are being devolved towards the broader applicability of ML/AI, specifically deep learning across multiple fields [36–39]. GPU-based accelerators also speed up the execution about hundreds times for ML/AI algorithms. GPUs are intended for high-performance scalar and parallel computations, which enhances the applicability of GPUs in a more meaningful manner [40].

Another platform that is used to implement the ML/AI algorithms is ASIC implementations. Since these are the chips that are designed for the specific purpose, many companies have started designing their AI chips for various applications [41], and lot of research is in progress in this direction. It is also claimed that specialized AI chips would perform better than CPUs and GPUs [42].

Recently, researchers have started implementation ML/AI algorithms using small single-board computers, e.g. Raspberry Pi for many applications [43–45]. Because of their tiny size, low cost, and lower power needs, single-board computers are a popular choice for AI applications.

1.6   HARDWARE IMPLEMENTATION OF ML/AI ALGORITHMS

In this section, hardware implementation of ML/AI algorithms and their accelerators are discussed, which are classified based on their implementation types, as shown in Figure 1.2. These papers are categorized in three categories:

- FPGA-based implementation
- GPU-based implementation
- ASIC-based implementation
- Other implementation

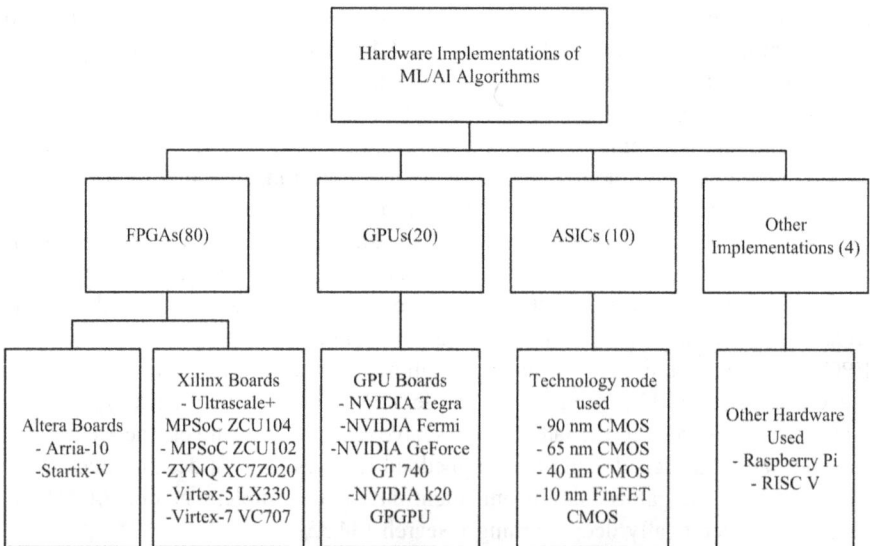

FIGURE 1.2   Summary of hardware implementations of ML/AI algorithms.

1.6.1   FPGA-Based Implementation

Many researchers have proposed hardware accelerators for ML/AI algorithms [21,27,46–79]. Many of them have reported FPGA-based accelerators for CNN architectures. All these research papers can be categorized into the following categories:

- Heterogeneous architectures (FPGA + CPU/ASIC)
- FPGA-embedded processors/FPGA-SoC
- FPGA

Some of the accelerators are presented where FPGAs are combined with CPUs to take advantage of both the architectures. Moss et al. [80] proposed the accelerator, which is designed in such a manner that the most intensive computational part of binarized neural network (BNN) is implemented on FPGA, whereas the rest of the implementation is done on Xeon^{TM} CPU. It is also claimed that the designed accelerator on FPGA along with Xeon give better performance and energy efficiency than the high-end GPU. Chi Zhang et al. [63] proposed a CPU-FPGA–based CNN accelerator, where they implemented the frequency domain algorithms on FPGA and used the data layout in shared memory to provide the essential communication to the data between FPGA and CPU. On the other hand, to design the hardware accelerator for GCN based on CPU-FPGA, Zeng and Prasanna [81] proposed the heterogeneous platform, where they performed the pre-processing on CPU and the rest of the computation was done on FPGA. It was also claimed that the magnitude of the training speeded up, with almost no accuracy loss compared to the existing implementations on multi-core platforms.

Other types of heterogeneous architectures that combine ASICs with FPGAs were proposed in Nurvitadhi et al. [82] for persistent recurrent neural networks (RNNs). Nurvitadhi et al. also discussed the integration of FPGA with ASIC chiplet TensorRAM to boost memory capacity and bandwidth. It also provides necessary throughput for matching bandwidth. Nurvitadhi et al. [83] focused on the deep learning (DL) domain with efficient tensor matrix/vector operations by proposing TensorTile ASICs for Stratix10.

A few researchers have presented the implementation of hardware accelerators of ML/AI algorithms on FPGA-embedded processors, such as NIOS-II by Santos et al. [84], in that the accelerator is designed for a pre-trained feedforward artificial neural network using the NIOS II processor. Jin Hee Kim et al. [85] presented the synthesized accelerator, which uses the Intel Arria 10 SoC FPGA with embedded ARM processor. Heekyung Kim et al. presented the FPGA-SoC design for CNN-based hardware implementation with respect to power consumption aspects. The authors have applied the proposed low-power scheme to Processing System (PS) and Programmable Logic (PL) architecture of the design to reduce power consumption effectively. Angelos Kyriakos et al. [86] focused on designing CNN accelerators based on Myriad2 and FPGA architecture.

FPGAs allow you to choose any precision that is appropriate for the intended CNN implementation. FPGA mappings, on the other hand, often rely on DSP blocks for maximum efficiency. Only fixed-precision DSP blocks are available. As a result, employing fixed-point quantization can boost speed while also conserving hardware resources [66,87].

A binarized CNN (BCNN) with weights and activation constrained to 2-values (+1/-1) has recently been presented. In well-optimized software on the CPU and GPU, BCNN can lead to huge increases in performance and power efficiency. Some research publications focused on using FPGAs to accelerate these BCNNs [53,54,57,57,57,60,88]. Dong Nguyen et al. [66] showed that CNNs are resilient up to 8-bit precision and proposed double MAC, which can increase throughput computationally of a CNN layer at a significant level. As a result, some implementations used 8-bit fixed point, which outperformed the 16-bit fixed-point implementation by more than 50% [89].

Some researchers have presented the ML algorithm implementations on FPGAs [90–93]. Jokic [93] presented an FPGA-based streaming camera, which can be classified as ROI with a BNN in real-time mode. Results show the energy savings of three times with respect to the external image processing.

1.6.2 GPU-BASED IMPLEMENTATION

GPUs can speed up a wide range of applications, including image recognition, gaming, data analytics, weather forecasting, and multimedia. GPUs have been very important in ML and deep learning accelerations, also. Many researchers have presented and discussed the GPU-based accelerator and ML/AI algorithm implementations [37,94–101]. Table 1.1 summarizes the work done on GPU-based implementation of ML/AI algorithms and hardware accelerators based on GPUs.

1.6.3 ASICS-BASED IMPLEMENTATIONS

Recently, several researchers have published papers related to the chip implementation of ML/AI algorithms related to many applications [103–109]. Conti et al. [103] proposed PULP architecture for faster performance and energy scalability. An application of this architecture is also presented, which is based on ConvNet detector for smart surveillance and works at ultra low energy on a 128 × 128 image. Joen et al. [104] presents an energy-efficient face-detection and recognition processor, which is aimed for mobile applications. The fabricated processor completes face detection and recognition in real time as 5.5 fps throughput with very little power consumption. Zheng et al. [105] proposed a hardware architecture that is an event-triggered architecture for a spike-time-dependent plasticity (STDP) learning algorithm. This architecture is implemented in a 65-nm technology. Chen et al. [106] designed 10-nm FinFET CMOS-based 1M-synapse chip to accelerate spiking neural network (SNN) learning. The designed chip reduces memory up to 16 times, with less than 2% overhead of memory. Frenkel et al. [107] presented the MorphIC, which is a quad-core binary weight digital neuro morphic processor embedding a S-SDSP learning rule. MorphIC demonstrates a promising improvement in area-accuracy tradeoff on the MNIST classification. Kim et al. [108] presented an energy-efficient STDP-based sparse coding processor, which is designed using a 65-nm CMOS process. In the inference mode of operations, the SNN hardware achieves the throughput of 374 Mps

TABLE 1.1

Summary of Work Done on GPU-Based Implementations of ML/AI Algorithms

References	Algorithms	Implementations
[37]	Immune convolutional neural network (ICNN)	NVIDIA GPU-based implementation results show higher recognition rate, faster computing speed, and more stable performance.
[94]	Feature-detection and segmentation algorithms	Implementation on multi-core SoC, which includes DSP-based TI66AK2H12 SoC and the GPU-based NIVIDIA Tegra K1/X1/X2 SoC.
[95]	HOG for feature extraction and SVM for classification	Fast vehicle classification using GPU computation. Implementation of GPU programming is in OpenCL.
[96]	LBP-based face detection algorithm	29 fps for full HD inputs is achieved on Tegra K1 platform, where Cortex-A15 CPU and 192-core GPU supported by CUDA are integrated.
[97]	Haar features and AdaBoost-based face detection	GPU acceleration is done to boost the performance and efficiency of the designed system.
[98]	Scanning-window technique, integral image, and local binary pattern (LBP)	Implemented on GPU for processing images of the speed up to 287 fps with the image dimensions of 640 x 480 pixels.
[99]	Viola-Jones face detection algorithm	CPU-GPU–based accelerator is designed to enhance performance.
[102]	Neural network acceleration (NNA)	Low overhead neurally accelerated architecture for GPUs.
[101]	Recurrent neural network language models (RNNLMs)	Implementation of RNN on multiple GPUs with the speed up of 3.4 times.

and 840.2 GSOP/s with the energy-efficiency of 781.52 pJ/pixels and 0.35 pJ/SOP. Chuang et al. [109] implemented a low-power and low-cost BW-SNN ASIC using 90-nm CMOS technology for image classification. The designed ASIC was demonstrated in real-time for bottled-drink recognition. It also provided very good accuracy and outperformed its efficiency.

1.6.4 OTHER IMPLEMENTATIONS

Recently, hardware implementations of ML/AI algorithms other than those based on FPGA, ASIC, and GPUs have been reported. They use other platforms like Raspberry Pi and RISC-V-based implementations [110–113]. Nagpal [110] implemented an LBPH algorithm using OpenCV on Raspberry Pi. This system uses a Haar cascade classifier for face detection in the image, and it follows LBPH for face recognition. Daryanavard et al. [111] presented the Haar cascade classifier algorithm implementation for face detection using the OpenCV tool on Raspberry Pi. The implementation is evaluated using a specific dataset of UAV facial images. Results are promising for face detection up to 5 meters height of the camera from the ground. Lage et al. [113] present the low cost IoT surveillance system. The system is designed and tested using Raspberry Pi and Up Squared Board devices. In addition to the person detection with an average precision of 0.48, the designed system is also tested for the benefits of hardware-acceleration by Intel® Movidius™ Neural Compute Stick (NCS). Zhang et al. [112] presented the hardware accelerator of YOLO using open source RISC-V core ROCKET as its controller. The hardware designed for real-time object detection system was verified using Xilinx Virtex-7 FPGA VC709.

1.6.5 SLR DISCUSSIONS AND RECOMMENDATIONS

Various hardware implementation approaches are used for ML/AI algorithms using ASICs, FPGAs, or GPUs. These approaches are presented here in terms of established figures of merits. Considered figures of merits for our survey are presented in Table 1.2.

It is a well known fact about ASIC designs that they are the most suitable approach for applications where the main objective of the design is to achieve low power with minimum area. These designs are also well suited in cases where the ICs produce in bulk for the given applications but no alteration is required after the IC is fabricated. ASICs are explored in the literature for image and video processing accelerators in mobile devices [104].

With recent devices like SoC-FPGAs, these devices enhance the design time ultimately with the on-board available embedded devices. For prototyping purposes, the real-time implementations of various applications of the latest FPGAs become the obvious choice of many researchers. Image processing applications, such as real-time object identification and recognition, are generally implemented on FPGAs. FPGAs are seen to be a simple and appropriate option for a real-time processing system.

FPGAs and ASICs give a hardware-level solution, whereas GPUs provide a software-level solution. In comparison to GPUs, FPGAs and ASICs offer more flexibility during the implementation process. As a result, GPU implementation is constrained by the underlying hardware, whereas FPGAs and ASICs rely only on design optimization performed during the development process.

TABLE 1.2
Comparisons of ML/AI Algorithms' Implementation Approaches

	ASIC	FPGA	GPU
Time to Market (ToM)	High	Medium	Medium
Design Flexibility	Low	High	High
Performance	High	Medium	Medium to low
Cost	High	Medium	Medium to low
Essential Skills	HDL, CAD tools for custom design	HDL, CAD tools to for programmable designs	High-Level Programming
Implementation Level	Hardware	Hardware	Software

TABLE 1.3

Research Question Findings from the Survey

Research Question	Question Description	Survey Findings
Q1	What are the most common applications for which ML/AI methods are used in hardware implementation?	Major applications are related to image processing by applying hardware implementations of ML/AI algorithms. Robotics, autonomous/semi-autonomous vehicles, navigation systems, and cyber security are other applications where ML/AI are implemented at the hardware level and used.
Q2	Identify the ML/AI algorithms and techniques produced between 2011 and 2020 that are most frequently used for hardware implementation.	Algorithms and techniques include Haar-Classifier, SURF, SVM, AdaBoost, computer vision algorithms, ANN, CNN, DNN, spike neural network, ICNN, YOLO, SLAM, BCNN, LSTM, YOLOv3, FP-DNN, etc.
Q3	What are the advantages and disadvantages of employing multiple hardware platforms for ML/AI acceleration?	Different types of hardware platforms used have their own advantages and disadvantages, and these are given in Table 1.2 in detail.

Hardware accelerators, whether they are based on FPGAs, ASICs, or GPUs, are becoming increasingly popular. It's noticeable in the deployment of AI and ML algorithms. These algorithms' superiority necessitates extremely high computing power and memory use, which hardware accelerators may provide. AI and ML are critical technologies that support daily social life, economic operations, and even medical applications. As a result of the ongoing development of these algorithms, new applications with larger resource demands will emerge. Implementing these algorithms on hardware will result in faster, more efficient, and more accurate AI processing, which will benefit all industries. Finally, Table 1.3 summarizes the research study's main findings.

1.7 CONCLUSIONS

In this chapter, we present an SLR study about VLSI and hardware implementations of ML/AI algorithms over the period between 2011 and 2020. The main objective of this SLR was to answer three research questions that address the survey from an application and implementation point of view. Only journal and conference papers were considered in this chapter from relevant search databases. Research papers that focused only on the VLSI and hardware implementations of ML/AI algorithms were selected. Hardware implementation selected here for the study were based in

FPGAs, ASICs, and GPUs. All the papers were selected in such a manner so that they were helpful to answer our set search questions. The work presented in this chapter gives a thorough analysis of different types of implementation platforms and approaches used in the last decade. This chapter would be useful for readers to understand the various approaches reported to implement ML/AI algorithms using ASICs, FPGAs, and GPUs, and it will also give a good understanding of hardware selection in terms of development time, performance results, and cost, also.

REFERENCES

[1] G. Meenalochini and S. Ramkumar, "Survey of machine learning algorithms for breast cancer detection using mammogram images," *Mater. Today Proc.*, vol. 37, pp. 2738–2743, 2021, doi: 10.1016/j.matpr.2020.08.543.

[2] P. Wang, E. Fan, and P. Wang, "Comparative analysis of image classification algorithms based on traditional machine learning and deep learning," *Pattern Recognit. Lett.*, vol. 141, pp. 61–67, 2021, doi: 10.1016/j.patrec.2020.07.042.

[3] M. Leo, S. Sharma, and K. Maddulety, "Machine learning in banking risk management: A literature review," *Risks*, vol. 7, no. 1, 2019, doi: 10.3390/risks7010029.

[4] A. Mustafa and M. Rahimi Azghadi, "Automated machine learning for healthcare and clinical notes analysis," *Computers*, vol. 10, no. 2, 2021, doi: 10.3390/computers10020024.

[5] M. Zekić-Sušac, S. Mitrović, and A. Has, "Machine learning based system for managing energy efficiency of public sector as an approach towards smart cities," *Int. J. Inf. Manage.*, vol. 58, p. 102074, 2021, doi: 10.1016/j.ijinfomgt.2020.102074.

[6] D. Van Aken, A. Pavlo, G. J. Gordon, and B. Zhang, "Automatic database management system tuning through large-scale machine learning," in *Proceedings of the 2017 ACM International Conference on Management of Data*, 2017, pp. 1009–1024, doi: 10.1145/3035918.3064029.

[7] S. Baee, E. Pakdamanian, V. Ordonez, I. Kim, L. Feng, and L. Barnes, "EyeCar: Modeling the visual attention allocation of drivers in semi-autonomous vehicles," *arXiv Prepr. arXiv1912.07773*, 2019.

[8] A. P. Sligar, "Machine learning-based radar perception for autonomous vehicles using full physics simulation," *IEEE Access*, vol. 8, pp. 51470–51476, 2020, doi: 10.1109/ACCESS.2020.2977922.

[9] A. L. Buczak and E. Guven, "A survey of data mining and machine learning methods for cyber security intrusion detection," *IEEE Commun. Surv. Tutorials*, vol. 18, no. 2, pp. 1153–1176, 2016, doi: 10.1109/COMST.2015.2494502.

[10] K. Shaukat, S. Luo, V. Varadharajan, I. A. Hameed, and M. Xu, "A survey on machine learning techniques for cyber security in the last decade," *IEEE Access*, vol. 8, pp. 222310–222354, 2020.

[11] J. Stilgoe, "Machine learning, social learning and the governance of self-driving cars," *Soc. Stud. Sci.*, vol. 48, no. 1, pp. 25–56, 2018.

[12] Z. Li, K. Xu, H. Wang, Y. Zhao, X. Wang, and M. Shen, "Machine-learning-based positioning: A survey and future directions," *IEEE Netw.*, vol. 33, no. 3, pp. 96–101, 2019, doi: 10.1109/MNET.2019.1800366.

[13] J. Misra and I. Saha, "Artificial neural networks in hardware: A survey of two decades of progress," *Neurocomputing*, vol. 74, no. 1–3, pp. 239–255, 2010.

[14] T. Baji, "Evolution of the GPU device widely used in AI and massive parallel processing," in *Proceedings of the 2018 IEEE 2nd Electron Devices Technology and Manufacturing Conference (EDTM)*, 2018, pp. 7–9.

[15] D. Budgen and P. Brereton, "Performing systematic literature reviews in software engineering," in *Proceedings of the 28th international conference on Software engineering*, 2006, pp. 1051–1052.

[16] R. Zhao, W. Luk, X. Niu, H. Shi, and H. Wang, "Hardware acceleration for machine learning," in *Proceedings of the 2017 IEEE Computer Society Annual Symposium on VLSI (ISVLSI)*, 2017, pp. 645–650.

[17] K. Tajiri and T. Maruyama, "FPGA acceleration of a supervised learning method for hyperspectral image classification," in *Proceedings of the 2018 International Conference on Field-Programmable Technology (FPT)*, 2018, pp. 270–273.

[18] A. I. Solis, "Dedicated hardware for machine/deep learning: Domain specific architectures," Masters' Thesis, 2019.

[19] S. Mittal, "A survey of FPGA-based accelerators for convolutional neural networks," *Neural Comput. Appl.*, vol. 32, no. 4, pp. 1109–1139, 2020.

[20] K. Guo, S. Zeng, J. Yu, Y. Wang, and H. Yang, "A survey of FPGA-based neural network accelerator," *arXiv Prepr. arXiv1712.08934*, 2017.

[21] A. Reuther, P. Michaleas, M. Jones, V. Gadepally, S. Samsi, and J. Kepner, "Survey and benchmarking of machine learning accelerators," in *Proceedings of the 2019 IEEE High Performance Extreme Computing Conference (HPEC)*, 2019, pp. 1–9.

[22] A. Reuther, P. Michaleas, M. Jones, V. Gadepally, S. Samsi, and J. Kepner, "Survey of machine learning accelerators," in *Proceedings of the 2020 IEEE High Performance Extreme Computing Conference (HPEC)*, 2020, pp. 1–12.

[23] Z. Li, Y. Wang, T. Zhi, and T. Chen, "A survey of neural network accelerators," *Front. Comput. Sci.*, vol. 11, no. 5, pp. 746–761, 2017, doi: 10.1007/s11704-016-6159-1.

[24] W. Zhang, "A Survey of FPGA Based CNNs Accelerators," EasyChair Preprint No. 3633, 2020.

[25] E. Koromilas, I. Stamelos, C. Kachris, and D. Soudris, "Spark acceleration on FPGAs: A use case on machine learning in Pynq," in *Proceedings of the 2017 6th International Conference on Modern Circuits and Systems Technologies (MOCAST)*, 2017, pp. 1–4, doi: 10.1109/MOCAST.2017.7937637.

[26] A. HajiRassouliha, A. J. Taberner, M. P. Nash, and P. M. F. Nielsen, "Suitability of recent hardware accelerators (DSPs, FPGAs, and GPUs) for computer vision and image processing algorithms," *Signal Process. Image Commun.*, vol. 68, pp. 101–119, 2018, doi: 10.1016/j.image.2018.07.007.

[27] Y. Ma, Y. Cao, S. Vrudhula, and J. Seo, "Optimizing the convolution operation to accelerate deep neural networks on FPGA," *IEEE Trans. Very Large Scale Integr. Syst.*, vol. 26, no. 7, pp. 1354–1367, 2018, doi: 10.1109/TVLSI.2018.2815603.

[28] Y. Ma, Y. Cao, S. Vrudhula, and J. Seo, "An automatic RTL compiler for high-throughput FPGA implementation of diverse deep convolutional neural networks," in *Proceedings of the 2017 27th International Conference on Field Programmable Logic and Applications (FPL)*, 2017, pp. 1–8.

[29] S. I. Venieris and C.-S. Bouganis, "fpgaConvNet: Mapping regular and irregular convolutional neural networks on FPGAs," *IEEE Trans. Neural Networks Learn. Syst.*, vol. 30, no. 2, pp. 326–342, 2019, doi: 10.1109/TNNLS.2018.2844093.

[30] A. Shawahna, S. M. Sait, and A. El-Maleh, "FPGA-based accelerators of deep learning networks for learning and classification: A review," *IEEE Access*, vol. 7, pp. 7823–7859, 2018.

[31] M. A. Dias and D. A. P. Ferreira, "Deep learning in reconfigurable hardware: A survey," in *Proceedings of the 2019 IEEE International Parallel and Distributed Processing Symposium Workshops (IPDPSW)*, 2019, pp. 95–98, doi: 10.1109/IPDPSW.2019.00026.

[32] E. Nurvitadhi *et al.*, "Can fpgas beat gpus in accelerating next-generation deep neural networks?," in *Proceedings of the 2017 ACM/SIGDA International Symposium on Field-Programmable Gate Arrays*, 2017, pp. 5–14.

[33] T. Ben-Nun and T. Hoefler, "Demystifying parallel and distributed deep learning: An in-depth concurrency analysis," *ACM Comput. Surv.*, vol. 52, no. 4, pp. 1–43, 2019.

[34] Y. LeCun, "1.1 Deep learning hardware: Past, present, and future," in *Proceedings of the 2019 IEEE International Solid- State Circuits Conference - (ISSCC)*, 2019, pp. 12–19, doi: 10.1109/ISSCC.2019.8662396.

[35] Y. Chen, B. Zheng, Z. Zhang, Q. Wang, C. Shen, and Q. Zhang, "Deep learning on mobile and embedded devices: State-of-the-art, challenges, and future directions," *ACM Comput. Surv.*, vol. 53, no. 4, pp. 1–37, 2020.

[36] J. Lemley, S. Bazrafkan, and P. Corcoran, "Deep learning for consumer devices and services: Pushing the limits for machine learning, artificial intelligence, and computer vision," *IEEE Consum. Electron. Mag.*, vol. 6, no. 2, pp. 48–56, 2017, doi: 10.1109/MCE.2016.2640698.

[37] T. Gong, T. Fan, J. Guo, and Z. Cai, "GPU-based parallel optimization of immune convolutional neural network and embedded system," *Eng. Appl. Artif. Intell.*, vol. 62, pp. 384–395, 2017.

[38] L. N. Huynh, Y. Lee, and R. K. Balan, "Deepmon: Mobile gpu-based deep learning framework for continuous vision applications," in *Proceedings of the 15th Annual International Conference on Mobile Systems, Applications, and Services*, 2017, pp. 82–95.

[39] M. A. Raihan, N. Goli, and T. M. Aamodt, "Modeling deep learning accelerator enabled GPUs," in *Proceedings of the 2019 IEEE International Symposium on Performance Analysis of Systems and Software (ISPASS)*, 2019, pp. 79–92, doi: 10.1109/ISPASS.2019.00016.

[40] N. Singh and S. P. Panda, "Enhancing the proficiency of artificial neural network on prediction with GPU," in *Proceedings of the 2019 International Conference on Machine Learning, Big Data, Cloud and Parallel Computing (COMITCon)*, 2019, pp. 67–71, doi: 10.1109/COMITCon.2019.8862440.

[41] D. Monroe, "Chips for artificial intelligence," *Commun. ACM*, vol. 61, no. 4, pp. 15–17, 2018.

[42] S. Greengard, "Making chips smarter," *Commun. ACM*, vol. 60, no. 5, pp. 13–15, 2017.

[43] D. K. Dewangan and S. P. Sahu, "Deep learning-based speed bump detection model for intelligent vehicle system using raspberry Pi," *IEEE Sens. J.*, vol. 21, no. 3, pp. 3570–3578, 2020.

[44] A. A. S. Zen, B. Duman, and B. Sen, "Benchmark analysis of Jetson TX2, Jetson Nano and Raspberry PI using deep-CNN," in *Proceedings of the 2020 International Congress on Human-Computer Interaction, Optimization and Robotic Applications (HORA)*, 2020, pp. 1–5, doi: 10.1109/HORA49412.2020.9152915.

[45] H. A. Shiddieqy, F. I. Hariadi, and T. Adiono, "Implementation of deep-learning based image classification on single board computer," in *Proceedings of the 2017 International Symposium on Electronics and Smart Devices (ISESD)*, 2017, pp. 133–137, doi: 10.1109/ISESD.2017.8253319.

[46] M. Motamedi, P. Gysel, V. Akella, and S. Ghiasi, "Design space exploration of FPGA-based deep convolutional neural networks," in *Proceedings of the 2016 21st Asia and South Pacific Design Automation Conference (ASP-DAC)*, 2016, pp. 575–580.

[47] A. Rahman, J. Lee, and K. Choi, "Efficient FPGA acceleration of convolutional neural networks using logical-3D compute array," in *Proceedings of the 2016 Design, Automation & Test in Europe Conference & Exhibition (DATE)*, 2016, pp. 1393–1398.

[48] Q. Xiao, Y. Liang, L. Lu, S. Yan, and Y.-W. Tai, "Exploring heterogeneous algorithms for accelerating deep convolutional neural networks on FPGAs," in *Proceedings of the 54th Annual Design Automation Conference 2017*, 2017, pp. 1–6.

[49] Y. Ma, Y. Cao, S. Vrudhula, and J. Seo, "Optimizing loop operation and dataflow in FPGA acceleration of deep convolutional neural networks," in *Proceedings of the 2017 ACM/SIGDA International Symposium on Field-Programmable Gate Arrays*, 2017, pp. 45–54.

[50] S. I. Venieris and C.-S. Bouganis, "Latency-driven design for FPGA-based convolutional neural networks," in *Proceedings of the 2017 27th International Conference on Field Programmable Logic and Applications (FPL)*, 2017, pp. 1–8, doi: 10.23919/FPL.2017.8056828.

[51] F. Sun *et al.*, "A high-performance accelerator for large-scale convolutional neural networks," in *Proceedings of the 2017 IEEE International Symposium on Parallel and Distributed Processing with Applications and 2017 IEEE International Conference on Ubiquitous Computing and Communications (ISPA/IUCC)*, 2017, pp. 622–629.

[52] J. Zhang and J. Li, "Improving the performance of OpenCL-based FPGA accelerator for convolutional neural network," in *Proceedings of the 2017 ACM/SIGDA International Symposium on Field-Programmable Gate Arrays*, 2017, pp. 25–34.

[53] H. Yonekawa and H. Nakahara, "On-chip memory based binarized convolutional deep neural network applying batch normalization free technique on an FPGA," in *Proceedings of the 2017 IEEE International Parallel and Distributed Processing Symposium Workshops (IPDPSW)*, 2017, pp. 98–105.

[54] H. Nakahara, H. Yonekawa, and S. Sato, "An object detector based on multiscale sliding window search using a fully pipelined binarized CNN on an FPGA," in *Proceedings of the 2017 International Conference on Field Programmable Technology (ICFPT)*, 2017, pp. 168–175.

[55] H. Nakahara, T. Fujii, and S. Sato, "A fully connected layer elimination for a binarizec convolutional neural network on an FPGA," in *Proceedings of the 2017 27th International Conference on Field Programmable Logic and Applications (FPL)*, 2017, pp. 1–4.

[56] L.-W. Kim, "DeepX: Deep learning accelerator for restricted boltzmann machine artificial neural networks," *IEEE Trans. Neural Networks Learn. Syst.*, vol. 29, no. 5, pp. 1441–1453, 2017.

[57] R. Zhao *et al.*, "Accelerating binarized convolutional neural networks with software-programmable fpgas," in *Proceedings of the 2017 ACM/SIGDA International Symposium on Field-Programmable Gate Arrays*, 2017, pp. 15–24.

[58] U. Aydonat, S. O'Connell, D. Capalija, A. C. Ling, and G. R. Chiu, "An opencl: Deep learning accelerator on arria 10," in *Proceedings of the 2017 ACM/SIGDA International Symposium on Field-Programmable Gate Arrays*, 2017, pp. 55–64.

[59] X. Wei *et al.*, "Automated systolic array architecture synthesis for high throughput CNN inference on FPGAs," in *Proceedings of the 54th Annual Design Automation Conference 2017*, 2017, pp. 1–6.

[60] M. Shimoda, S. Sato, and H. Nakahara, "All binarized convolutional neural network and its implementation on an FPGA," in *Proceedings of the 2017 International Conference on Field Programmable Technology (ICFPT)*, 2017, pp. 291–294.

[61] A. X. M. Chang and E. Culurciello, "Hardware accelerators for recurrent neural networks on FPGA," in *Proceedings of the 2017 IEEE International Symposium on Circuits and Systems (ISCAS)*, 2017, pp. 1–4.

[62] J. Guo, S. Yin, P. Ouyang, L. Liu, and S. Wei, "Bit-width based resource partitioning for CNN acceleration on FPGA," in *Proceedings of the 2017 IEEE 25th*

Annual International Symposium on Field-Programmable Custom Computing Machines (FCCM), 2017, p. 31.

[63] C. Zhang and V. Prasanna, "Frequency domain acceleration of convolutional neural networks on CPU-FPGA shared memory system," in *Proceedings of the 2017 ACM/SIGDA International Symposium on Field-Programmable Gate Arrays*, 2017, pp. 35–44.

[64] L. Lu, Y. Liang, Q. Xiao, and S. Yan, "Evaluating fast algorithms for convolutional neural networks on FPGAs," in *Proceedings of the 2017 IEEE 25th Annual International Symposium on Field-Programmable Custom Computing Machines (FCCM)*, 2017, pp. 101–108.

[65] L. Gong, C. Wang, X. Li, H. Chen, and X. Zhou, "A power-efficient and high performance FPGA accelerator for convolutional neural networks: Work-in-progress," in *Proceedings of the Twelfth IEEE/ACM/IFIP International Conference on Hardware/Software Codesign and System Synthesis Companion*, 2017, pp. 1–2.

[66] D. Nguyen, D. Kim, and J. Lee, "Double MAC: Doubling the performance of convolutional neural networks on modern FPGAs," in *Proceedings of the Design, Automation & Test in Europe Conference & Exhibition (DATE), 2017*, 2017, pp. 890–893.

[67] K. Abdelouahab, M. Pelcat, J. Serot, and F. Berry, "Accelerating CNN inference on FPGAs: A survey," *arXiv Prepr. arXiv1806.01683*, 2018.

[68] S. Yin *et al.*, "A high throughput acceleration for hybrid neural networks with efficient resource management on FPGA," *IEEE Trans. Comput. Des. Integr. Circuits Syst.*, vol. 38, no. 4, pp. 678–691, 2018.

[69] N. Shah, P. Chaudhari, and K. Varghese, "Runtime programmable and memory bandwidth optimized FPGA-based coprocessor for deep convolutional neural network," *IEEE Trans. Neural Networks Learn. Syst.*, vol. 29, no. 12, pp. 5922–5934, 2018.

[70] C. Zhang, G. Sun, Z. Fang, P. Zhou, P. Pan, and J. Cong, "Caffeine: Toward uniformed representation and acceleration for deep convolutional neural networks," *IEEE Trans. Comput. Des. Integr. Circuits Syst.*, vol. 38, no. 11, pp. 2072–2085, 2018.

[71] L. Wei, B. Luo, Y. Li, Y. Liu, and Q. Xu, "I know what you see: Power side-channel attack on convolutional neural network accelerators," in *Proceedings of the 34th Annual Computer Security Applications Conference*, 2018, pp. 393–406.

[72] L. Lu and Y. Liang, "SpWA: An efficient sparse winograd convolutional neural networks accelerator on FPGAs," in *Proceedings of the 55th Annual Design Automation Conference*, 2018, pp. 1–6.

[73] L. Gong, C. Wang, X. Li, H. Chen, and X. Zhou, "MALOC: A fully pipelined FPGA accelerator for convolutional neural networks with all layers mapped on chip," *IEEE Trans. Comput. Des. Integr. Circuits Syst.*, vol. 37, no. 11, pp. 2601–2612, 2018.

[74] J.-W. Chang, K.-W. Kang, and S.-J. Kang, "An energy-efficient FPGA-based de-convolutional neural networks accelerator for single image super-resolution," *IEEE Trans. Circuits Syst. Video Technol.*, vol. 30, no. 1, pp. 281–295, 2018.

[75] H. Kim and K. Choi, "Low power FPGA-SoC design techniques for CNN-based object detection accelerator," in *Proceedings of the 2019 IEEE 10th Annual Ubiquitous Computing, Electronics Mobile Communication Conference (UEMCON)*, 2019, pp. 1130–1134, doi: 10.1109/UEMCON47517.2019.8992929.

[76] A. Shawahna, S. M. Sait, and A. El-Maleh, "FPGA-based accelerators of deep learning networks for learning and classification: A review," *IEEE Access*, vol. 7, pp. 7823–7859, 2019, doi: 10.1109/ACCESS.2018.2890150.

[77] X. Lian, Z. Liu, Z. Song, J. Dai, W. Zhou, and X. Ji, "High-performance FPGA-based CNN accelerator with block-floating-point arithmetic," *IEEE Trans. Very Large Scale Integr. Syst.*, vol. 27, no. 8, pp. 1874–1885, 2019.

[78] Y. Ma, Y. Cao, S. Vrudhula, and J.-S. Seo, "Performance modeling for CNN inference accelerators on FPGA," *IEEE Trans. Comput. Des. Integr. Circuits Syst.*, vol. 39, no. 4, pp. 843–856, 2019.

[79] M. P. Véstias, R. P. Duarte, J. T. de Sousa, and H. C. Neto, "A fast and scalable architecture to run convolutional neural networks in low density FPGAs," *Microprocess. Microsyst.*, vol. 77, p. 103136, 2020.

[80] D. J. M. Moss *et al.*, "High performance binary neural networks on the Xeon +FPGA platform," in *Proceedings of the 2017 27th International Conference on Field Programmable Logic and Applications (FPL)*, 2017, pp. 1–4, doi: 10.23919/FPL.2017.8056823.

[81] H. Zeng and V. Prasanna, "GraphACT: Accelerating GCN training on CPU-FPGA heterogeneous platforms," in *Proceedings of the 2020 ACM/SIGDA International Symposium on Field-Programmable Gate Arrays*, 2020, pp. 255–265.

[82] E. Nurvitadhi *et al.*, "Why compete when you can work together: FPGA-ASIC integration for persistent RNNs," in *Proceedings of the 2019 IEEE 27th Annual International Symposium on Field-Programmable Custom Computing Machines (FCCM)*, 2019, pp. 199–207, doi: 10.1109/FCCM.2019.00035.

[83] E. Nurvitadhi *et al.*, "In-package domain-specific asics for intel®stratix®10 fpgas: A case study of accelerating deep learning using tensortile asic," in *Proceedings of the 2018 28th International Conference on Field Programmable Logic and Applications (FPL)*, 2018, pp. 106–1064.

[84] P. Santos, D. Ouellet-Poulin, D. Shapiro, and M. Bolic, "Artificial neural network acceleration on FPGA using custom instruction," in *Proceedings of the 2011 24th Canadian Conference on Electrical and Computer Engineering(CCECE)*, 2011, pp. 450–455, doi: 10.1109/CCECE.2011.6030491.

[85] J. H. Kim, B. Grady, R. Lian, J. Brothers, and J. H. Anderson, "FPGA-based CNN inference accelerator synthesized from multi-threaded C software," in *Proceedings of the 2017 30th IEEE International System-on-Chip Conference (SOCC)*, 2017, pp. 268–273, doi: 10.1109/SOCC.2017.8226056.

[86] A. Kyriakos, E.-A. Papatheofanous, B. Charalampos, E. Petrongonas, D. Soudris, and D. Reisis, "Design and performance comparison of CNN accelerators based on the Intel Movidius Myriad2 SoC and FPGA embedded prototype," in *Proceedings of the 2019 International Conference on Control, Artificial Intelligence, Robotics Optimization (ICCAIRO)*, 2019, pp. 142–147, doi: 10.1109/ICCAIRO47923.2019.00030.

[87] G. Feng, Z. Hu, S. Chen, and F. Wu, "Energy-efficient and high-throughput FPGA-based accelerator for Convolutional Neural Networks," in *Proceedings of the 2016 13th IEEE International Conference on Solid-State and Integrated Circuit Technology (ICSICT)*, 2016, pp. 624–626.

[88] Y. Yoshimoto, D. Shuto, and H. Tamukoh, "FPGA-enabled binarized convolutional neural networks toward real-time embedded object recognition system for service robots," in *Proceedings of the 2019 IEEE International Circuits and Systems Symposium (ICSyS)*, 2019, pp. 1–5, doi: 10.1109/ICSyS47076.2019.8982469.

[89] K. Guo *et al.*, "Angel-eye: A complete design flow for mapping cnn onto embedded fpga," *IEEE Trans. Comput. Des. Integr. circuits Syst.*, vol. 37, no. 1, pp. 35–47, 2017.

[90] N. Paulino, J. C. Ferreira, and J. M. P. Cardoso, "Optimizing OpenCL code for performance on FPGA: k-Means case study with integer data sets," *IEEE Access*, vol. 8, pp. 152286–152304, 2020, doi: 10.1109/ACCESS.2020.3017552.

[91] L. A. Dias, J. C. Ferreira, and M. A. C. Fernandes, "Parallel implementation of K-means algorithm on FPGA," *IEEE Access*, vol. 8, pp. 41071–41084, 2020, doi: 10.1109/ACCESS.2020.2976900.

[92] T. Aoki, E. Hosoya, T. Otsuka, and A. Onozawa, "A novel hardware algorithm for real-time image recognition based on real AdaBoost classification," in *Proceedings*

of the 2012 IEEE International Symposium on Circuits and Systems (ISCAS), 2012, pp. 1119–1122, doi: 10.1109/ISCAS.2012.6271427.

[93] P. Jokic, S. Emery, and L. Benini, "BinaryEye: A 20 kfps streaming camera system on FPGA with real-time on-device image recognition using binary neural networks," in *Proceedings of the 2018 IEEE 13th International Symposium on Industrial Embedded Systems (SIES)*, 2018, pp. 1–7, doi: 10.1109/SIES.2018.8442108.

[94] B. Ramesh, E. Shea, and A. D. George, "Investigation of multicore SoCs for on-board feature detection and segmentation of images," in *Proceedings of the NAECON 2018 - IEEE National Aerospace and Electronics Conference*, 2018, pp. 375–381, doi: 10.1109/NAECON.2018.8556637.

[95] S. Prabhu, V. Khopkar, S. Nivendkar, O. Satpute, and V. Jyotinagar, "Object detection and classification using GPU acceleration," in *Proceedings of the International Conference On Computational Vision and Bio Inspired Computing*, 2019, pp. 161–170.

[96] C. Oh, S. Yi, and Y. Yi, "Real-time face detection in Full HD images exploiting both embedded CPU and GPU," in *Proceedings of the 2015 IEEE International Conference on Multimedia and Expo (ICME)*, 2015, pp. 1–6, doi: 10.1109/ICME.2015.7177522.

[97] V. Mutneja and S. Singh, "GPU accelerated face detection from low resolution surveillance videos using motion and skin color segmentation," *Optik (Stuttg).*, vol. 157, pp. 1155–1165, 2018.

[98] M. R. Ikbal, M. Fayez, M. M. Fouad, and I. Katib, "Fast implementation of face detection using LPB classifier on GPGPUs," in *Proceedings of the Intelligent Computing-Proceedings of the Computing Conference*, 2019, pp. 1036–1047.

[99] Y. Lee, C. Jang, and H. Kim, "Accelerating a computer vision algorithm on a mobile SoC using CPU-GPU co-processing: A case study on face detection," in *Proceedings of the International Conference on Mobile Software Engineering and Systems*, 2016, pp. 70–76.

[100] A. Yazdanbakhsh, J. Park, H. Sharma, P. Lotfi-Kamran, and H. Esmaeilzadeh, "Neural acceleration for gpu throughput processors," in *Proceedings of the 48th international symposium on microarchitecture*, 2015, pp. 482–493.

[101] X. Zhang, N. Gu, and H. Ye, "Multi-gpu based recurrent neural network language model training," in *Proceedings of the International Conference of Pioneering Computer Scientists, Engineers and Educators*, 2016, pp. 484–493.

[102] A. Prakash, N. Ramakrishnan, K. Garg, and T. Srikanthan, "Accelerating computer vision algorithms on heterogeneous edge computing platforms," in *Proceedings of the 2020 IEEE Workshop on Signal Processing Systems (SiPS)*, 2020, pp. 1–6, doi: 10.1109/SiPS50750.2020.9195221.

[103] F. Conti, D. Rossi, A. Pullini, I. Loi, and L. Benini, "PULP: A ultra-low power parallel accelerator for energy-efficient and flexible embedded vision," *J. Signal Process. Syst.*, vol. 84, no. 3, pp. 339–354, 2016.

[104] D. Jeon *et al.*, "A 23-mW face recognition processor with mostly-read 5T memory in 40-nm CMOS," *IEEE J. Solid-State Circuits*, vol. 52, no. 6, pp. 1628–1642, 2017, doi: 10.1109/JSSC.2017.2661838.

[105] N. Zheng and P. Mazumder, "A low-power hardware architecture for on-line supervised learning in multi-layer spiking neural networks," in *Proceedings of the 2018 IEEE International Symposium on Circuits and Systems (ISCAS)*, 2018, pp. 1–5, doi: 10.1109/ISCAS.2018.8351516.

[106] G. K. Chen, R. Kumar, H. E. Sumbul, P. C. Knag, and R. K. Krishnamurthy, "A 4096-Neuron 1M-Synapse 3.8-pJ/SOP spiking neural network with on-chip STDP learning and sparse weights in 10-nm FinFET CMOS," *IEEE J. Solid-State Circuits*, vol. 54, no. 4, pp. 992–1002, 2019, doi: 10.1109/JSSC.2018.2884901.

[107] C. Frenkel, J.-D. Legat, and D. Bol, "MorphIC: A 65-nm 738k-Synapse/mm^2 quad-core binary-weight digital neuromorphic processor with stochastic spike-

driven online learning," *IEEE Trans. Biomed. Circuits Syst.*, vol. 13, no. 5, pp. 999–1010, 2019, doi: 10.1109/TBCAS.2019.2928793.

[108] H. Kim, H. Tang, W. Choi, and J. Park, "An energy-quality scalable stdp based sparse coding processor with on-chip learning capability," *IEEE Trans. Biomed. Circuits Syst.*, vol. 14, no. 1, pp. 125–137, 2020, doi: 10.1109/TBCAS.2019.2963676.

[109] P.-Y. Chuang, P.-Y. Tan, C.-W. Wu, and J.-M. Lu, "A 90nm 103.14 TOPS/W binary-weight spiking neural network CMOS ASIC for real-time object classification," in *Proceedings of the 2020 57th ACM/IEEE Design Automation Conference (DAC)*, 2020, pp. 1–6, doi: 10.1109/DAC18072.2020.9218714.

[110] G. S. Nagpal, G. Singh, J. Singh, and N. Yadav, "Facial detection and recognition using OpenCV on Raspberry Pi Zero," in *Proceedings of the 2018 International Conference on Advances in Computing, Communication Control and Networking (ICACCCN)*, 2018, pp. 945–950, doi: 10.1109/ICACCCN.2018.8748389.

[111] H. Daryanavard and A. Harifi, "Implementing face detection system on UAV using Raspberry Pi platform," in *Proceedings of the Iranian Conference on Electrical Engineering (ICEE)*, 2018, pp. 1720–1723, doi: 10.1109/ICEE.2018.8472476.

[112] G. Zhang, K. Zhao, B. Wu, Y. Sun, L. Sun, and F. Liang, "A RISC-V based hardware accelerator designed for Yolo object detection system," in *Proceedings of the 2019 IEEE International Conference of Intelligent Applied Systems on Engineering (ICIASE)*, 2019, pp. 9–11, doi: 10.1109/ICIASE45644.2019.9074051.

[113] E. S. Lage, R. L. Santos, S. M. T. Junior, and F. Andreotti, "Low-cost IoT surveillance system using hardware-acceleration and convolutional neural networks," in *Proceedings of the 2019 IEEE 5th World Forum on Internet of Things (WF-IoT)*, 2019, pp. 931–936, doi: 10.1109/WF-IoT.2019.8767325.

2 Machine Learning for Testing of VLSI Circuit

Abhishek Choubey and Shruti Bhargava Choubey
Sreenidhi Institute of Science and Technology, Hyderabad, India

CONTENTS

2.1   INTRODUCTION

Digital and analog devices and circuits are basic electronic components in the widest type of electronic devices. In addition to consumer electronics markets, the IC industry more than ever is pressed by the enormous demand for medical, healthcare, automotive, or security electronics [1]. Analog/random forest (RF) components are already present in more than 50% of the total IC shipments yearly; thus, their design, test, and validation are fundamental tasks to meet the stringent time-to-market constraints and production costs [2].

The complexity of components and scaling of a device has led to much effort in testing Very Large Scale Integrated (VLSI). To solve the gradually prominent test cost issue of analog and digital devices, the machine learning (ML) approach has received a lot of consideration [3]. ML algorithms are presently incorporated in solutions to numerous VLSI testing issues. ML stances are at the height of current technological development, enhanced by the influence of digital cloud computing and the availability of vast data and storage. In modern manufacturing, ML-based approaches are now commonly used as they can provide effective results to complex issues that were thought insoluble a decade earlier. Nearly any domain spans the spectrum of their implementations and retains commitments that are restricted only by human imagination. In recent years, there have been various developments in ML and artificial intelligence (AI), including the emergence of deep learning

DOI: 10.1201/9781003201038-2

(DL) as useful to fields from system-level to logic-level programming, physical design, and testing and evaluation throughout electronic design [4].

ML methods offer reasonably decent concepts and quality of result, which find them appropriate to achieve detection of faults, as shown by numerous current studies. The performance of ML in all these areas can be due to the rapid solutions it offers since the decision-making method for predicting the parameters decreases to a simple feature estimation once the model is trained from the results [5]. The accessibility of graphic processing units and high-speed hardware that accelerate the needed computation is the additional substantial explanation behind the upthrust in ML, especially DL. In this chapter, we emphasize the issues of testing digital and analog hardware logic and deliberate the part of ML in resolving such issues. There are enough opportunities to explore applications of these methods and to create new solutions, while ML is revolutionary in this field.

The technology for electronics is around 50 years old. The architecture types of complementary metal-oxide-semiconductor (CMOS) are widely used to incorporate digital circuits with very high levels of integration on a broad scale (VLSI) [6]. IC-chips today consist of trillions of transistors on a single die. In addition to designing, checking for manufacturing flaws is a critical phase in the digital IC-chip development cycle as it impacts reliability, expense, and time to delivery. In order to approximate the chip's yield and to shed light on process variations, successful testing is also important. In the past three decades, various aspects of fault modeling, identification, diagnosis, built-in self-test, fault simulation, and design-for-testability (DFT) have been extensively deliberate, contributing to successful testable prototypes, fault-diagnosis algorithms, and test generation. A range of industrial tools for analog and digital circuit and device testing have been developed over the years [7].

In the design of analog and digital ICs, computer-aided design (CAD) tools are important. The massification balances the design effort of the digital and analog/RF circuits in consumer electronics. The lack of EDA, however, challenges the design of the custom ICs needed to generate state-of-the-art tailored equipment and creates obstacles to innovation. The implementation of automation mechanisms will reduce their development time dramatically while improving their efficiency at the same time [8].

However, design automation in analog IC design flow is far from being the norm, despite the enormous efforts made by the EDA community over the past few decades. Analog IC design is in sharp contrast to the digital IC design flow, where plenty of EDA tools are available and established. Analog ICs' nonlinear behavior, the increasing complexity observed in current applications, and the challenges in deep nanometer integration technologies only further increase the difficulties faced on analog/RF IC design and tests, placing additional pressure on analog/RF IC designers and EDA development teams [5]. Figure 2.1 illustrates the traditional design flow. When following the traditional flow, the designer repeats the flow for each targeted specification, even for the same issue. The experience, expertise, and instincts of the designer are of the utmost importance, but still, the existence of formalization greatly limits the distribution and reuse of knowledge.

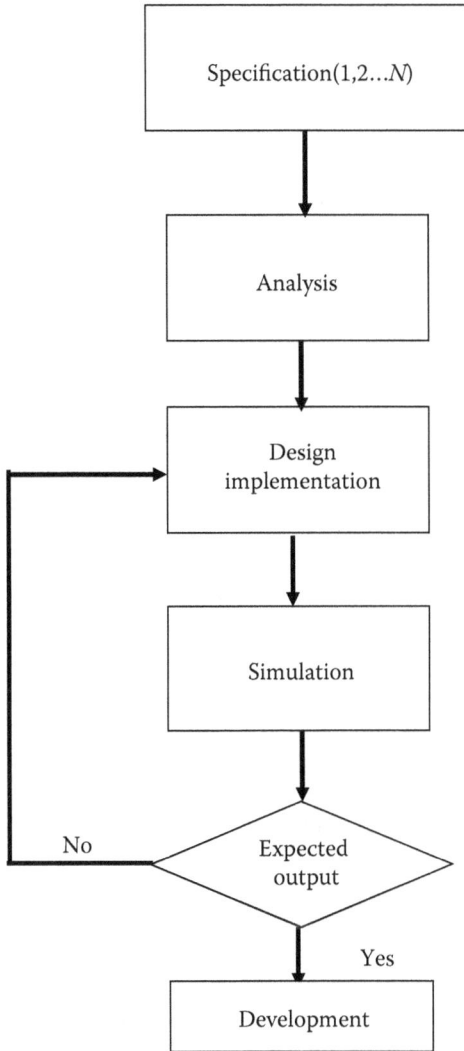

FIGURE 2.1   Traditional design flow.

2.2   MACHINE LEARNING OVERVIEW

As the term suggests, ML is an area that attempts to "learn" data and data information using the computational capacity of the new machine," namely computers. While its basic purpose, i.e. to derive from the data, is close to that of
mathematical learning, the combination of multiple procedures and the form of data
on which it is applicable is more sensitive. Its approach ranges over mathematical
learning, from the application of geometry and computational science to the mere
imitation of neural network biological mechanisms that have yet to be thoroughly
understood [9]. The data set can differ from plain tabular, organized data to

structured data such as videos and images that are more complex, to unstructured data such as graphs. The basic procedure for the implementation of ML algorithms is as follows:

i. In the preprocessing step, the appropriate features are first chosen and the data with these characteristics are then extracted from the raw data so that they can be used to discriminate between the various values of the target outputs. After that, multiple data cleansing and engineering features, such as the preference of key features, scaling, and generating the sample dataset for learning. Dimensionality reduction is performed.

ii. In the learning process, to extract the models from the training dataset, the required learning algorithms are chosen and executed. These models are associated with a certain underlying data principle. Cross-validation, assessment of performance, and hyper-parameters are performed to achieve the final versions.

iii. In the assessment phase, the final models are evaluated to measure their performance, the test dataset. In practice, it is possible to pick and configure assessment parameters according to various scenarios.

iv. The final models are analyzed in the simulation process to infer the predicted target output values for the new input results.

In compliance with the essence of the data types that ML processes, there are two main types of categories for learning that are digital and analog areas that are commonly explored: supervised learning and unsupervised learning [10]. The supervised learning approach utilizes the labeled knowledge to carry out model training. The final models to estimate the target classes are chosen and provide new input data; for unsupervised learning, on the other hand, techniques concentrate mainly on identifying the possible relations, characterizing the data when the data labels for all area groups are inaccessible through learning from them. In fact, the data marks being open is the main distinction between supervised learning and unsupervised learning.

Figure 2.2 shows a description of the most popular ML algorithms in digital and analog/RF testing. The prominent approaches used for unsupervised ML are Bayesian induction, clustering of k-means, and spectral clustering. Artificial neural networks (ANNs), decision trees, support vector machines (SVMs), RFs, and Bayesian networks are widespread methods that are used for supervised learning. Due to the availability of standard instruments, particularly ANNs and SVMs, supervised learning is more popular. The ML-based design produces responsive solutions [11]. The caveat is how many effective models are obtained. The application of the principles and methods of ML to digital and analog devices should come as no surprise, especially in the semiconductor industry, which produces billions of devices each year. The ML-based design flow is shown in Figure 2.3.

Many problems that can be solved using ML methods are common in semiconductor manufacturing, such as feature learning (to determine the optimal circuit that produces a certain output given a certain input), prediction (to determine chip efficiency indirectly), classification (to classify a device as functional or defective, given a set of test results and data from previous training). In numerous processes spanning the entire chip design and

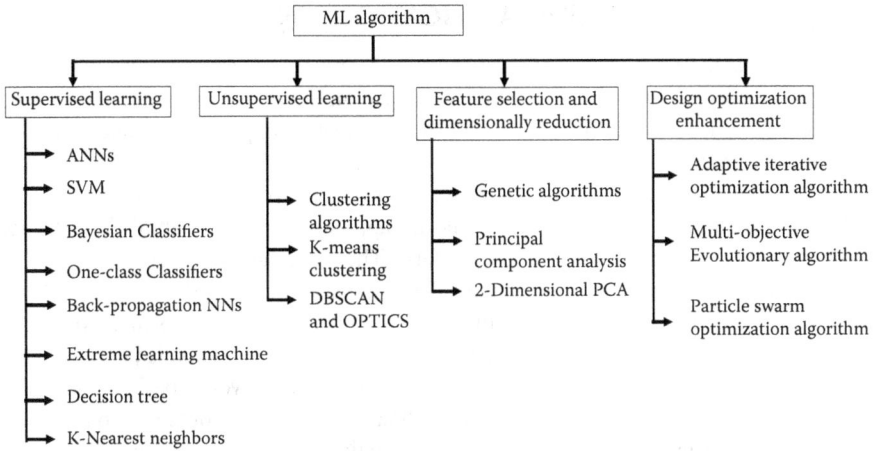

FIGURE 2.2   Summary of ML algorithm.

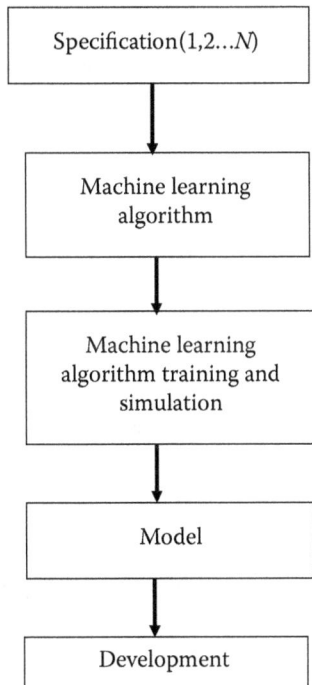

FIGURE 2.3   ML-based design flow.

production period, these issues appear, including physical design, routing, optimization of design, performance representation testing for fault detection, validation, health estimation, and security, to name only the most representative [12].

2.3 MACHINE LEARNING APPLICATIONS IN IC TESTING

The best procedure for analyzing analog, digital, and RF circuits is a specification-based test consisting of measuring the performance offered in the datasheet one by one and comparing it to its specifications. In order to greatly simplify and reduce the test cost and test process, an alternative test paradigm was proposed [13] in addition to the specification tests. The set of test responses is greatly simplified by using a common test setup (exploiting a low-cost ATE) and by applying the same stimuli in order to simultaneously evaluate different parameters. The alternate test flow principle is shown in Figure 2.4. As shown in Figure 2.4, it consists of two defect filters such as lenient and strict defect filters. Due to the correlation between variations in the process, performance characteristics, and test response, performance prediction is possible. The alternative test exploits the notion that the resulting reaction would be influenced by the same variety of circuit system parameters that often influence output with a proper stimulus range. Statistical research, in such a situation, would be necessary to expose the association between response and output from a wide sample of circuits, offering a prediction framework for the latter that avoids direct measuring.

Alternate studies demonstrate substantial benefits. The most important examples are the following

1. Allows, based on a single circuit reaction, prediction of multiple performances.
2. Using a standard and reasonably basic test setup for all output characteristics and also a common trigger, resulting in a substantial reduction in test time price and cost.
3. Both circuit stimulation and reaction tests requires less complicated and less inefficient devices.

The summary of ML in IC testing is shown in Table 2.1.

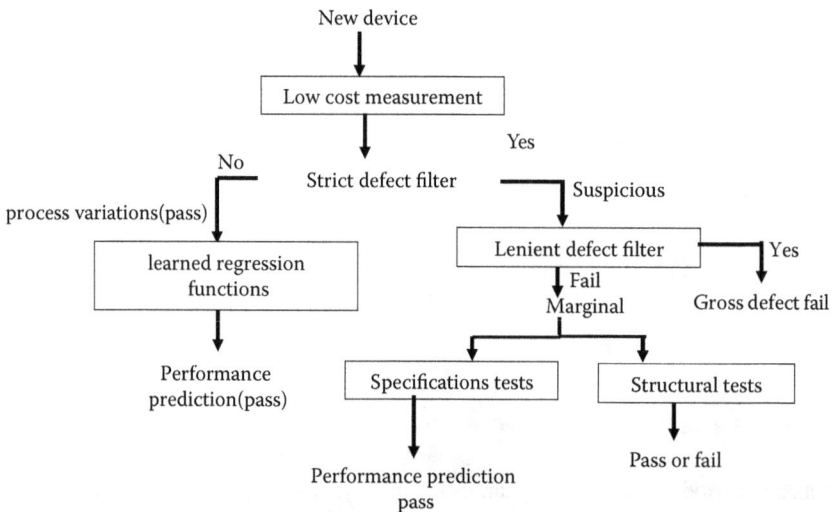

FIGURE 2.4 Alternate test flow principle.

TABLE 2.1

Summary ML in IC Testing

S.No.	Applications of ML in Test	Contribution	Reference
1	Alternate test	It is the replacement of the specification test with low cost.	[14]
2	Test compaction	PCA can classify to predict a fail or pass result of the tests.	[15]
3	Fault diagnosis and yield learning	Multi-class classifiers can be explored to map a set of features extracted on test data to one defect.	[16]
4	Post-manufacturing tuning	ML method can explore the mapping of a set of features extracted from test data.	[17]
5	Outlier detection	Clustering and class classifier algorithms can be explored to perform robust and quick outlier screening.	[18]
6	Adaptive test	Adaptive test based on fixed test process for die and wafer.	[19]
7	Test cost reduction through wafer-level spatial correlation modeling	ANN model can explore untested design in other places of the wafer.	[20]
8	Analog Test Metrics Estimation	ML model can explore analog test metrics estimation. A classifier is performed to separate a set of circuit instances in the space of process parameters.	[21]
9	Neuromorphic on-chip testers	It can be used to predict a learning machine on-chip or in the digital signal processor (DSP)	[22]
10	Other test-related tasks	ML method can be used for physical verification, functional verification, or speeding up IC laser trimming.	[23]

2.4 ML IN DIGITAL TESTING

The block diagram of digital circuit testing is shown in Figure 2.5. As shown in Figure 2.5, digital circuit testing consists of the combinational circuit (CUT) and sequential circuit, i.e. flip-flop and both reverse and directions, and output/input ports, which can feed the signal. In testing, there are two parts: test application and test generation. The block diagram of automatic test pattern generation (ATPG) is shown in Figure 2.6. As shown in Figure 2.6, it consists of CUT, input vector space is $2n$, and comparator. Two popular fault models are commonly used: transition faults and stuck-at faults. To determine whether they fail or pass the test, they are then compared to the predicted circuit outputs. The performance response is further evaluated if a test fails, the output response is additionally studied for diagnosis.

SVMs, ANNs, decision trees, Bayesian networks, and RFs are common methods used for supervised learning. Due to the accessibility of usual instruments,

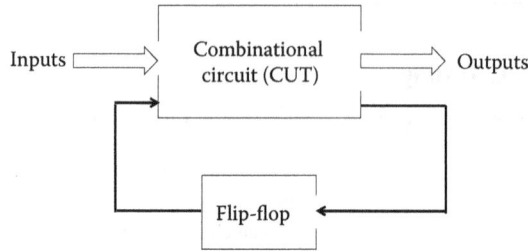

FIGURE 2.5 Block diagram of digital circuit testing.

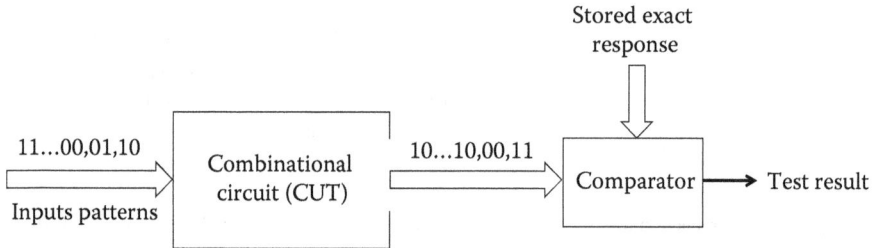

FIGURE 2.6 Block diagram of automatic test pattern generation (ATPG).

particularly SVMs and ANNs [24], supervised learning is more recent. While supervised learning is often chosen over unsupervised methods, labels are often not present or tough to obtain. Therefore, the method must rely on the type of available data. There are frequent prospects for the application of supervised learning in the field of digital logic testing. Table 2.2 shows several data sources that have been used or may be used for ML applications in the field of digital electronic testing.

   The technique of defect location uses a feedback mechanism called diagnosis. If IC technology decreases and the level of integration increases, the number and variety of bugs eventually increase, and the diversity and number of defects certainly escalate. Conventionally, defects are found using a physical-level approach called physical failure analysis (PFA). Defects can affect both memory and combinational components. In scan-chain diagnosis, the flip-flops are tested first and diagnosed for defects during the manufacturing test [11,31–35]. Thereafter, the faults in the rest of the circuit are diagnosed. Diagnosis is hierarchically performed. ML methods that have been used for diagnosis at different levels of circuit hierarchy are shown in Table 2.3.

2.5 ML IN ANALOG CIRCUIT TESTING

ANNs have recently become a viable alternative to techniques for numerical simulation, empirical methods, and empirical models. Those models will instantly be available for a pre-trained issue or to construct the solution; therefore, the designer may circumvent multiple costly simulations [38]. ML-based modeling has been used over the years at various levels (from a single unit to a large system) and for

TABLE 2.2

Summary of Digital Electronic Testing and ATPG with ML Techniques

S.No.		Contribution	Reference
1	Manufacturing test response	Identification of patterns of failure response of a range of defective ICs	[25,26]
2	Simulation	Testing of data with CAD tools	[27]
3	Historical data on diagnosis	The analysis of test failure response/syndromes	[28,29]
4	Circuit parameters	Number of output and input port	[5]
5	Circuit structure	The study of the structure of logical networks	[12,30]
6	Physical layout	Testing of layout	[24]
7	Test compression	In this technique, test cost is measured using test time and test data volume	[5]
8	Circuit testability	The analysis of test-point insertion and evaluation of the loss of fault coverage	[1]
9	Analysis of X- sensitivity	The analysis of loss of fault coverage in digital circuit	[12]
10	Test-point insertion	Handling a test problem	[30]
11	Timing analysis	Determining frequency of the clock	[5]

various purposes (analog, RF, and heterogeneous). The summary of ML approaches in analog and RF circuits is shown in Figure 2.7.

In order to achieve the functional versions of analog circuits, SVMs and ANN-based methods are widely used. In analog circuit modeling, SVMs are typically preferred because they are not easily stuck at the local minimum and suffer from the curse of dimensionality as the data points are calculated taking the measurements into account. A separate ANN-based modeling application is presented in which analog circuit power consumption is modeled and then calculated using empirical-based ANN instead of achieving output using input parameters [39]. The principle behind this study is to estimate the mathematical definition of energy consumption as a function of varied energy consumption. First, through a measurement set-up involving a PC, analog circuit voltage measurements are conducted to produce various input patterns and save the power data. Second, the data collected is used to train the ANN to obtain the power consumption of a continuous mathematical function. Three layers are used in the neural networks: one input, one secret, and one output layer. Linear and sigmoid are the activation functions for the output layer and hidden layer, respectively. Training based on backpropagation is used. It is paired with a data-flow–based generalized functional model of the circuit until the power model is obtained [40]. Therefore, the outputs of all circuits and the instantaneous power usage are collected, which makes it possible to predict circuit output without any analytical tests being carried out. Any analog circuit inputs parameters using neural networks. The method suggested is general and also sufficient for heterogeneous systems. In addition, through the proposed strategy, online power consumption calculations can be carried out.

TABLE 2.3

Summary of ML Techniques for a Different Levels of Circuits

S.No.	Test	Article	Main Research Contribution
1	Wafer-level diagnosis	[13,36,37]	i. Applying SVM approach using historical data set to find automated die inking ii. Identifying defect clusters using failure dataset and clustering approach iii. Applying statistical correlation method to the correction of failure and parameters using the failure dataset
2	Scan-chain diagnosis	[10,27]	i. Using Bayesian approach with failure dataset to target hard-to-model faults
3	Fault diagnosis: pre-processing	[31,32]	i. Using classification technique with failure dataset to the regulation of test data volume ii. Using random forest method with failure dataset to find inferring diagnostic efficiency
4	Fault diagnosis: Defect postprocessing identification	[24,33]	i. Defect classification using ANN approach with the simulated dataset ii. Using decision tree approach with simulated and failure dataset to identifying bridging defects
	Improving diagnostic resolution	[25,26]	i. Transient and intermittent faults identification using Bayesian network with simulated dataset ii. Improving diagnostic resolution fault detection using the simulated dataset
5	Volume diagnosis	[11,25]	i. Volume diagnosis of unmodeled faults by SVM approach with the simulated dataset ii. Volume diagnosis for root cause identification by MLE and Bayesian network algorithm with simulated dataset iii. Identification of systematic defects with failure dataset using clustering technique
6	Board-level diagnosis	[28,29]	i. Fault isolation with past data using ANN/SVM decision tree technique ii. The computation of missing syndrome using the Bayes technique
7	Test compression	[5]	Test cost optimization with simulated dataset using SVR algorithm
8	Circuit testability	[12,30]	i. Prediction of X-sensitivity with structural feature and simulated dataset using SVR approach ii. Test point insertion with structural feature and simulated dataset using GCN technique
9	Timing analysis	[34]	Based on PSN with simulated dataset using multiple tools

```
                        ┌─────────────────┐
                        │  ML-in circuit  │
                        │    modeling     │
                        └────────┬────────┘
                                 │
    ┌──────────┬─────────────────┼─────────────────┬──────────────┐
    │          │                 │                 │              │
┌─────────┐┌─────────┐     ┌──────────┐      ┌──────────┐  ┌──────────┐
│ Analog  ││   RF    │     │    IC    │      │Analog/RF │  │Analog IC │
│ circuit ││ circuit │     │ circuit  │      │ layout   │  │  fault   │
│modeling ││modeling │     │synthesis │      │synthesis │  │ testing  │
└─────────┘└─────────┘     └────┬─────┘      └──────────┘  │   and    │
                               │                          │diagnosis │
                               │                          └──────────┘
                 ┌───────┬─────┼──────┬──────────┐
            ┌─────────┐┌─────────┐┌──────────┐┌──────────┐
            │  Fault  ││  Pre-   ││  Post-   ││  Volume  │
            │diagnosis││diagnosis││diagnosis ││diagnosis │
            └─────────┘└─────────┘└──────────┘└──────────┘
```

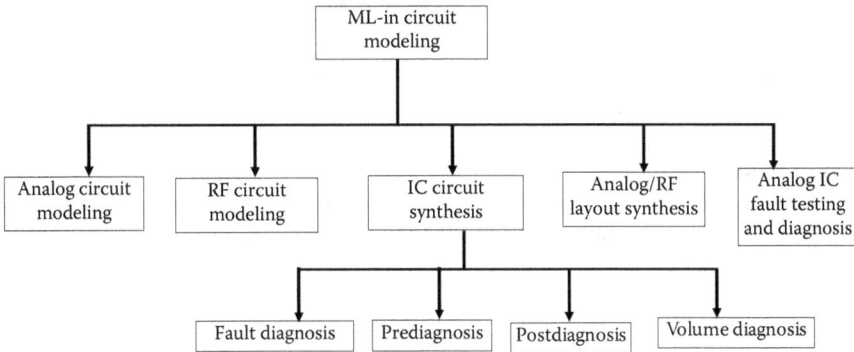

FIGURE 2.7   ML in analog circuit modeling.

For RF and microwave modeling and architecture, neural networks have been used, where passive/active component/circuit models based on ANN are then used at higher levels of design. Therefore, in contrast to the costly system, a reliable response of the whole system can be achieved within a shorter period. From theory to implementation, modeling and architecture was discussed. The authors noted that neural networks are desirable alternatives to traditional approaches, such as computationally costly numerical simulation methods or analytical methods that may be hard to achieve for new devices [41]. They included examples where neural networks are used in printed circuit boards (PCBs), coplanar waveguide (CPW) discontinuities, and MESFETs to model signal propagation delays of a VLSI interconnect network, all from previous works in the literature.

To address the problem of time efficiency, ML-based synthesis approaches have become popular. The concept behind the use of ML in circuit synthesis is to replace simulations with practical simulations. Model(s) were developed with ML techniques; therefore, during the synthesis process, the excessive number of simulations can be prevented. The concept behind the use of ML in circuit synthesis is to replace the simulations with the functional model(s) produced by ML techniques, so during the synthesis process, the unnecessary number of simulations can be avoided. Optimization-based circuit synthesis, which uses an optimization approach to explore the design space, is the most known method for automating circuit synthesis. The design time in which many nature-inspired algorithms (evolutionary, particle swarm, reinforcement learning, etc.) are used to scan the design space to find an optimal solution for a given circuit problem is probably accelerated by analog/RF circuit optimization tools. One use of ML-based optimization techniques is to leverage optimization tools for dataset creation [40,42,43].

The widespread application of ML to numerous fields, including the automation of analog/RF IC layout, opens new perspectives for the creation of push-button technologies that combine legacy data or expert design insights simultaneously in a way that was not feasible in previous generations of EDA software. These recent layout automation ML technologies vary from placement software to drafters for routing, but also pre- and post-placement analysis. The several ML approaches, analog and RF circuits, are summarized in Table 2.4.

TABLE 2.4
Summary of ML Techniques for Analog and RF Circuits

	Article	Main Research Contribution
Analog and RF application device		
RF-CPW components	[42]	ANN-EM can classify modeling of CPW components.
RF-microwave components, MESFET and HMT	[43]	ANN (several) can be explored for CAD-based microwave design and nonlinear modeling of microwave design.
RF-MESFET	[11,43]	ANN (WNN-MLP) can classify the large non-linear power circuits and transistor.
Analog and AMS circuit CMOS	[40,44]	SVMs and ANN (ITDN) can be employed for the model of AMS circuit and feasible design space selection.
IC circuit synthesis applications		
Analog circuit synthesis	[45,46]	i. RL(L2DC) can explore sizing of analog circuits. ii. ANN can classify and evaluate the hyperparameters of analog circuit synthesis. iii. ANN model can generate better FOMs for Op-Amps. iv. ANN (GRP+MLP) can explore sizing of analog building blocks. v. DL+RELU model can estimate multiple performance estimation of Op-amps.
Analog circuit optimization	[8,41,47]	i. LCB+GP+NSGA-II+MOEA/D+BNN model can classify accurate and fast optimization to acquire better PFs. ii. GA+ANN can explore accurate and fast layout-aware Op-Amp synthesis. iii. Large-scale data mining can explore KNN with boosted regression. iv. SPEA2 and ANN can classify efficient optimization.
Performance space exploration	[6,38]	i. Automatic generation of POFs is done by polynomial regression. ii. Sparse regression and ANN-based text mining can explore on the Internet via knowledge harvesting. iii. Bayesian regression can explore large-scale design space.
Layout automation		
Routing	[48]	i. Assist A* search using ANN is used as VAE with the semi-supervised training model. ii. ANN is adopted as the classifier due to the complexity of data.
Building block identification	[49]	This model alternates to subgraph isomorphism.
Well definition	[50]	Based on supervised training dataset with GAN, model reproduces legacy data patterns.
Placement	[3]	i. ANN with four hidden layers model using unsupervised training for sizing data. ii. Weights assigned by hill-climbing training data using ANN for discrete WxH layout plane. iii. Supervised training with ANN used to reproduce legacy data patterns.

TABLE 2.4 (Continued)
Summary of ML Techniques for Analog and RF Circuits

Article		Main Research Contribution
Analog IC fault calibration, diagnosis and testing		
One-shot calibration	[51]	ANN model used as post-fabrication calibration in a single calibration phase to counter output deviation due to fabrication.
Test set compression	[52]	ONN technique used NSGA optimization to diagnose the CUT.
Post-layout modeling	[1]	BMF approach can explore for efficient post-silicon modeling.
Fault diagnosis	[7,53]	i. DBN can classify the feature engineering with end-to-end learning. ii. LDA to enhance class separability while compressing the space of the features using the Fisher DT LDA approach. iii. Haarwavelet+MLP+kPCA to reduce dimensionally of features. iv. ANN+ Wavelet+ Dictionary+PCA to reduce the dimensionality of features. v. SVM approach with defect filter identifies soft and hard faults.
Remaining useful performance	[54]	Kernel RVM model can classify the trajectories of the circuit's health.

2.6 ML IN MASK SYNTHESIS AND PHYSICAL PLACEMENT

The effectiveness of hotspot detection is highly dependent on removing the layout function and choosing the model. The layout function reflects the layout attributes that are important for identification in order to determine hotspots and non-hotspots. The hotspot identification challenge in the physical design and verification phases is to identify hotspots with the quick turnaround time on a given architecture. Using dynamic models of lithography, traditional lithography simulation obtains pattern images [22]. The precise simulation of full-chip lithography is computationally costly and therefore does not provide fast input to lead the early physical design phases. In bridging the broad distance between modeling and process-aware physical design methods, hotspot detection plays an important role. A lot of ML-based hotspot detection work has been completed. ML methods create a regression. A model is based on a collection of data from preparation. This methodology can recognize previously unknown hotspots naturally. It will produce false alerts, however, which are not true hotspots.

Principle component analysis (PCA) is useful for data reduction and feature extraction. Adding SVM with PCA can significantly expand the accuracy of detection. Recently, a classification model based on boosting was proposed. The use of the weakly nonlinear learning model was identified to correctly detect hotspots with a low false alert via a simpler interface function.

A new placement flow with automated data-path abstraction was suggested in which all first-order, relevant data paths were evaluated and rated, and configured

TABLE 2.5
Summary of ML Method for Mask Synthesis

	Article	Main Research Contribution
Sub-Resolution Assist Features (SRAFs)	[55]	Supervised learning can classify efficiently approximate model-based SRAF and predict whether SRAFs can cover pixels.
Optical Proximity Correction	[56]	Bayes model (HBM) and generalized linear mixed model (GLMM) can explore OPC problem with CCAS feature like convex, line-end edge, different edge types, etc.
Clock optimization	[57]	A learning-based model that can be classified for latch optimization is proposed.
Lithography hotspot detection	[58]	The deep convolutional neural network (CNN) can classify the feature of lithography hotspot detection.

together with general-purpose wire-length guided placement. In the training stage, in the original netlist, to classify and evaluate the datapath patterns, SVM and ANN techniques were combined to build compact and run-time-efficient models. An error-tolerant procedure was combined with a special selection of working sets in the SVM model. ANN works to achieve a high-dimensional decision diagram-like data structure by configuring complex neuron networks given training samples and decision hints. For both ANN and SVM, the optimization goal was to maximize the accuracy of datapath and non-datapath pattern assessments.

Several resolution enhancement methods (RETs), such as source mask co-optimization, optical proximity correction (OPC), and sub-resolution assist functions (SRAFs), become a requirement when the technology nodes scale to the limit of light wavelength. Supervised learning promises to effectively produce estimated model-based SRAF to retain high quality and maximize turn-around time. Training knowledge comes from the model-based generation of SRAF [55]. The model is trained to predict whether SRAFs can cover a pixel. With the guidance of the model, the actual SRAFs are created, according to SRAF laws. Focused on learning, the generation of SRAF is formulated into a classification problem where feature vectors are extracted with CCAS and both logistic regression and SVM are adopted by the kernel models. Summary of ML method for mask synthesis is shown in Table 2.5.

2.7   CONCLUSION

We also looked at numerous issues that emerge in the research and diagnosis of VLSI circuits where ML has been applied. In managing the complexities of the problem, they have outperformed conventional heuristic-based methods and given practical solutions much faster. ML-based approaches have recently been used successfully in many applications, where increased learning capacity makes them unique in solving any complex/nonlinear problem. IC architecture has also bene-fited from ML techniques at various stages of design, from device simulation to fabricated research. The attempts in device/circuit/system modeling have been aimed at generating precise models at various abstraction levels and replacing the simulator with these models especially in RF applications; thus, it is possible to minimize human effort and design time.

In the future, solutions to automatic feature and data generation innovation problems will set off further implementation of ML methods to other chip testing issues. Useless to mention, ample scope remnants for data generation and the development of digital circuit illustration procedures that will enrich industrial as well as an academic study in the field of ML-guided testing.

ACKNOWLEDGMENT

The authors would like to thank the Sreenidhi Institute of Science and Technology Hyderabad for providing the infrastructure to conduct the research.

REFERENCES

[1] Wang, F. et al., "Bayesian model fusion: Large-scale performance modeling of analog and mixed-signal circuits by reusing early-stage data," *IEEE Trans. Comput.-Aided Design Integr. Circuits Syst.*, vol. 35, no. 8, pp. 1255–1268, 2016.

[2] Afacan, E., Lourenço, N., Martins, R., & Dündar, G., "Review: Machine learning techniques in analog/RF integrated circuit design, synthesis, layout, and test," *Integration*, vol. 77, pp. 113–130, 2020.

[3] Gusmao, A. et al., "Semi-supervised artificial neural networks towards analog IC placement recommender," in Proc. IEEE International Symposium on Circuits and Systems (ISCAS), 2020, pp. 1–5.

[4] Pradhan, M., & Bhattacharya, B. B., "A survey of digital circuit testing in the light of machine Learning," *WIREs Data Mining Knowl. Discov.*, vol. 11, 2020.

[5] Li, Z., Colburn, J. E., Pagalone, V., Narayanun, K., & Chakrabarty, K., "Test-cost optimization in a scan-compression architecture using support-vector regression," in Proc. VTS, 2017, pp. 1–6.

[6] Pan, P.-C., Huang, C.-C., & Chen, H.-M., "Late breaking results: An efficient learning-based approach for performance exploration on analog and RF circuit synthesis," in Proc. 56th ACM/IEEE Design Automation Conference (DAC), 2019, pp. 1–2.

[7] Xiao, Y., & He, Y., "A novel approach for analog fault diagnosis based on neural networks and improved kernel PCA," *Neurocomputing*, vol. 74, no. 7, pp. 1102–1115, 2011.

[8] Islamoglu, G. et al., "Artificial neural network assisted analog IC sizing tool," in Proc. 16th International Conference on Synthesis, Modeling, Analysis and Simulation Methods and Applications to Circuit Design (SMACD), IEEE, 2019, pp. 9–12.

[9] Murphy, K. P., "Machine learning a probabilistic perspective (adaptive computation and machine learning series)," *MIT Press*, 2012.

[10] Huang, Y., Benware, B., Klingenberg, R., Tang, H., Dsouza, J., & Cheng, W., "Scan chain diagnosis based on unsupervised machine learning," in Proc. ATS, 2017, pp. 225–230.

[11] Cheng, W., Tian, Y., & Reddy, S. M., "Volume diagnosis data mining," In Proc. Proc. ETS, 2017, pp. 1–10.

[12] Pradhan, M., Bhattacharya, B. B., Chakrabarty, K., & Bhattacharya, B. B., "Predicting X-sensitivity of circuit-inputs on testcoverage: A machine-learning

approach," *IEEE Trans. Comput.-Aided Design*, vol. 38, no. 12, pp. 2343–2356, December 2019.

[13] Tikkanen, J., Siatkowski, S., Sumikawa, N., Wang, L., & Abadir, M. S., "Yield optimization using advanced statistical correlation methods," in Proc. ITC, 2014, pp. 1–10.

[14] Hsiao, S.-W. et al., "Analog sensor-based testing of phase-locked loop dynamic performance parameters," in Proc. IEEE Asian Test Symp., 2013, pp. 50–55.

[15] Sumikawa, N. et al., "An experiment of burn-in time reduction based on parametric test analysis," in Proc. Proc. IEEE Int. Test Conf., 2012.

[16] Vasan, A. S. S. et al., "Diagnostics and prognostics method for analog electronic circuits," *IEEE Trans. Ind. Electron.*, vol. 60, no. 11, pp. 5277–5291, 2013.

[17] Andraud, M. et al., "One-shot non-intrusive calibration against process variations for analog/RF circuits," *IEEE Trans. Circuits Syst. I, Reg. Papers*, vol. 63, no. 11, pp. 2022–2035, 2016.

[18] Lin, F. et al., "AdaTest: An efficient statistical test framework for test escape screening," in Proc. IEEE Int. Test Conf., 2015.

[19] Stratigopoulos, H.-G., & Streitwieser, C., "Adaptive test with test escape estimation for mixed-signal ICs," *IEEE Trans. Computer.-Aided Design Integr. Circuits Syst.*, vol, 37, no. 10, pp. 2125–2138, 2018.

[20] Huang, K. et al., "Low-cost analog/RF IC testing through combined intra-and inter-die correlation models," *IEEE Des. Test. Comput.*, vol. 32, no. 1, pp. 53–60, 2015.

[21] Stratigopoulos, H.-G., "Test metrics model for analog test development," *IEEE Trans. Comput.-Aided Design Integr. Circuits Syst.*, vol. 31, no. 7, pp. 1116–1128, 2012.

[22] Banerjee, D. et al., "Real-time use-aware adaptive RF transceiver systems for energy efficiency under BER constraints," *IEEE Trans.Comput.-Aided Design Integr. Circuits Syst.*, vol. 34, no. 8, pp. 1209–1222, 2015.

[23] Wang, L.-C., "Experience of data analytics in EDA and test principles, promises, and challenges," *IEEE Trans. Comput.-Aided Design Integr. Circuits Syst.*, vol. 36, no. 6, pp. 885–898, 2017.

[24] Nelson, J. E., Tam, W. C., & Blanton, R. D., "Automatic classification of bridge defects," in Proc. ITC, 2010, pp. 1–10.

[25] Huisman, L. M., Kassab, M., & Pastel, L., "Data mining integrated circuit fails with fail commonalities," in Proc. ITC, 2004.

[26] Xue, Y., Poku, O., Li, X., & Blanton, R. D., "Padre: Physically-aware diagnostic resolution enhancement," in Proc. ITC, 2013, pp. 1–10.

[27] Chern, M., Lee, S.-W., Huang, S.-Y., Huang, Y., Veda, G., Tsai, K.-H. H., & Cheng, W.-T., "Improving scan chain diagnostic accuracy using multi-stage artificial neural networks," in Proc. ASPDAC, 2019, pp. 341–346. New York, NY: ACM.

[28] Sun, Z., Jiang, L., Xu, Q., Zhang, Z., Wang, Z., & Gu, X., "Agentdiag: An agent-assisted diagnostic framework for boardlevel functional failures," in Proc. ITC, 2013, pp. 1–8.

[29] Ye, F., Chakrabarty, K., Zhang, Z., & Gu, X., "Self-learning and adaptive board-level functional fault diagnosis," in Proc. ASPDAC, 2015, pp. 294–301.

[30] Ma, Y., Ren, H., Khailany, B., Sikka, H., Luo, L., Natarajan, K., & Yu, B., "High performance graph convolutional networks with applications in testability analysis," in Proc. DAC, 2019, pp. 1–6.

[31] Wang, H., Poku, O., Yu, X., Liu, S., Komara, I., & Blanton, R. D., "Test-data volume optimization for diagnosis," in Proc. DAC, 2012, pp. 567–572.

[32] Huang, Q., Fang, C., Mittal, S., & Blanton, R. D. S., "Improving diagnosis efficiency via machine learning," in Proc. ITC, 2018, pp. 1–10.

[33] Gómez, L. R., & Wunderlich, H., "A neural-network-based fault classifier," in Proc. ATS, 2016, pp. 144–149.

[34] Ye, F., Firouzi, F., Yang, Y., Chakrabarty, K., & Tahoori, M. B., "On-chip droop-induced circuit delay prediction based on support-vector machines," *IEEE Trans. Comput.-Aided Design*, vol. 35, no. 4, pp. 665–678, April 2016.

[35] Zhang, Q.-J., Gupta, K. C., & Devabhaktuni, V. K., "Artificial neural networks for RF and microwave design-from theory to practice," *IEEE Trans. Microw. Theory Tech.*, vol. 51, no. 4, pp. 1339–1350, 2003.

[36] Sumikawa, N., Nero, M., & Wang, L., "Kernel based clustering for quality improvement and excursion detection," in Proc. ITC, 2017, pp. 1–10.

[37] Xanthopoulos, C., Sarson, P., Reiter, H., & Makris, Y., "Automated die inking: A pattern recognition-based approach," in Proc. ITC, 2017, pp. 1–6.

[38] Kaya, E., Afacan, E., & Dundar, G., "An analog/RF circuit synthesis and design assistant tool for analog IP: DATA-IP," in Proc. 15th International Conference on Synthesis, Modeling, Analysis and Simulation Methods and Applications to Circuit Design (SMACD), IEEE, 2018, pp. 1–9.

[39] Stratigopoulos, H.-G., "Machine learning pplications in IC testing," in Proc. 2018 23rd IEEE ETS, 2018.

[40] Grabmann, M., Feldhoff, F., & Gläser, G., "Power to the model: Generating energy-aware mixed-signal models using machine learning," in Proc. 16th International Conference on Synthesis, Modeling, Analysis and Simulation Methods and Applications to Circuit Design (SMACD), IEEE, 2019, pp. 5–8.

[41] Liu, H. et al., "Remembrance of circuits past: Macro modeling by data mining in large analog design spaces," in Proc. 39th Annual Design Automation Conference, 2002, pp. 437–442.

[42] Watson, P. M., & Gupta, K. C., "Design and optimization of CPW circuits using EM-ANN models for CPW components," *IEEE Trans. Microw. Theory Tech.*, vol. 45, no. 12, pp. 2515–2523, 1997.

[43] Ceperic, V., & Baric, A., "Modeling of analog circuits by using support vector regressionmachines," in Proc. 2004 11th IEEE International Conference on Electronics, Circuits and Systems (ICECS), IEEE, 2004, pp. 391–394.

[44] Harkouss, Y. et al., "The use of artificial neural networks in nonlinear microwave devices and circuits modeling: An application to telecommunication systemdesign (invited article)," *Int. J. RF and Microw. Comput.-Aided Eng.*, vol. 9, no. 3, pp. 198–215, 1999.

[45] Vural, R. et al., "Process independent automated sizing methodology for current steering dac," *Int. J. Electron.*, vol. 102, no. 10, pp. 1713–1734, 2015.

[46] Bhatia, V., Pandey, N., & Bhattacharyya, A., "Modelling and design of inverter threshold quantization based current comparator using artificial neural networks," *Int. J. Electr. Comput. Eng.*, vol. 6, no. 1, pp. 2088–8708, 2016.

[47] Wolfe, G., & Vemuri, R., "Extraction and use of neural network models in automated synthesis of operational amplifiers," *IEEE Trans. Comput.-Aided Design Integr. Circuits Syst.*, vol. 22, no. 2, pp. 198–212, 2003.

[48] Zhu, K., et al., "Geniusroute: A new analog routing paradigm using generative neural network guidance," in Proc. International Conference on Computer Aided Design (ICCAD), 2019.

[49] Kunal, K. et al., "Align: Open-source analog layout automation from the ground up," in Proc. 56th Annual Design Automation Conference (DAC), 2019, pp. 1–4.

[50] Xu, B. et al., "Wellgan: Generative-adversarial-network-guided well generation for analog/mixed-signal circuit layout," in Proc. 56th ACM/IEEE Design Automation Conference (DAC), IEEE, 2019, pp. 1–6.

[51] Andraud, M., Stratigopoulos, H., & Simeu, E., "One-shot non-intrusive calibration against process variations for analog/RF circuits," *IEEE Trans. Circuits Syst. I Regul. Pap.*, vol. 63, no. 11, pp. 2022–2035, 2016.

[52] Stratigopoulos, H. et al., "RF specification test compaction using learning machines," IEEE Trans. Very Large Scale Integr. VLSI Syst., vol. 18, no. 6, pp. 998–1002, 2010.

[53] Binu, D., & Kariyappa, B. S., "RideNN: A new rider optimization algorithm-based neural network for fault diagnosis in analog circuits," *IEEE Trans. Instrum. Meas.*, vol. 68, no. 1, pp. 2–26, 2019.

[54] Zhang, C. et al., "A multiple heterogeneous kernel RVM approach for analog circuit fault prognostic," *Cluster Comput.*, vol. 22, no. 2, pp. 3849–3861, 2019.

[55] Xu, X., Matsunawa, T., Nojima, S., Kodama, C., Kotani, T., & Pan, D. Z., "A machine learning based framework for sub-resolution assist feature generation," in Proc. ACM International Symposium on Physical Design (ISPD), 2016, pp. 161–168.

[56] Matsunawa, T., Yu, B., & Pan, D. Z., "Optical proximity correction with hierarchical bayes model," *J. Micro Nanolithogr. MEMS MOEMS*, vol. 15, no. 2, pp. 021,009–021,009, 2016.

[57] Ward, S. I., Viswanathan, N., Zhou, N. Y., Sze, C. C., Li, Z., Alpert, C. J., & Pan, D. Z., "Clock power minimization using structured latch templates and decision tree induction," in Proc. IEEE/ACM International Conference on Computer-Aided Design (ICCAD), 2013, pp. 599–606.

[58] Yu, Y. T., Chan, Y. C., Sinha, S., Jiang, I. H. R., & Chiang, C., "Accurate process-hotspot detection using critical design rule extraction," in Proc. ACM/IEEE Design Automation Conference (DAC), 2012.

3 Online Checkers to Detect Hardware Trojans in AES Hardware Accelerators

Sree Ranjani Rajendran[1] and
Rajat Subhra Chakraborty[2]
[1]RISE Lab, Indian Institute of Technology Madras, Chennai,
Tamil Nadu, India
[2]Indian Institute of Technology Kharagpur, Kharagpur,
West Bengal, India

CONTENTS

3.1 INTRODUCTION: BACKGROUND AND DRIVING FORCES

ICs embedded in crypto-devices should be trustworthy in nature. However, these ICs may themselves be compromised. IC manufacturing industries have to spend millions of dollars to upgrade their fabrication unit; hence, they outsource the fabrication process to a third party. The attacker in the untrustworthy fabrication units can easily access the design by reverse engineering and malicious circuit (a.k.a hardware Trojan, or HT) to the original design [1]. It is reported that hardware attacks cause a loss of $4 billion annually to the semiconductor industry. These hardware-related security issues directly spoil the efficiency of the architectures where hardware plays a major role in implementation, such as the Internet of Medical Things (IoMT) applications [2]. Trojan taxonomy, operation of HTs, the emerging methods of Trojan detection and prevention techniques and their major challenges are discussed in Ranjani and Nirmala Devi, and Nixon et al. [3,4].

DOI: 10.1201/9781003201038-3

The design stages of the entire IC design flow is insecure for various vulnerabilities. The adversaries in the design house will have access to the design and might sabotage the design to serve other interests. The mitigation of these security threats, with minimal overhead on design, fabrication, and testing of ICs, is a primary motivation of widespread research. The security threats in various stages of IC design flow and their corresponding countermeasures are described in Figure 3.1.

Strategies either to simplify detection of HT or prevention of HT insertion become a prominent research. HT prevention schemes, with significant design overhead, modify the circuit at various levels of design abstraction by adding additional locking circuitry [5–10]. Thus, reverse engineering to the adversary at the foundry is difficult in the obfuscated circuit. These DFTr techniques can be broadly categorized, as shown in Figure 3.2.

Sametinger et al. [11] discussed security challenges for medical devices. The authors addressed security risk in medical devices due to HT. HT implanted in

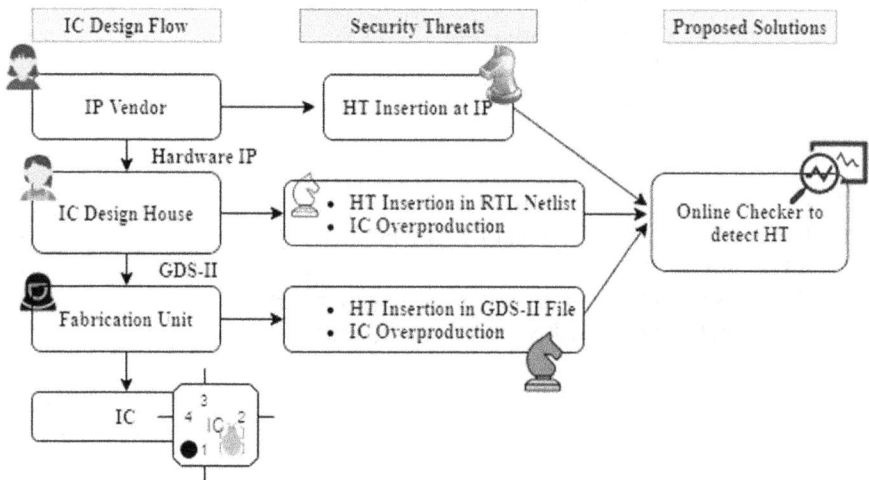

FIGURE 3.1 Modern IC life cycle and its security threats with countermeasures.

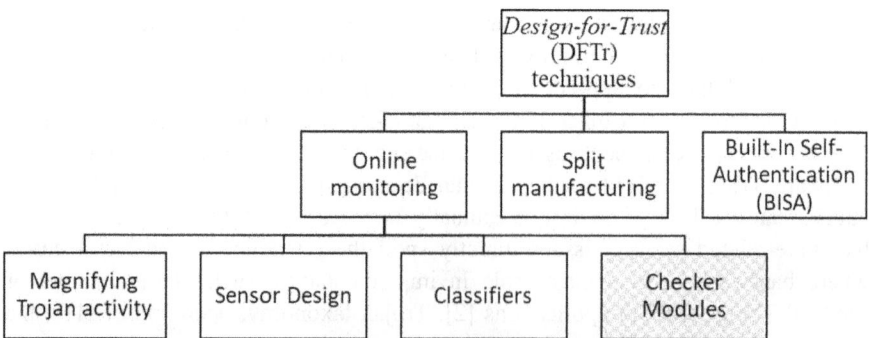

FIGURE 3.2 Classification of design-for-trust schemes.

pacemakers installed in the human heart is a real-time example. Dong et al. [12] developed a framework to protect Internet of Things chips from multi-layer HT called RG-Secure, which combines the third-party intellectual property trusted design strategy with the scan-chain netlist feature analysis technology. The RG-Secure detects HTs by analyzing multilayer rather than directly analyzing the functionality based on netlists. However, the design overhead is very high in RG-Secure framework. With this motivation and background, we propose an online monitoring system of HT detection for smart healthcare devices during runtime of the device. The main objective of the proposed work is to establish a secured hardware adopting a detection scheme, through the development of low hardware and computational overhead, which can counteract the security threats at various stages of modern IC life-cycle. The EDA tools are used to implement the proposed schemes with automated design methodology. The main goal of this research is to provide a comprehensive solution for all the stages of the design flow. This is achieved by the promising online checking technique [13], to detect hard-to-detect HTs with acceptable and controlled hardware overhead. Detection and prevention of HTs at the circuit netlist-level have thus emerged as a significance research area along with the investigation of newer threats. This work includes detection of HT at the design phase with minimal hardware overhead with high accuracy due to the node selection by reliability analysis.

3.1.1 THREAT MODEL

A tiny combinational HT is considered in this work, which creates malfunction whenever triggered. It is assumed that the HT instance satisfies the following constraints:

- The HT instances are inserted within a radius of two logic gates of the inserted checker.
- The HT instances are inserted at a radius of ten or more logic gates away from the inserted checker.

This is the more probable scenario with respect to HT insertion.

There are various attack models outlined in the literature, from which the following attack models are chosen for this work; 1) by an attacker in the third-party IP (3PIP) vendor, or at design house and as well as in the foundry. Table 3.1 illustrates the threat model of HT attack scenarios. In this scenario, an attacker may add, delete, or modify any gate in the foundry, such that an HT has inserted in the design.

TABLE 3.1
Threat Model

Scenarios	IP Vendor	Design House	Foundry	End-user
i	☻	☻	☻	☺

☻ – An attacker; ☺ – A trustworthy entity

3.2 PROPOSED METHODOLOGY: ONLINE MONITORING FOR HT DETECTION

The core idea of the proposed online monitoring technique is to monitor the internal logic values at the probable HT insertion sites of IC. The focus is on the expected circuit functionality and to monitor the correlation between the sets of logic values of the selected sites to that of the neighbor internal nodes. If their internal logics are uncorrelated, then the checker will detect the presence of an HT.

The following are the primary objectives of the proposed technique to enhance HT detectability:

- For a given netlist, the internal nodes are identified to which the HTs are inserted.
- The low-overhead checkers are inserted at a selected subset of internal nodes.
- The logic malfunctions are identified and reported by the online checkers.
- The logic errors are propagated to the primary outputs, and they are independent to the test vectors generated.
- The numbers of checkers inserted are restricted to have a low hardware overhead with maximum HT detection accuracy.

The proposed technique is based on the following main observation: that there exists a logical correlation between the HT-infected node to its nearby nodes. By neighborhood of a node N which is input to c different gates G1; G2;...; GC, we mean the input and output nodes of these c logic gates. Hence, the checkers are inserted at well-chosen specific nodes, to monitor and report logic malfunctions in its proximity.

Figure 3.3 shows an example of both single-rail and double-rail checkers are included in this work. The checker does not add any functionality to the circuit; that is, circuit functionality and as well as the triggering condition of the HT will not be altered by the proposed checker.

3.2.1 RELIABILITY-BASED NODE SELECTION TO INSERT CHECKER

Randomly selected nodes to insert the proposed checker module would incur unacceptable overhead and may not detect HTs effectively. Hence, the proposed online checking methodology should rely on selection of important nodes identified by the algorithm, which uses an analytical tool called signal probability reliability analysis (SPRA) [14]. For all possible input scenarios, the overall circuit reliability is calculated by SPRA in execution time, which is a linear function of the circuit size, and hence is much more efficient than gate-level circuit simulation by random vectors. An attack model is considered, where the attack causes a logical malfunction on HT activation, that renders the chip completely useless. The HTs inserted are minimal in number, and hence their impact on manifestational parameters (such as power consumption) are negligible. This make an ineffective HT detection based on the side-channel analysis. In this case, the adversary will select the gates with lower reliability for HT insertion, which is considered in this work.

Figure 3.4 demonstrates the complete automated design flow of the proposed online monitoring methodology for HT detection. A circuit netlist was given, and its

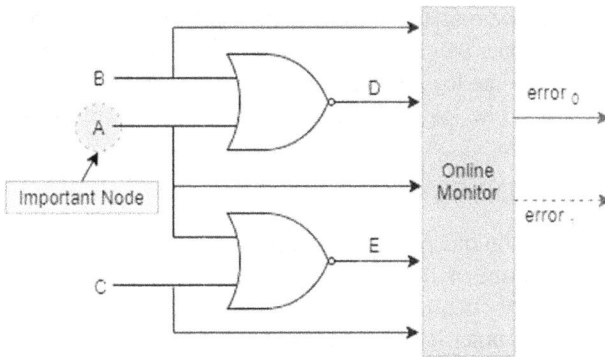

Single-Rail Checker Logic

If (A==1)
$error_0$= (D^~B) | (E^~C);
else if (A==0)
$error_0$= ((D, E)! = 2'b11);

Double-Rail Checker Logic

If (A==1)
{$error_0$, $error_1$} = ((D^~B) | (E^~C))?2'b10:2'b01;
else if (A==0)
{$error_0$, $error_1$} = ((D, E)! = 2'b11) ?2'b10:2'b01;

FIGURE 3.3 Single-rail and double-rail checker at the probable HT insertion sites to check the logical correlation of important node with the neighborhood node [15].

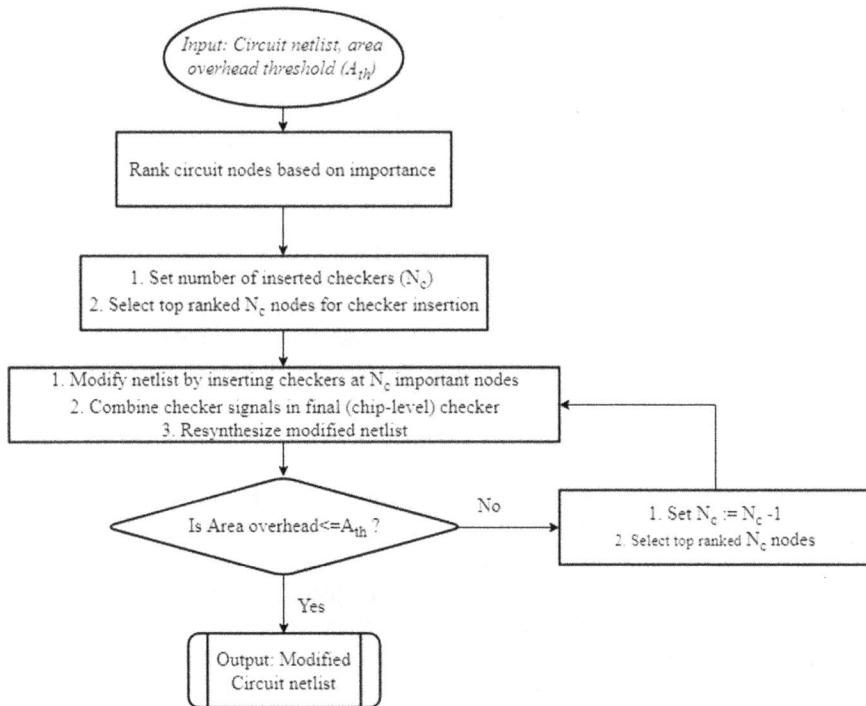

FIGURE 3.4 An automated design flow of the proposed HT detection methodology [15].

important internal nodes were first identified by the SPRA algorithm [15]. Then, for a given area threshold (upper bound) Ath on the area overhead, a trial Nc number of checkers with the appropriate logic were inserted in the netlist, the checker outputs were combined in a final checker, and the modified netlist was resynthesized.

3.3 RESULTS AND DISCUSSION

The online monitoring technique proposed is applied and validated on the collective Verilog netlists of a subset of the ISCAS'85 combinational circuits, ISCAS'89 sequential circuits, ITC'99 circuits [16] and a prototype AES hardware accelerator implementation [17], a symmetric block cipher. The original circuit netlists and the modified checker embedded netlists are synthesized using the Synopsys Design Compiler (DC) tool, using a 45 nm standard cell library. The input test patterns are generated exhaustively for the smallest benchmark circuit (c17) considered, and randomly for the other complex benchmark circuits. C++ was used to implement the entire automated design methodology.

3.3.1 RESULTS OF BENCHMARK CIRCUITS

The number of locations in which the checker modules are inserted is to be considered such that there should be 10% of overhead of the modified design to that of the original design, including the final checker. Figures 3.5 and 3.6 show the improvement in HT detection coverage by inserting HT near and away from the checker locations, for benchmark circuits.

It is clearly demonstrated from the experimental results that the proposed technique substantially increases in detection coverage. As expected, for the HTs inserted near the checkers, this improvement is more pronounced. Table 3.1 describes the comparison of the proposed node selection-based checker insertion with the random insertion of checker modules. That which clears the node selection improves the detection accuracy to an average of 47.47% than the randomly inserted checker design. Figure 3.7 presents the design overheads of the modified designs with respect to the original designs, along with the target area overhead threshold Ath. In some circuits, the inserted checker logic and the critical path do not have the same path, and hence, path delay in some of the circuits is zero.

3.3.2 RESULTS FOR AES ENCRYPTION UNIT

Simulation results corresponding to the application of the proposed online checking technique to enhance HT detectability on AES encryption are now presented. AES is widely deployed, and a RTL or gate-level IP of AES is widely available.

In this simulation, the open-source Verilog Register Transfer Level (RTL) implementation of 128-bit AES core available at Opencores [17] was used. The cores were integrated into a single encrypted/decrypted core and controlled by mode-control signal. The three sub-modules of AES module – the key-expand, the S-Box, and the inverse S-Box – were separately subjected to the design flow with an area

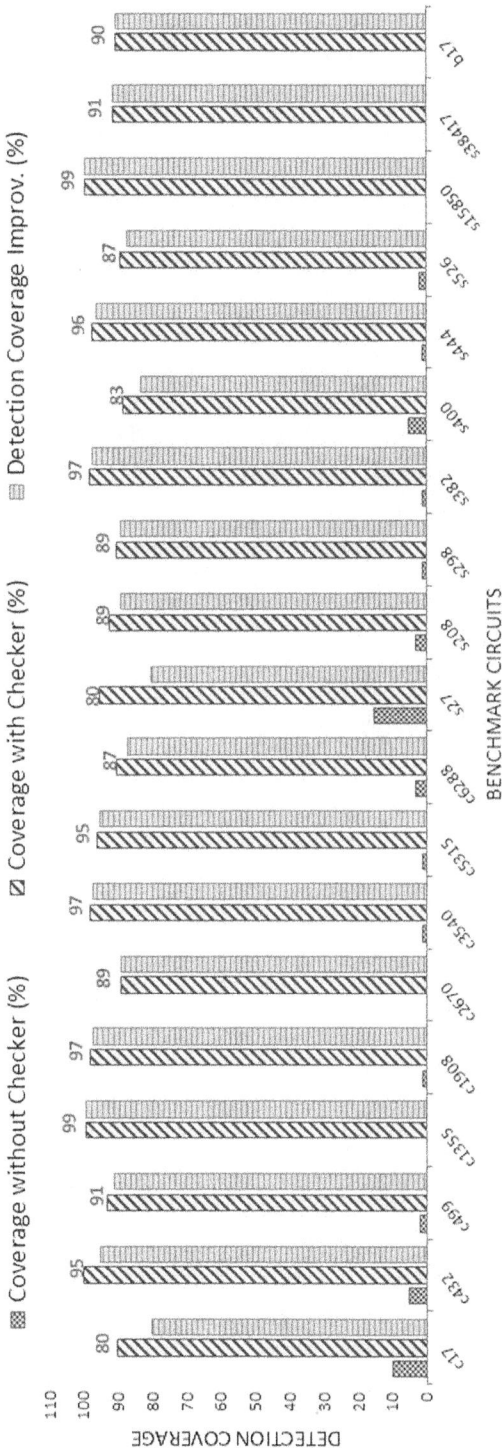

FIGURE 3.5 Improvement in HT detection coverage by inserting HT near to the checker.

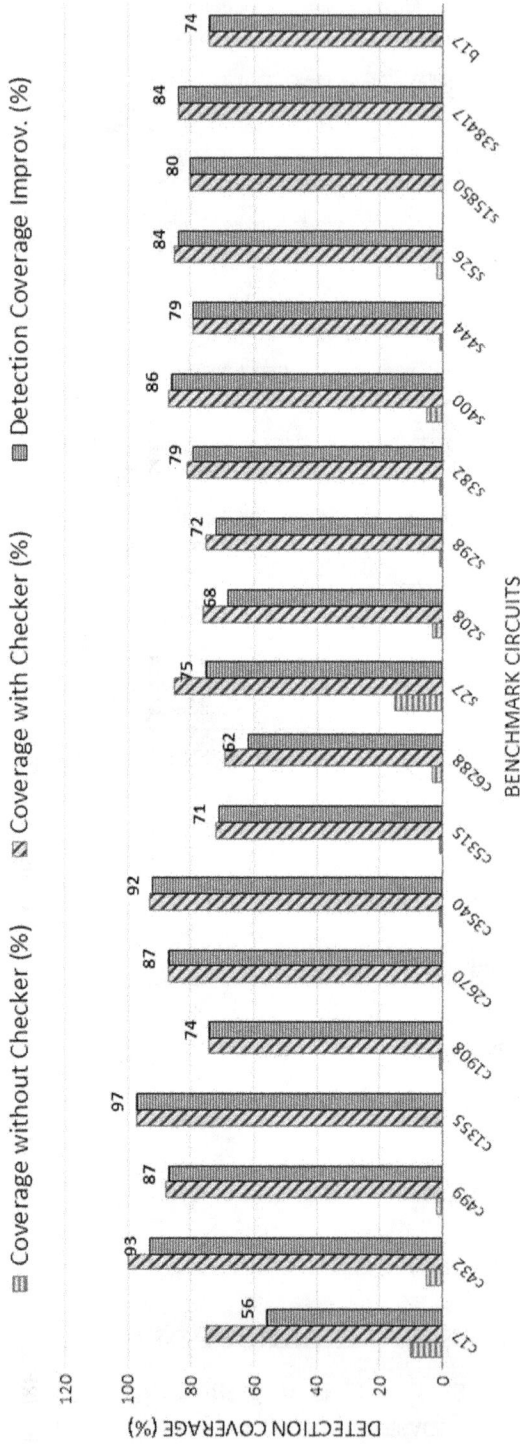

FIGURE 3.6 Improvement in HT detection coverage when HT is inserted away from checker.

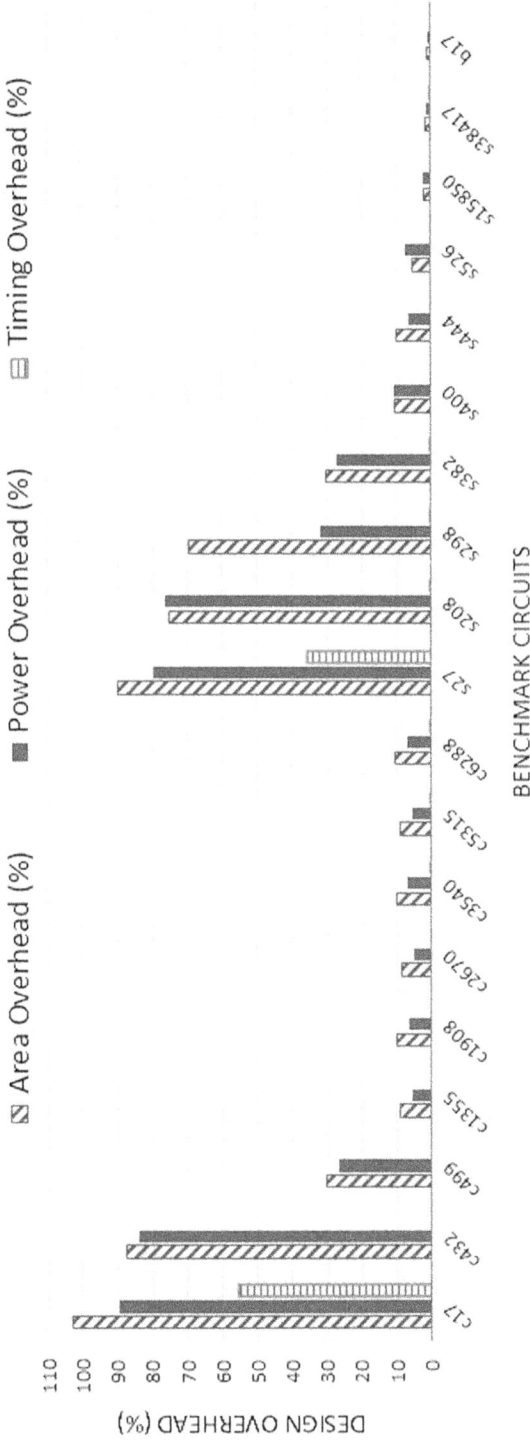

FIGURE 3.7 Design overhead results of proposed online monitoring technique for enhanced HT detection.

TABLE 3.2

Detection Coverage Improvement When Checkers Are at Important Nodes

Circuit	# Checkers Inserted	Coverage with Checker (%)		Improved Detection Coverage (%)
		Random Nodes	Important Nodes	
c17	1	50	90	40
c432	1	40	100	60
c499	8	53	93	40
c1355	11	66	99	33
c1908	10	35	98	63
c2670	12	24	89	65
c3540	18	32	98	66
c5315	33	30	96	66
c6288	4	26	90	64
s27	1	58	95	37
s208	2	54	92	38
s298	2	81	90	9
s382	7	79	98	19
s400	12	68	88	20
s444	12	75	97	22
s526	15	28	89	61
s15850	45	31	99	68
s38417	65	33	91	58
b17	80	20	93	73
Avg.	17.84	46.47	93.94	47.47

overhead below 5%. Table 3.2 shows the performance and design overheads for the three AES core designs integrated with checker module. The actual area overhead was less than 5%, and the power and timing overheads were within the acceptable limits. The overall design overhead of the proposed scheme is shown in Table 3.3.

TABLE 3.3

Checker Efficiency and Design Overheads of the AES Modules

AES Modules	Nodes Modified (%)	Design Overhead (%)		
		Area	Delay	Power
Key Expand	0.8	3.44	0	3.45
S-Box	0.8	3.63	1.24	4.32
Inverse S-Box	0.8	4.72	1.5	5.51

TABLE 3.4
Overall Design Overheads of the AES Core (%)

Parameter	AES	AES+Checker	Overhead (%)
Area(μm)^2	2732061	2840250	4
Power (mW)	564	590.03	4.43
Detection Coverage (%)	0.8	97	96.2

Again, the area and power overheads of the checker integrated AES core were less than 5% with a 96.2% of detection coverage overhead (Table 3.4).

3.4 CONCLUSION

Physical devices used in the IoMT to monitor and guide patients remotely are not trustworthy, due to the insertion of HTs in the ICs embedded in them. Detection of HTs is important, since an undetected HT in a deployed electronic system can be a serious threat to privacy, security, and safety. In this context, we have developed an automated low overhead online monitoring technique to detect HT at the gate-level netlist. The proposed technique was applied to develop a secured system architecture directed towards design-for-trust. The effectiveness of the proposed scheme was experimentally validated, where detection coverage was close to 100% when controlled design overhead was 10%.

The proposed technique is resistant to cell replacement attack, since reverse engineering is not possible in the modified design and can be extended easily to secure the system designed. The proposed online monitoring scheme is validated on ISCAS'85, ISCAS'95, ITC'99, and AES encryption modules to detect HTs with high detection coverage. Thus, the proposed online checkers could possibly provide secured hardware for smart healthcare devices at the design level.

An extension of the current work is to develop online checkers for all smart healthcare devices that also include data and software security along with the proposed hardware security.

REFERENCES

[1] Rajendran, Sree Ranjani, Rijoy Mukherjee, and Rajat Subhra Chakraborty. "SoK: Physical and logic testing techniques for hardware Trojan detection." In *Proceedings of the 4th ACM Workshop on Attacks and Solutions in Hardware Security*, pp. 103–116, 2020.
[2] Ranjani, Rajendran Sree. "Machine learning applications for a real-time monitoring of arrhythmia patients using IoT." In *Internet of Things for Healthcare Technologies*, pp. 93–107, Springer, Singapore, 2020.
[3] Ranjani, R. Sree, and M. Nirmala Devi. "Malicious hardware detection and design for trust: an analysis." *Elektrotehniski Vestnik* 84, no. 1/2 (2017): 7.

[4] Nixon, Patrick, Waleed Wagealla, Colin English, and Sotirios Terzis. "Security, privacy, and trust issues in smart environments." In *Smart Environments: Technology, Protocols and Applications*, pp. 220–240, Wiley-Blackwell, USA, 2005.

[5] Kanuparthi, Arun, Ramesh Karri, and Sateesh Addepalli. "Hardware and embedded security in the context of internet of things." In *Proceedings of the 2013 ACM workshop on Security, privacy & dependability for cyber vehicles*, pp. 61–64, 2013.

[6] Colins, D. "Trust in integrated circuits (tic)." *DARPA Solicitation BAA07-24* (2007).

[7] Tehranipoor, Mohammad, and Farinaz Koushanfar. "A survey of hardware trojan taxonomy and detection." *IEEE Annals of the History of Computing* 01 (1900): 10–25.

[8] Rajendran, Sree Ranjani. "KARNA for a trustable hardware." In *Proceedings of the 2020 24th International Symposium on VLSI Design and Test (VDAT)*, pp. 1–4, IEEE, 2020.

[9] Ranjani, R. Sree, and M. Nirmala Devi. "Enhanced logical locking for a secured hardware IP against key-guessing attacks." In *Proceedings of the International Symposium on VLSI Design and Test*, pp. 186–197, Springer, Singapore, 2018.

[10] Ranjani, R. Sree, and M. Nirmala Devi. "Secured hardware design with locker-box against a key-guessing attacks." *Journal of Low Power Electronics* 15, no. 2 (2019): 246–255.

[11] Sametinger, Johannes, Jerzy Rozenblit, Roman Lysecky, and Peter Ott. "Security challenges for medical devices." *Communications of the ACM* 58, no. 4 (2015): 74–82.

[12] Dong, Chen, Guorong He, Ximeng Liu, Yang Yang, and Wenzhong Guo. "A multi-layer hardware trojan protection framework for IoT chips." *IEEE Access* 7 (2019): 23628–23639.

[13] Nicolaidis, Michael, and Yervant Zorian. "On-line testing for VLSI—a compendium of approaches." *Journal of Electronic Testing* 12, no. 1–2 (1998): 7–20.

[14] Pagliarini, Samuel Nascimento. "Reliability analysis methods and improvement techniques applicable to digital circuits." PhD diss., 2013.

[15] Chakraborty, Rajat Subhra, Samuel Pagliarini, Jimson Mathew, Sree Ranjani Rajendran, and M. Nirmala Devi. "A flexible online checking technique to enhance hardware trojan horse detectability by reliability analysis." *IEEE Transactions on Emerging Topics in Computing* 5, no. 2 (2017): 260–270.

[16] Benchmarkcircuits. "The iscas85, iscas89 and itc99 benchmark circuits." http:// pld.ttu.ee/ maksim/benchmarks/.

[17] Opencores. "The 128-bit advanced encryption standard ip core." 2009. www.opencores. org/projects.cgi/web/aescore.

4 Machine Learning Methods for Hardware Security

Soma Saha[1] and Bodhisatwa Mazumdar[2]
[1]Department of Computer Engineering, SGSITS Indore,
Indore, Madhya Pradesh, India
[2]Department of Computer Science and Engineering, IIT
Indore, Indore, Madhya Pradesh, India

CONTENTS

DOI: 10.1201/9781003201038-4

4.1   INTRODUCTION

The areas of artificial intelligence (AI) and the real-life applications of machine learning (ML) are jointly thriving, and have permeated in almost every aspect of human lives in the present day. With the advent of both AI and ML-induced computing systems, society is struggling to absorb and understand this faster growing technology with pace. Therefore, instead of making the world safer and affordable, this technology exploits security challenges that may endanger public and private life. In another aspect, the presence of an enormous number of off-shore vendors in integrated circuit (IC) design, fabrication and manufacturing process for cost-effective production of ICs, the insertion of malfunctioning hardware by malicious agents in the entire IC supply chain and design for even the smallest computing systems has turned out to be one of severe personal as well as national threat. Additionally, with the increasing count and power of hardware-based attacks, the urge to address and resolve hardware security challenges has become evident in the context of VLSI design as well as in the context of data confidentiality, availability, and integrity.

In another aspect, the full potential of this thriving technology is yet to be realized in various fields, such as hardware security and network security. Exceptional success of applying ML in a variety of research domains has encouraged researchers from the hardware security domain to explore its potential to address a variety of security challenges. In recent years, ML methods have been used in a variety of hardware-related security challenges with the aim of providing powerful defense mechanisms by (i) understanding and mitigating vulnerabilities in the IC supply chain, counterfeiting, overbuilding, (ii) constructing defence mechanisms against hardware Trojans (HTs), (iii) detecting HTs through reverse engineering (RE) efforts, (iv) inserting logic locking mechanisms, and so on. Research works in hardware security in terms of attack mechanisms also incorporated ML algorithms in (i) side-channel analysis (SCA), and (ii) physical unclonable function (PUF) attack.

4.2   PRELIMINARIES

In this section, we provide an ensemble of ML models that have been used extensively in hardware security research. The ML models comprise different types of supervised and unsupervised ML approaches that have been applied in various implementations of hardware security attacks and countermeasures.

4.2.1 MACHINE LEARNING MODELS USED IN HARDWARE SECURITY

Primarily, the ML field has been focused on generating the computational "models", also known as learning algorithms that learn from data-centered past experiences [1–3]. These trained models are utilized for classification, or prediction on new data. As depicted in Figure 4.1, the general procedure for applying ML algorithms requires the following steps:

1. In the pre-processing phase, (i) relevant features are extracted to distinguish between the different values of the target output or new features are created from already-existing original features; this sub-phase is known as the feature extraction phase, (ii) features with the highest information content are ranked and selected, which is termed the feature selection phase, (iii) high-dimensional features are mapped to low-dimensional space in the dimensionality reduction sub-phase. In summary, the pre-processing phase extracts, chooses, and reduces the number of relevant features for a particular model based on available data.
2. In the training or learning phase, first suitable learning algorithms are chosen and executed to attain the models from the training dataset. These ML models correlate to the underlying relation between the input features and target output.
3. In the model validation or evaluation phase, several operations such as result evaluation, cross-validation, and hyper-parameter optimizations are implemented on training models to select the final models with best performance based on test dataset.
4. In the testing or prediction phase, the selected final models are utilized to derive the predicted values of the target output for previously unseen data.

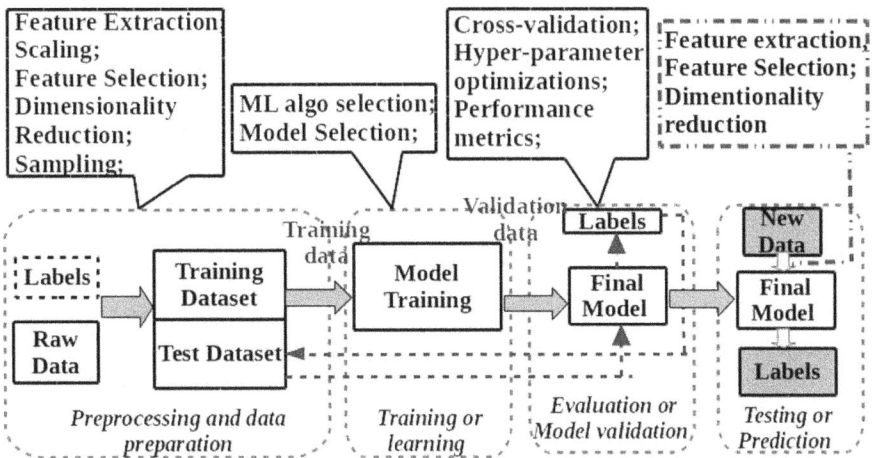

FIGURE 4.1   The general flow of ML, primarily supervised learning framework.

According to the nature of data available and processed, the learning tasks in ML algorithms can be categorized into supervised learning and unsupervised learning. The above-mentioned steps in the last paragraph for applying ML algorithms are aligned to supervised learning. In the hardware security domain, both supervised and unsupervised learning are widely explored based on the nature of problem and available data types. When input data are available with their corresponding paired correct target output, data are labeled. The supervised learning techniques use labeled data for training the models and select best performed models to predict the target output classes for newly obtained input data. In contrast, the unsupervised learning techniques deal with unlabeled data, and focus on learning from the underlying structures of available labeled data along with the data for which labeling is missing. In the following paragraphs, we present an overview of most widely explored ML techniques, feature selection methods, dimensionality reduction methods, and several optimization and model enhancement techniques in hardware security.

4.2.1.1 Supervised Learning

Supervised learning algorithms employ labeled training data to construct attack models, e.g. DTs, neural networks. These models are first trained for parametric tuning and then used to detect threats and vulnerabilities. The rest of this section presents an overview of supervised learning algorithms that have been used by both attackers and defenders in the domain of hardware security.

4.2.1.1.1 Support Vector Machines

Support vector machine (SVM) is a supervised learning technique that focuses on solving an optimization problem by constructing a hyperplane or a set of hyperplanes to separate data samples into different target classes with maximum intervals/margins in between [4,5]. As shown in Figure 4.2a, sample data points are separated into two target classes by the learned hyperplane with maximum intervals in between when only two features are considered.

    a. Classification using two-class SVM
    b. Classification using one-class SVM

If the data are linearly non-separable, a kernel trick is used with specific functions to map the data samples to a higher dimensional space, where they can be linearly separated. A variety of kernels is used in SVMs, for example, polynomial function/kernel, wavelet kernel, and Gaussian radial basis function (RBF) kernel. SVMs are efficient in high-dimensional space. SVMs have different kernel-dependent hyper parameters and a soft margin parameter or a slack variable that are required to be fine-tuned along with the selection of appropriate kernels to enhance the model prediction accuracy. In SVM, a trade-off between maximizing the interval/margin and minimizing the training error can be controlled by tuning the slack variable value. Increasing the slack variable value may cause overfitting or poor generalized performance on new data; however, it provides better accuracy over training data. Extended versions of basic SVMs for binary classifications are utilized to support multi-class classification.

(a)

(b)

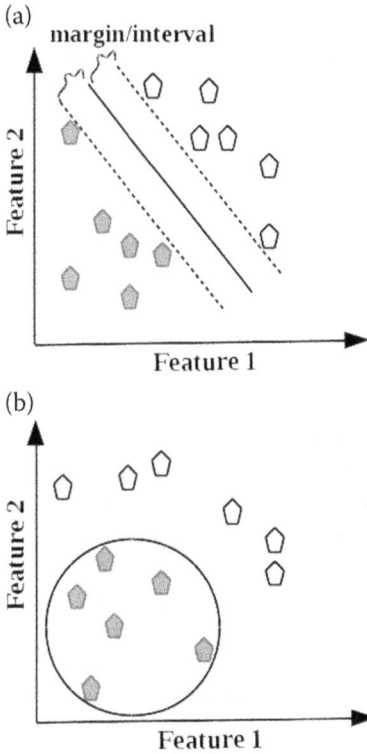

FIGURE 4.2 Examples of ML techniques, (a) support vector machine (SVM), and (b) one class classifier (OCC).

4.2.1.1.2 One-Class Classifiers

One-class classifiers (OCCs) aim to construct supervised classification models when the negative class in which new training data originally belong to is missing, poorly sampled, or not well defined. OCCs are well known in the fields of novelty and anomaly detection. Therefore, they are used extensively in hardware security when data from attacked systems are not known a priori. One-class SVM and one-class neural networks are two variants of OCCs. Generally, the one-class SVM is used as an optimization approach that minimizes the volume around a hyper-sphere enclosing most of the training data in the feature space, also known as one-class, as depicted in Figure 4.2b. Data points that do not belong within the hyper-sphere during the testing phase are known as anomalous data points.

4.2.1.1.3 Bayesian Classifiers

Bayesian classifiers address the classification problem by learning the distribution of instances when different class values are given. Instead of learning an explicit decision rule, Bayesian classification learning reduces to estimating probabilities. These classifiers can achieve a minimum error rate. They can choose the best possible classifier if a sufficient number of training instances are provided without considering the computational limitation.

4.2.1.1.4 Linear Regression

Linear regression (LR) is a supervised model that aims to predict a continuous quantity for the given input variables using a linear function. This statistical technique is used for investigating and modeling the relationship between the target output and independent input variables.

4.2.1.1.5 Multivariate Adaptive Regression Splines (MARS)

MARS technique aims to find a flexible nonparametric regression model that works on high-dimensional data and represents the non-linear relations between variables [6,7].

4.2.1.1.6 Decision Tree (DT)

Decision tree (DT) is a tree-based supervised learning model used for classification and regression. DT comprises a root node and several internal and external/leaf nodes. Each leaf node in DT represents a target class, whereas each internal node corresponds to attribute test conditions to split the training data points that pose different characteristics. In the decision-learning process, a data point traverses the tree starting at root node according to the attribute test criteria at each node till the target class at leaf node is reached. Depending on different splitting or attribute test criteria, for example, Gini score or information gain, at each internal node, different DT learning algorithms, such as classification and regression trees (CART), ID3, and C4.5, are used for training data. Figure 4.3a depicts an example of DT on two-dimensional data, where decision variables are x and y.

4.2.1.1.7 Random Forest (RF)

Random forest (RF) learning model focuses on building a predictive ensemble with a set of DTs. These DTs grow in randomly selected subspaces of data. RF scheme is applicable for both classification and regression to improve prediction accuracy. Here, the training set for each tree is drawn by sampling with replacement, or bootstrap sampling from original data. After the tree construction phase, all data are tracked down the tree, and proximities are computed for each pair of cases. The proximity of two cases is increased by one if two cases occupy the same terminal node. In the next run, proximities are normalized by dividing the number of trees, and these normalized proximities are used in replacing missing data and tracing outliers. The final class decision is the most voted class among different DTs or the average of predicted real values of all individual trees in the ensemble. Figure 4.3b shows an example of classification using a RF of three trees with three decision variables, x, y, and z.

4.2.1.1.8 Logistic Regression (LR)

LR technique classifies an observation into one of two classes or one of many classes. This regression model is well-suited to discover the association between features.

(a)

(b)

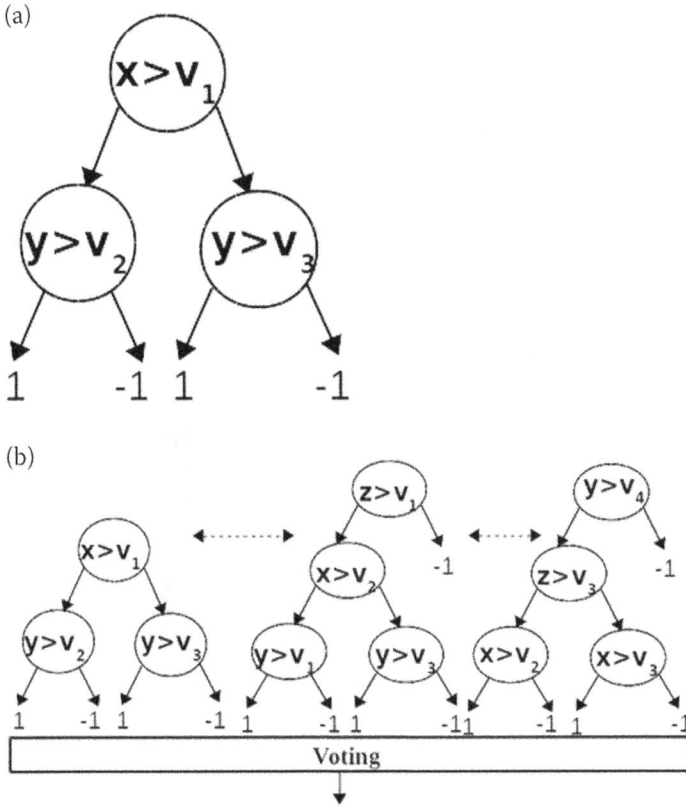

FIGURE 4.3 Example of ML techniques. Classification using (a) decision tree on a two-dimensional dataset, and (b) random forest of three decision trees.

4.2.1.1.9 AdaBoost or Adaptive Boosting

Adaptive boosting is well-known as the best out-of-the-box classifier due to its ability of combining multiple weak learning algorithms to generate a strong classifier. Initially, a weak learning algorithm learns the training data. Then, in each iteration, another weak learning algorithm is combined to reduce the training error of previously applied weak learning algorithms.

4.2.1.1.10 Artificial Neural Networks

Artificial neural networks (ANNs) are inspired by the biological learning systems of the human brain that are made of complex webs of interconnected neurons. An ANN can be viewed as a mathematical model consisting of an input layer, an output layer, and several hidden layers in between, as shown in Figure 4.4.

Each layer comprises neurons with activation functions. Neurons in one layer are connected to neurons in the following layer using synapses with weights. Depending on different connection patterns and how the synaptic weights are learned during the training phase, various ANNs are constructed. For example, in

(a)

(b)

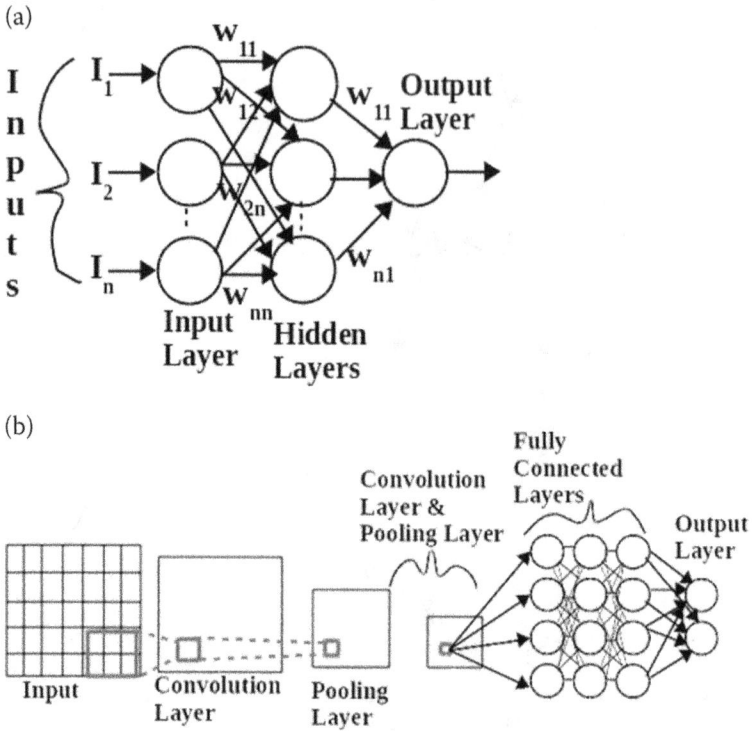

FIGURE 4.4 Examples of ML techniques. Classification using (a) artificial neural networks (ANNs) and (b) convolutional neural networks (CNN).

backpropagation neural networks (BPNN) [8], the difference between generated output and the expected output, known as training error, is used to update the synaptic weights. ANNs have been successfully applied to diverse applications of classification, including HT detection and recycled and counterfeit IC detection.

4.2.1.1.11 Convolutional Neural Network

Convolutionl neural network (CNN) is a variant of neural networks where the hidden layers consist of a series of convolutional layers that convolve with multiplication or other dot products, pooling layers, and fully connected layers [9]. A pooling layer sub-samples its input feature vector to reduce its dimension that subsequently decreases the required amount of computational time and weights. Figure 4.4 shows the general structure of a CNN.

4.2.1.1.12 AutoEncoder

AutoEncoder is a neural network with layers that work as encoders and decoders. The encoder aims at learning a representation (or encoding) for a set of data for dimensionality reduction by training the underlying network to ignore "noise" [10]. The decoder aims at mapping the encoded output obtained from the encoder to the original input through reconstruction. Figure 4.5 depicts the general structure of an

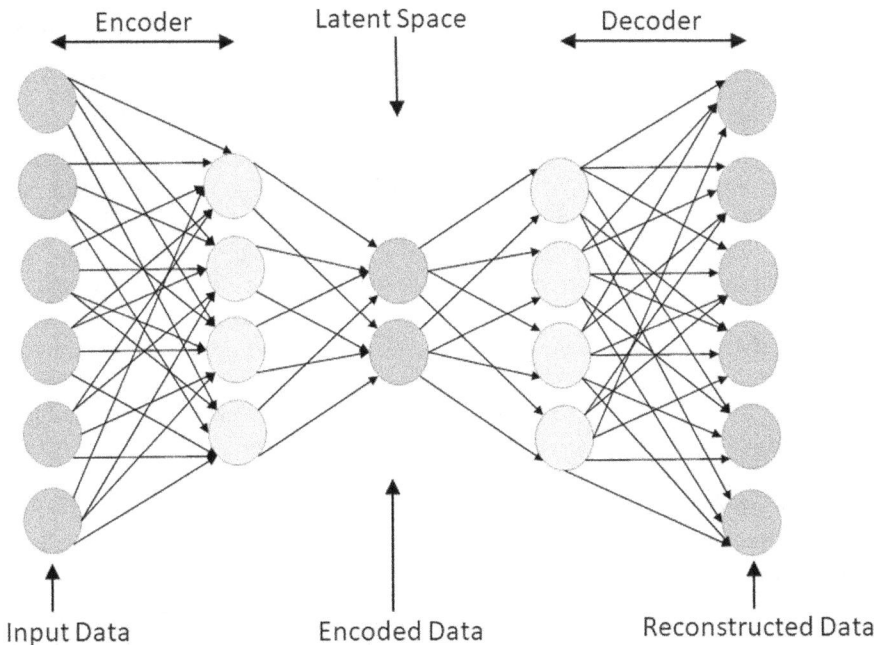

FIGURE 4.5 Taxonomy of different machine learning algorithms used in attacks and defense approaches in hardware security.

autoencoder. During the training phase of backpropagation, the difference between the reconstructed input and original input is computed, which is known as reconstruction error. It is used to estimate the optimal weights that are needed to minimize the reconstruction error.

4.2.1.1.13 Recurrent Neural Network

Recurrent neural network (RNN) is a class of neural networks with the ability to use their internal state (memory) to process variable length sequences of inputs [11]. In RNN, connections between nodes form a directed graph along with a temporal sequence that is allowing the network to exhibit temporal dynamic behavior. Thus, RNN is suitable for modeling the temporal dependencies in time series and data streams.

4.2.1.1.14 Extreme Learning Machine

Extreme learning machine (ELM) is a single hidden layer feedforward neural network (SLFN) that obtains the corresponding output weight by randomly initializing the input weight and bias. Due to its faster learning speed and fine generalization ability, ELM has been extensively used in many fields, including hardware security [12].

4.2.1.1.15 Long Short-Term Memory

Long short-term memory (LSTM) is a variant of the RNN model used in the field of deep learning. This model is well-suited for classification, processing, and making predictions as it deals with data streams and time-series with arbitrary time-lags between instances [13].

4.2.1.1.16 Half-Space Trees

Half-space trees (HS-trees) are useful models for detecting anomalies in streaming or near continuous data [14]. Initially, an anomaly score is assigned to each incoming data point in the streaming data. Then, each half-space tree splits the feature space into two halves that ensembles the left and right child of a binary tree. The splitting is done at each node by picking a randomly selected dimension. Node expansion continues until all nodes reach their maximum depth, which will be the same for all nodes. Each node in the HS-tree records the mass profile of the training data that traverses through the tree. In the testing phase, an anomaly score for the input data point that traverses each tree is computed depending on the mass profile of the terminating node.

4.2.1.1.17 K-Nearest Neighbors (KNN)

K-nearest neighbors (KNN) is a well-known supervised learning model for both classification and regression. This model ensembles the concept of proximity by measuring the Euclidean distance between different data points in training data. Depending on this measured value, k most similar samples in a feature space are chosen for one category or class.

4.2.2 Unsupervised Learning

The main aim of unsupervised learning is to identify hidden patterns in unlabeled data. In unsupervised learning, the model is not provided with correct results during the training. It can be used to cluster the input data in classes only on the basis of their statistical properties. Such techniques facilitate classification and analysis of raw datasets, thereby helping to generate information from unlabeled data. In the recent past, advances in hierarchical learning, clustering algorithms, outlier detection, etc., have significantly improved advancements in state-of-the-art unsupervised ML techniques. In this section, we provide a summary of existing unsupervised ML techniques that have been used in the domain of hardware security.

4.2.2.1 Clustering Algorithms

Clustering algorithms (CAs) aim at grouping unlabeled data into clusters [15]. A cluster is a class of data items that have similarity between them and are dissimilar to data items in other clusters. Clustering comprises the following components:

  i. Proximity measure, either a similarity measure or a dissimilarity measure, for instance, Euclidean distance or Manhattan distance
  ii. Criterion function to evaluate a clustering
  iii. Algorithm to compute clustering, for instance, optimizing the criterion function

CAs can work without a priori knowledge about the input data. Therefore, CAs can be used in the HT detection and protection field, where features of ICs/golden designs are not accessible.

4.2.2.2 K-means Clustering Algorithm

K-means clustering is a variant of the clustering algorithm that intends to partition the dataset into k predefined distinct non-overlapping clusters in which each data point belongs to a single cluster only [16]. This clustering intends to retain,

    i. the distance metric values between data points within each intra-cluster minimal
    ii. the distance metric values within inter-clusters as maximum as possible

Initially, cluster centroids are chosen randomly. Then, data points are allocated to the nearest clusters. Subsequently, the centroid of each cluster is recomputed to reduce intra-cluster distance metric values. These steps are repeated until no changes in the centroids are possible or any other explicit stopping criteria is reached.

4.2.2.3 Partitioning Around Medoids (PAM)

Partitioning around Medoids (PAM) is similar to K-means clustering algorithm with the difference in initial centroid selection and re-computation strategy [17]. Here, the initial centroids are chosen as one of the real data points from the dataset, known as the medoids of a cluster. After the allocation of data points to each cluster, centroids are reevaluated based on overall squared error or most similarities with other data points within the cluster. These two features allow PAM to be more robust to noise than K-clustering algorithms.

4.2.2.4 Density-Based Spatial Clustering (DBSCAN) and Ordering Points to Identify the Clustering Structure (OPTICS)

DBSCAN and OPTICS are density-based clustering algorithms [18,19]. They find the clusters by beginning with the estimated density distribution of corresponding nodes. These clustering algorithms form clusters with various sizes and shapes. However, these variations in clusters do not affect the accuracy of clustering.

4.2.3 FEATURE SELECTION AND DIMENSIONALITY REDUCTION

The feature selection process comprises selection of subset features that are relevant that can be used for model construction. In other words, for an input vector x, and corresponding output vector y, feature selection comprises finding a subset of selection features that preserves most of the output prediction capabilities. This comprises filtering approaches that filter out features with small potential to enable correct prediction of outputs, and wrapper approaches that involve selecting features to optimize the accuracy of the chosen classifier. Dimensionality reduction is a classification problem that comprises input data $\{x_1, x_2,..., x_n\}$, such that $x_i = \{x_i^1, x_i^2, ..., x_i^d\}$, and a corresponding set of output labels $\{y_1, y_2,...., y_n\}$, wherein the dimension d of the datapoint x is very large. The problem is to classify x. With high dimensional input vectors, it requires a large number of parameters to run. Given this condition, a small-sized input dataset can result in large variance of estimates and overfit. In this section, we present the algorithms that are associated with feature selection and dimensionality reduction.

4.2.3.1   Genetic Algorithms

Genetic algorithms are well-known natural evolution inspired heuristic algorithms [15]. In classification problems, genetic algorithms can be used for feature selection that aims at revealing a small subset of variables from a dataset to provide maximum classification accuracy.

4.2.3.2   Pearson's Correlation Coefficient

The Pearson's correlation coefficient is widely used as a feature selection metric. This correlation is computed between each feature and the target output. Features having the largest Pearson's correlation coefficient with the target output are selected for model training and testing.

4.2.3.3   Minimum Redundancy Maximum Relevance (mRMR)

The Minimum Redundancy Maximum Relevance (mRMR) feature selection method chooses features with minimum correlation with other input features and maximum correlation with the target output [20].

4.2.3.4   Principal Component Analysis

Principal component analysis (PCA) is a well-known feature dimensionality reduction technique in which one-dimensional features are mapped into k-dimensional features [21,22]. These features are also known as principal components that reflect orthogonal property. Each of the principal components can reflect most of the original variables, thus underlying information is not repeated and the dimension of original features is reduced. The dimension k is usually defined by the user in advance and has a value much lower than n.

4.2.3.5   Two-Dimensional Principal Component Analysis

Two-dimensional PCA (2DPCA) is a variant of PCA technique that can overcome the deficiencies of PCA, such as the requirement of high computational time due to computational complexity [18]. In 2DPCA, the covariance matrix is constructed based on 2D matrices rather than converted into 1D vector in PCA.

4.2.3.6   Self-Organizing Maps (SOMs)

Self-organizing maps (SOMs) are special variants of neural networks where it maps high-dimensional feature vectors into a single neuron in the grid [23]. Thus, SOMs can be used for dimensionality reduction.

In addition, numerous techniques exist for optimizing the design of an ML model and enhancing the model, for example, Multi-Objective Evolutionary Algorithm (MOEA), Adaptive Iterative Optimization Algorithm (AIOA), and Particle Swarm Optimization (PSO) Algorithm. Due to space limitation, we are unable to include the details and references of these techniques.

4.3 HARDWARE SECURITY CHALLENGES ADDRESSED BY MACHINE LEARNING

The taxonomy of hardware security threats that use different ML algorithms are shown in Figure 4.5. As shown, there are different variants of RE attacks that are employed in both attack and defense strategies of hardware strategies. For instance, there has been algorithmic structural and functional analysis of underlying circuit topology resulting in IP piracy acts. However, similar algorithmic approaches can be used in detection of HTs and malicious design changes resulting in untrusted design components in an IC. In the next section, we enumerate the hardware security threats that exist in the IC supply chain.

4.3.1 HARDWARE TROJANS

HTs are malicious alterations incorporated in ICs during design or fabrication in an untrusted design house or foundry by untrusted people or design tools [1]. HTs can also be used to mount denial-of-service attacks by tampering the design functionality. These malicious components are hard to detect, which remain inactive for most of the time IC execution is in process. HTs are activated and are triggered in corner case conditions or rare signal events occurring during an IC operation.

HTs are usually inserted in System-on-Chips (SoCs) while integrating untrustworthy IP circuits obtained from either third parties or internally by a rogue employee. As shown in Figure 4.6a, HTs can be inserted as a combinational circuit, or a sequential circuit shown in Figure 4.6b. HTs comprise two parts, a trigger which checks for the activation conditions, and a payload that propagates the activity effect of the HT when it is activated [24].

HTs can be activated with change in circuit functionality or change in physical parameters, such as temperature. These HTs exhibit stealthy behavior, i.e. they go undetected during functional testing or validation stages of the SoCs [25]. However, such HTs can be activated after a long exposure to in-field execution.

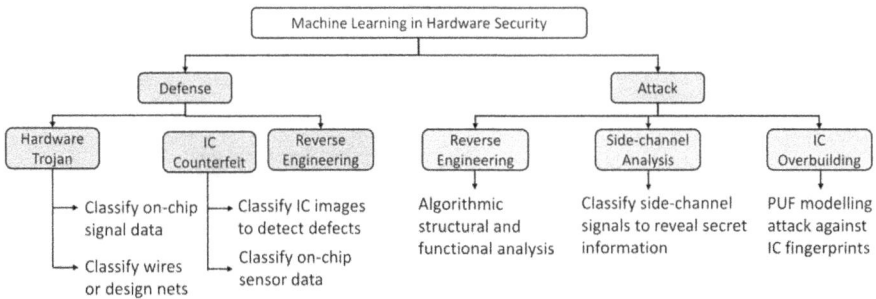

FIGURE 4.6 Examples of Hardware Trojans, (a) A combinational Trojan, which can be triggered with corner case condition, a = 0, b = 0, c = 0, (b) A sequential Trojan, which can be triggered when the corner case condition, a = 0, b = 1 occurs for 2^n times, where n is the length of the counter.

Few major objectives of HT attacks among many from an attacker's perspective are (i) altering major national infrastructure by incorporating malicious functionality in electrical components used in mission critical systems such as nuclear reactors and military weapons, (ii) defaming the impression of a company for gaining competitive advantages in the market, (iii) illegally accessing a secure system by leaking secret information from inside a chip, or (iv) prompting irreversible and fatal damage to the system [1,3]. In short, HT attacks may leak confidential data, degrade the system performance level, or cause denial-of-service and pose serious security threats to IC suppliers and end IC users. Therefore, it is challenging to address defensive strategies for diminishing the potential security threats posed by HT attacks [3].

4.3.2 Reverse Engineering

RE involves disassembling systems and devices and extracting their underlying design using destructive and nondestructive methods [26,27]. Such operations often lead to device cloning, duplication, and reproduction [28]. The motivation to mount RE analysis can be both honest, which can be legal as long as patent and copyright

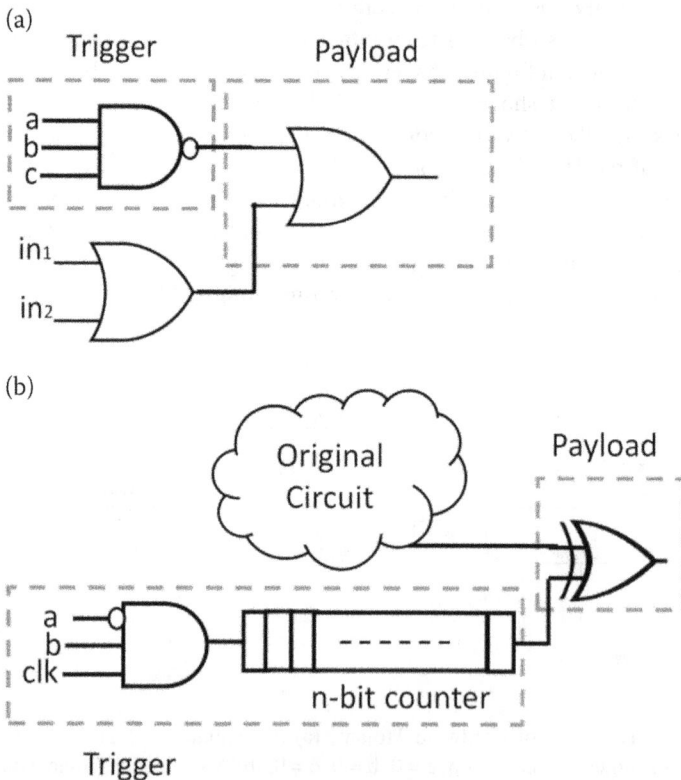

FIGURE 4.7   Classification of reverse engineering applications.

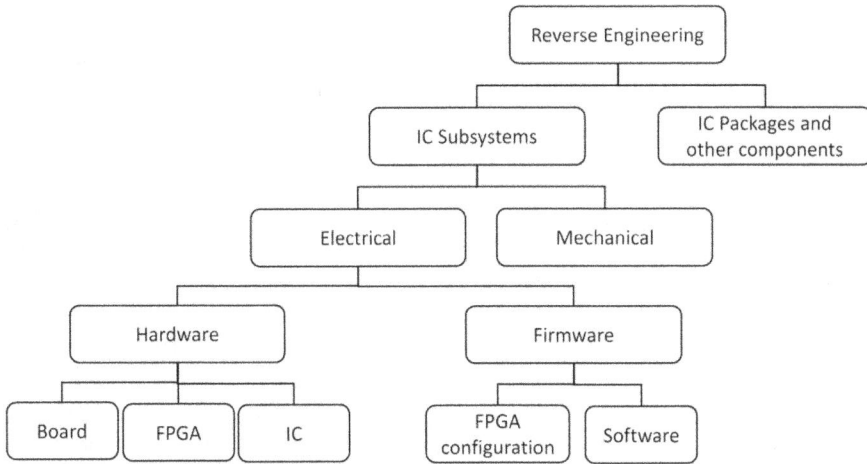

FIGURE 4.8 Interconnections in an integrated circuit (Source: https://semiengineering.com/all-about-interconnects/).

are not violated, and dishonest. Reasons to perform RE with honest motivation comprise functional verification, fault analysis, and understanding the working of deployed products. Dishonest intentions to mount RE comprise cloning, piracy, design counterfeiting, and insertion of HT. RE of electronic devices range from chip to system levels. A broad classification of RE application is shown in Figure 4.7.

An IC comprises multiple electronic devices fabricated using semiconductor materials. ICs constitute package material, bond wires, die, and lead frame. The die is composed of multiple metal layers, vias, and metal layer interconnections, as shown in Figure 4.8. X-ray tomography is one of the non-destructive RE methods that provide layer-wise images of the ICs for analysis of internal wire connections, vias, wire bonding, capacitors, etc. Destructive RE methods comprise etching and grinding each layer for analysis. In this process, images are captured with either *scanning electron microscope* (SEM) or *transmission electron microscope* (TEM).

In addition, printed circuit boards (PCBs) are reverse engineered first with identification of ICs, and other components mounted on them, and corresponding traces on the visible layers. Subsequently, X-ray imaging or delayering are used to further identify traces, connections, and vias of the internal layers in the PCBs. Furthermore, system-level RE targets the system's firmware that constitutes information about the system's operations and corresponding sequence of events. System firmware remains embedded in nonvolatile memories (NVMs), such as ROM, EEPROM, and flash memories. RE can provide insight into system functionality by analyzing the contents of such memories.

4.3.3 SIDE-CHANNEL ANALYSIS

SCAs are one of the strongest forms of passive attacks on implementations of cryptographic algorithms in electronic devices. An SCA adversary monitors physical observables, such as power, electromagnetic emanation, execution time, etc. A key-sensitive intermediate state of the implementation that affects the physical observable is modeled to infer the secret key or leak any information statistical parameter of the key.

4.3.4 IC COUNTERFEITING

IC counterfeiting is a longstanding problem that has grown in scope and magnitude over the past decade. With the advent of shrinking VLSI technology and the complexity of ICs used, they are assembled and fabricated globally over different geographical regions. This trend has led to a thriving illicit market that undercuts the market competition with counterfeit ICs and electronic systems. Some of the topmost counterfeited semiconductors comprise analog IC, microprocessor IC, memory IC, programmable logic IC, transistors, and others. In the recent past, several ML-based countermeasures have been employed for detecting counterfeit ICs. Automation of inspection procedures have recently used image processing algorithms and ANNs [29].

4.3.5 IC OVERPRODUCTION

IC overproduction involves the foundry producing more ICs than that required by an IC design house. Subsequently, the foundry sells the ICs in the market without authorization of the IC design house.

4.4 PRESENT PROTECTION MECHANISMS IN HARDWARE SECURITY

In existing literature, multiple ML-based countermeasures have been proposed that are classified as {\em design-for-security} techniques, as shown in Figure 4.9. Although a large portion of such countermeasures belong to HT defense domains, they are categorized into, (i) Trojan detection assistance, (ii) implantation prevention, and (iii) trusted libraries. The figure summarizes the research contributions of ML in design for security.

At present, there exist multiple hardware-based approaches for enhancing security, which comprise security primitives, such as PUF and crypto-processors, which form a component of an overall security architecture. ML algorithms intersect with these techniques constructively and destructively. Defenders can use ML along with hardware-based observations to construct models of IC operations for attack detection, in terms of software, hardware, and environmental conditions. On the other hand, attackers can use the same ML algorithms to extract sensitive data and process information from an IC, thus breaking corresponding trust assumptions in hardware security.

The role of ML towards existing hardware security challenges is summarized in Table 4.1. In the mentioned opportunities and threats offered by ML algorithms, some problems, such as model-building attacks in PUFs, have been well explored. On the other hand, problems such as analysis of implications of ML techniques in IC design flow towards hardware security threats, are yet to be explored.

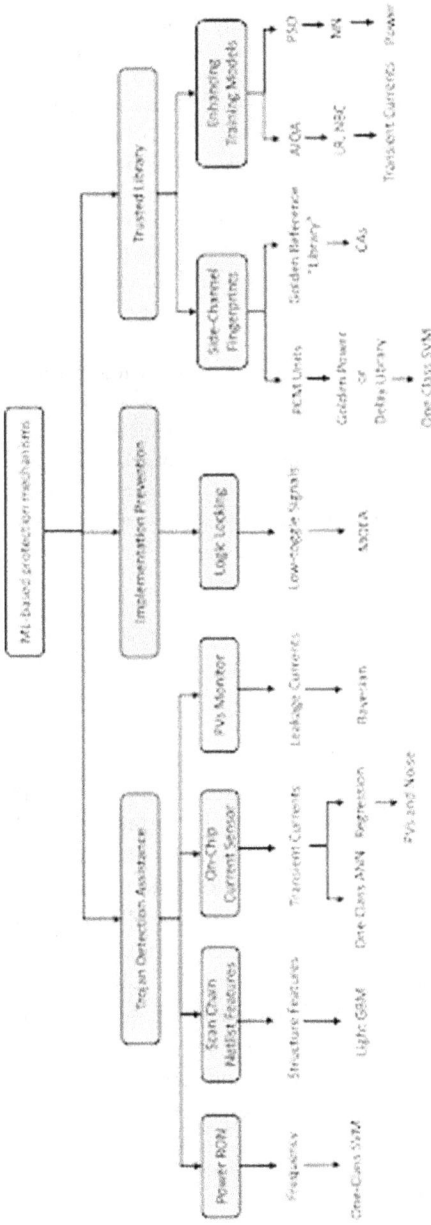

FIGURE 4.9 Classification of machine learning algorithms in design-for-security techniques.

TABLE 4.1

Role of Machine Learning in Hardware Security Challenges

	Challenges in Hardware Security	ML Opportunities in Hardware Security	Hardware Security Threats from ML	Challenges to ML Implementations
Hardware for Security	Power/area overhead, reliability, detecting and preventing bugs	Develop classification models for detecting functional anomalies, automation of security requirement extraction, attack detection and mitigation in real time	Model-building attacks on PUFs, exploit generation from automated bug detection processes, adaptive side-channel analysis	Construct exact artifact representation, producing high-quality training datasets, robustness against adversarial power, computational efficiency
Security of Hardware	Mitigate RE, IC piracy, IC overbuilding, IC counterfeiting, IC supply chain integrity, HW Trojan detection/ prevention, minimize side-channel leakage	Improve design flows, explore security techniques, automating HT detection, Automating bug detection, analysis, and patching	ML threats under adversarial settings, ML-based signal processing and decision making	

As hardware operates below software with operating system and hypervisor sandwiched in between, researchers have shown the path for behavior monitoring using hardware components, such as hardware performance counters used for detection of malware execution [30,31]. Such techniques depend on exact analysis and interpretation of sequence of instruction execution, and memory access events. ML techniques have been shown to learn characteristics of malicious behavior. Such techniques have been employed to interpret and make inferences by detecting patterns in multi-modal data [32].

In recent years, there have been multiple hardware-based primitives in the security architecture; however, with the advent of emerging threats, current research has focused on security of hardware design process and supply chain. One of the emerging potential attack vectors are the design tools themselves in the entire design flow [33].

4.4.1 HARDWARE TROJAN DETECTION

We hereby present the methods with which ML algorithms can be employed in the entire design flow for HT detection. HTs can be inserted in almost all stages of the design flow. They typically represent minor changes in area, timing delay, and other design metrics. Even after being stealthy, numerous ML-based approaches have been shown to have good success rates at HT detection in gate-level netlists [34–37]. Almost all these approaches are based on training and evaluation of ML models on benchmarks from Trust-Hub [38]. Efforts employing ML-techniques are non-trivial, as countermeasures first need to determine an appropriate input representation, such as multidimensional matrices, which can be easily represented but require significant effort to generate. For instance, in Hasegawa et al. [36], features were proposed based on net characteristics, gate count in the fan-in of a target net, and distance of a target net to the closest MUX in the design.

HT detection-based countermeasures can be classified in two categories: (i) countermeasures with availability of golden HT-free ICs, and (ii) countermeasures that assume HT-free ICs are unavailable. The countermeasures can be further classified in terms of extracted features, variants of trained models, and the functionality of HTs. The corresponding attacks can be of different variants, such as information leakage, denial-of-service, and functionality change for performance degradation. The HTs can be identified as abnormalities from on-chip sources, such as data streams from performance counters, and current measurements. The current measurements from on-chip sensors are converted to DC voltage. Subsequently, they are classified by one-class neural networks to pinpoint whether the chip has an embedded activated HT [39]. Furthermore, neural networks have been used to classify power consumption traces to detect HT-infected designs [40]. Process variations can be calibrated to facilitate HT detection by using Bayesian inference through analysis of leakage currents [41].

The countermeasures based on golden (HT free) ICs use the data extracted from the ICs to construct ML models that can categorize gate-level design nets, on-chip sensor data, or traffic between multiple cores to detect HTs. The summary of ML algorithms that have been employed for detecting HTs is shown in Figure 4.10.

ML-based methods have been applied to the domain of reverse engineering, circuit feature analysis, and SCA. In reverse engineering-based HT detection, one-class SVM or K-means can classify captured IC images [42,43]. Moreover, pattern mining techniques have been employed to reverse gate-level netlists to input-output tracking model [44].

In circuit feature analysis, dynamic cluster algorithms have been demonstrated to clusterless toggled signals (LTS) in the gate-level netlists in IP cores [45]. From the netlist of ICs, HT-related features have been extracted using two-class SVMs and multiple ANNs [46,47]. Furthermore, RF analysis has been shown to aid in selection of the best set of IC features that are HT infected. The HT detection accuracy can be increased by using tools, such as Voting Ensemble of RF, Bayes, and AdaBoot that can classify structural and functional features of ICs [48].

In SCA, extreme learning machines (ELM) have been shown to classify electrical current features in the design nets of ICs [12]. ML algorithms, such as BPNN, KNN, DT classification, naive Bayes classifier (NBC), PCA and Markov chain, and clustering algorithms have been used to classify power consumption traces, or profiling features of ICs [49,50]. Moreover, unsupervised approaches comprising one-class SVM with RBF kernel have been used for HT detection with better performance accuracy than template-based detection techniques [35].

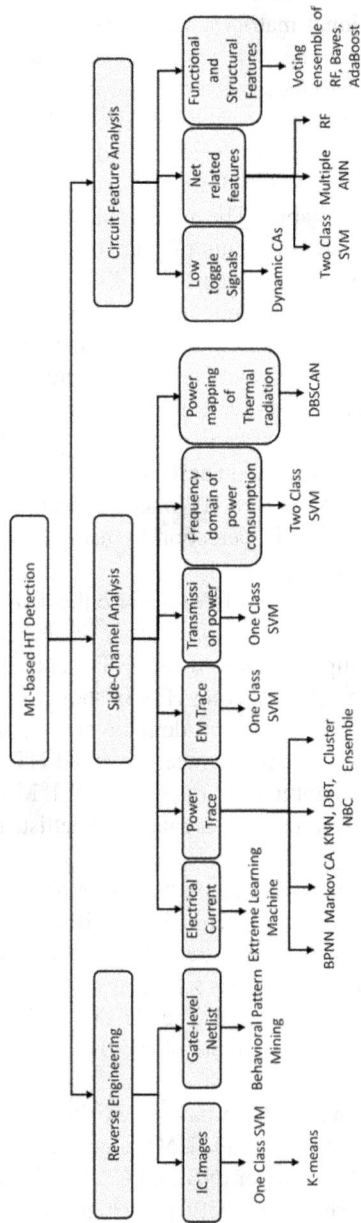

FIGURE 4.10 Summary of machine learning approaches in hardware Trojan detection.

4.4.2 IC COUNTERFEITING COUNTERMEASURES

With the advent of globalization in the IC manufacturing industry, IC counterfeiting is an emerging threat that has necessitated several ML-based countermeasures to identify such threats. Over the past decade, image processing techniques and ANNs have been proposed to detect physical defects that reveal IC wear out [29]. Recycled ICs were identified by analysis and classification of parametric measurements from on-chip sensors by using one-class SVMs [51]. Moreover, one-class SVMs have been used in detection of recycled ICs by classifying the frequencies and their degradation of ageing in ring oscillators, which find their applications in FPGA under test. These frequencies can be also clustered using a K-means clustering algorithm to detect recycled FPGAs.

4.4.3 REVERSE ENGINEERING APPROACH

ML-based RE countermeasures against a variety of IP piracy attacks and IC counterfeiting primarily comprise decapsulation, delayering, imaging, annotation, and reconstructing schematics. In the recent past, ML-based models have been trained to classify between HT-free and HT-infected ICs. The models are built with the assumption that golden physical layout is available since the HTs are added in the fabrication stage in the foundry. The image of the IC and its corresponding layout are partitioned into grids. These grids are then compared with the grids in the golden layout of the ICs. Subsequently, the difference between the corresponding grids in terms of area and centroid locations in the IC are employed for training ML models. For this purpose, one-class SVM with polynomial kernels are used [42]. Furthermore, unsupervised K-means clustering algorithm can cluster the IC grids to pinpoint the HT-infected grids [43].

4.5 MACHINE-LEARNING–BASED ATTACKS AND THREATS

In the previous section, ML-based techniques were mentioned that designers can employ to monitor and detect stealthy attacks by constructing models to pinpoint anomalies with hardware-enabled observation. However, adversaries can employ same techniques build models to extract sensitive information regarding circuit design in the IC. Furthermore, in the recent past, data input perturbations and training data poisoning by the adversary has put forth new issues in safeguarding the confidentiality and integrity of hardware design flow.

4.5.1 SIDE-CHANNEL ANALYSIS

SCA exploits information leakage from hardware and software implementations of cryptographic algorithms to extract secret information embedded in devices. In power-based SCA, power traces from circuits are captured, and by applying an appropriate power leakage model, the goal of the adversary is to identify the secret key that has been processed internally. Over the past decade, profiled SCA has emerged as the strongest variant as it has been shown to be effective even when only few traces or measurements are available [52].

Of the most commonly used profiled analysis, template attacks have emerged as the most powerful variant. However, there are many instances of worst case side-channel security threat wherein ML-based algorithms outperform template based attacks. Attackers are often provided with a sufficiently large number of power traces that aid in building precise leakage models. With a properly chosen algorithm and parameter tuning phase in ML algorithms, such attacks turn out to be more efficient. If the algorithm is properly tuned, it requires an even smaller number of features to achieve a high success rate of the attack. A measure called data confusion factor is proposed in [53] to differentiate between various ML methods. In supervised ML approach, SVMs, RF, rotation forest (RTF) [54], and multiboost algorithm [55] comprise classifiers that exhibited high accuracy on multiple datasets [56].

4.5.1.1   Side-Channel Analysis for Cryptographic Key Extraction

In this section, we present the state-of-the-art usage of ML algorithms in SCA to retrieve keys in cryptographic implementations. In profiling-based SCA, models are constructed that are based on the relation between the captured side-channel signals and the corresponding datapath intermediate values that are a function of the key.

In ML-based profiling attacks, profiling phase and attack form the two major phases. The attacker first captures side-channel signals for all possible target secret key values. For this capture, data points containing significant information of intermediate values in the datapath are chosen for profiling. Subsequently, a supervised ML model is trained to predict the value of secret keys embedded in the device.

4.5.1.2   Side-Channel Analysis for Instruction-Level Disassembly

Adversaries employing side-channel information can exploit the information to reveal executed instructions when programs run. For instance, pinpointing program blocks and corresponding operations in execution is possible with any captured side-channel information. This can be used to reverse engineer the programs in execution. Hidden Markov models can be used to extract assembly instructions from captured power traces. A block of program-in-execution can also be predicted using a KNN classifier based on a dynamic time warping metric from captured power races.

4.5.2   IC OVERBUILDING

IC overbuilding can be thwarted using techniques, such as hardware metering, logic obfuscation, watermarking, and fingerprinting, For IC fingerprinting, PUFs are used as the enforced challenge-response-pair (CRP) behavior to the system rely on certain physical traits of the system. In the recent past, ML models have been used to learn the CRP behavior of PUFs. Ring oscillator PUFs and arbiter PUFs have been efficiently modeled using ML algorithms.

4.6 EMERGING CHALLENGES AND NEW DIRECTIONS

In previous sections, ML-based algorithms have been shown to be both malevolent and benevolent in hardware security research. Defenders can use ML algorithms to identify stealthy attacks by building ML models to pinpoint anomalies. On the other hand, attackers can build similar models to infer secret information from the secured device. As these algorithms can be used for behavior monitoring, a pertinent question is, can ML algorithms enable decisions to guide a sequence of operations to preserve system integrity or thwart an attack? Moreover, if subjected to adversarial input perturbations and training data poisoning, how can the integrity of hardware design flows be safeguarded? In addition, can we render side-channel information deliberately biased so as to prevent ML-based modeling attacks? These challenges can usher new frontiers in ongoing research in ML algorithms for hardware security.

REFERENCES

[1] Tom M. Mitchell. *Machine Learning, International Edition. McGraw-Hill Series in Computer Science.* McGraw-Hill, Noida, India, 1997.

[2] Rana Elnaggar and Krishnendu Chakrabarty. Machine learning for hardware security: Opportunities and risks. *Journal of Electronic Testing,* 34(2):183–201, 2018.

[3] Z. Huang, Q. Wang, Y. Chen, and X. Jiang. A survey on machine learning against hardware trojan attacks: Recent advances and challenges. *IEEE Access,* 8:10796–10826, 2020.

[4] M.A. Hearst, S.T. Dumais, E. Osman, J. Platt, and B. Scholkopf. Support vector machines. *Intelligent Systems and Their Applications, IEEE,* 13(4):18–28, 1998.

[5] Gabriel Hospodar, Benedikt Gierlichs, Elke De Mulder, Ingrid Verbauwhede, and Joos Vandewalle. The investigation of neural networks performance in side-channel attacks. *Journal of Cryptographic Engineering,* 1(293):2190–8516, 2011.

[6] Jerome H. Friedman. Multivariate adaptive regression splines. *The Annals of Statistics,* 19(1):1–67, 1991.

[7] Y. Liu, G. Volanis, K. Huang, and Y. Makris. Concurrent hardware trojan detection in wireless cryptographic ics. In 2015 IEEE International Test Conference (ITC), pages 1–8, 2015.

[8] Yinan Kong and Ehsan Saeedi. The investigation of neural networks performance in side-channel attacks. *Artificial Intelligence Review,* 52(1):607–623, 2019.

[9] Yann LeCun and Yoshua Bengio. *Convolutional Networks for Images, Speech, and Time Series,* pages 255–258. MIT Press, Cambridge, MA, USA, 1998.

[10] Pierre Baldi. Autoencoders, unsupervised learning and deep architectures. In 2011 International Conference on Unsupervised and Transfer Learning Workshop - Volume 27, UTLW'11, pages 37–50. JMLR.org, 2011.

[11] Danilo P. Mandic and Jonathon Chambers. *Recurrent Neural Networks forPrediction: Learning Algorithms, Architectures and Stability.* John Wiley Sons, Inc., USA, 2001.

[12] Sixiang Wang, Xiuze Dong, Kewang Sun, Qi Cui, Dongxu Li, and Chunxiao He. Hardware trojan detection based on elm neural Network. In 2016 First IEEE International Conference on Computer Communication and the Internet (ICCCI), pages 400–403. IEEE, 2016.

[13] Felix A. Gers, Nicol N. Schraudolph, and J̈urgen Schmidhuber. Learning precise timing with LSTM recurrent networks. *The Journal of Machine Learning Research*, 3:115–143, 2002.

[14] Sweechuan Tan, Kaiming Ting, and Tonyfei Liu. Fast anomaly detection for streaming data. In Twenty-Second International Joint Conference on Artificial Intelligence, pages 1511–1516, 2011.

[15] N. Karimian, F. Tehranipoor, M. T. Rahman, S. Kelly, and D. Forte. Genetic algorithm for hardware trojan detection with ring oscillator network (ron). In 2015 IEEE International Symposium on Technologies for Homeland Security (HST), pages 1–6, 2015.

[16] Chongxi Bao, Domenic Forte, and Ankur Srivastava. On reverse engineering-based hardware trojan detection. *IEEE Transactions on Computer-Aided Design of Integrated Circuits and Systems*, 35(1):49–57, 2016.

[17] L. Kaufman and P. J. Rousseeuw. *Partitioning Around Medoids (Program PAM)*, chapter 2, pages 68–125. John Wiley Sons, Ltd., Hoboken, 1990.

[18] A.N. Nowroz, K. Hu, F. Koushanfar, and S. Reda. Novel techniques for high-sensitivity hardware trojan detection using thermal and power maps. *IEEE Transactions onComputer-Aided Design of Integrated Circuits and Systems*, 33(12):1792–1805, 2014.

[19] B. Cakır and S. Malik. Hardware trojan detection for gate-level ics using signal correlation based clustering. In 2015 Design, Automation Test in Europe Conference Exhibition (DATE), pages 471–476, 2015.

[20] Hanchuan Peng, Fuhui Long, and Chris Ding. Feature selection based on mutual information: Criteria of max-dependency, max-relevance, and min-redundancy. *IEEE Transactions on Pattern Analysis and Machine Intelligence*, 27(8):1226–1238, 2005.

[21] Svante Wold, Kim Esbensen, and Paul Geladi. Principal component analysis. *Chemometrics and Intelligent Laboratory Systems*, 2(1):37–52, 1987. Proceedings of the Multivariate Statistical Workshop for Geologists and Geochemists.

[22] Y. Liu, Y. Jin, A. Nosratinia, and Y. Makris. Silicon demonstration of hardware trojan design and detection in wireless cryptographic ICs. *IEEE Transactions on Very Large Scale Integration (VLSI) Systems*, 25(4):1506–1519, 2017.

[23] T. Kohonen. The self-organizing map. *Proceedings of the IEEE*, 78(9):1464–1480, 1990.

[24] Mohammad Tehranipoor and Farinaz Koushanfar. A survey of hardware trojan taxonomy and detection. *IEEE Design & Test of Computers*, 27(1):10–25, 2010.

[25] Swarup Bhunia, Michael S. Hsiao, Mainak Banga, and Seetharam Narasimhan. Hardware trojan attacks: Threat analysis and countermeasures. *Proceedings of the IEEE*, 102(8):1229–1247, 2014.

[26] Robert J. Abella, James M. Daschbach, and Roger J. McNichols. Reverse engineering industrial applications. *Computers & Industrial Engineering*, 26(2):381–385, 1994.

[27] Randy Torrance and Dick James. The state-of-the-art in ic reverse engineering. In International Workshop on Cryptographic Hardware and Embedded Systems, pages 363–381. Springer, 2009.

[28] Ian McLoughlin. Secure embedded systems: The threat of reverse engineering. In 2008 14th IEEE International Conference on Parallel and Distributed Systems, pages 729–736. IEEE, 2008.

[29] Navid Asadizanjani, Mark Tehranipoor, and Domenic Forte. Counterfeit electronics detection using image processing and machine learning. In *Journal of Physics: Conference Series*, volume 787, page 012023. IOP Publishing, 2017.

[30] Nisarg Patel, Avesta Sasan, and Houman Homayoun. Analyzing hardware based malware detectors. In 2017 54th ACM/EDAC/IEEE Design Automation Conference (DAC), pages 1–6. IEEE, 2017.

[31] Xueyang Wang, Charalambos Konstantinou, Michail Maniatakos, and Ramesh Karri. Confirm: Detecting firmware modifications in embedded systems using hardware performance counters. In 2015 IEEE/ACM International Conference on Computer-Aided Design (ICCAD), pages 544–551. IEEE, 2015.

[32] Adrien Facon, Sylvain Guilley, Xuan-Thuy Ngo, and Thomas Perianin. Hardware-enabled AI for embedded security: A new paradigm. In 2019 3rd International Conference on Recent Advances in Signal Processing, Telecommunications & Computing (SigTelCom), pages 80–84. IEEE, 2019.

[33] Kanad Basu, Samah Saeed, Christian Pilato, Mohammed Ashraf, Mohammed Nabeel, Krishnendu Chakrabarty, and Ramesh Karri. Cad-base: An attack vector into the electronics supply chain. *ACM Transactions on Design Automation of Electronic Systems*, 24:1–30, 2019.

[34] Konstantinos G. Liakos, Georgios K. Georgakilas, Serafeim Moustakidis, Patrik Karlsson, and Fotis C. Plessas. Machine learning for hardware trojan detection: A review. In 2019 Panhellenic Conference on Electronics & Telecommunications (PACET), pages 1–6. IEEE, 2019.

[35] Dirmanto Jap, Wei He, and Shivam Bhasin. Supervised and unsupervised machine learning for side-channel based trojan detection. In 2016 IEEE 27th International Conference on Application-specific Systems, Architectures and Processors (ASAP), pages 17–24. IEEE, 2016.

[36] Kento Hasegawa, Youhua Shi, and Nozomu Togawa. Hardware trojan detection utilizing machine learning approaches. In 2018 17th IEEE International Conference on Trust, Security and Privacy in Computing and Communications/12th IEEE International Conference on Big Data Science and Engineering (Trust-Com/ BigDataSE), pages 1891–1896. IEEE, 2018.

[37] Tao Han, Yuze Wang, and Peng Liu. Hardware trojans detection at register transfer level based on machine learning. In 2019 IEEE International Symposium on Circuits and Systems (ISCAS), pages 1–5. IEEE, 2019.

[38] Hassan Salmani, Mohammad Tehranipoor, and Ramesh Karri. On design vulnerability analysis and trust benchmarks development. In 2013 IEEE 31st International Conference on Computer Design (ICCD), pages 471–474. IEEE, 2013.

[39] Yier Jin, Dzmitry Maliuk, and Yiorgos Makris. Post-deployment trust evaluation in wireless cryptographic ics. In 2012 Design, Automation & Test in Europe Conference & Exhibition (DATE), pages 965–970. IEEE, 2012.

[40] Jun Li, Lin Ni, Jihua Chen, and Errui Zhou. A novel hardware trojan detection based on bp neural network. In 2016 2nd IEEE International Conference on Computer and Communications (ICCC), pages 2790–2794. IEEE, 2016.

[41] Xiaoming Chen, Lin Wang, Yu Wang, Yongpan Liu, and Huazhong Yang. A general framework for hardware trojan detection in digital circuits by statistical learning algorithms. *IEEE Transactions on Computer-Aided Design of Integrated Circuits and Systems*, 36(10):1633–1646, 2016.

[42] Chongxi Bao, Domenic Forte, and Ankur Srivastava. On application of one-class SVM to reverse engineering-based hardware trojan detection. In Fifteenth International Symposium on Quality Electronic Design, pages 47–54. IEEE, 2014.

[43] Chongxi Bao, Domenic Forte, and Ankur Srivastava. On reverse engineering-based hardware trojan detection. *IEEE Transactions on Computer-Aided Design of Integrated Circuits and Systems*, 35(1):49–57, 2015.

[44] Wenchao Li, Zach Wasson, and Sanjit A. Seshia. Reverse engineering circuits using behavioral pattern mining. In 2012 IEEE International Symposium on Hardware-Oriented Security and Trust, pages 83–88. IEEE, 2012.

[45] Er-Rui Zhou, Shao-Qing Li, Ji-Hua Chen, Lin Ni, Zhi-Xun Zhao, and Jun Li. A novel detection method for hardware trojan in third party ip cores. In 2016 International Conference on Information System and Artificial Intelligence (ISAI), pages 528–532. IEEE, 2016.

[46] Kento Hasegawa, Masao Yanagisawa, and Nozomu Togawa. A hardware-trojan classification method using machine learning at gate-level netlists based on trojan features. *IEICE Transactions on Fundamentals of Electronics, Communications and Computer Sciences*, 100(7):1427–1438, 2017.

[47] Kento Hasegawa, Masao Yanagisawa, and Nozomu Togawa. Trojan-feature extraction at gate-level netlists and its application to hardware-trojan detection using random forest classifier. In 2017 IEEE International Symposium on Circuits and Systems (ISCAS), pages 1–4. IEEE, 2017.

[48] Tamzidul Hoque, Jonathan Cruz, Prabuddha Chakraborty, and Swarup Bhunia. Hardware IP trust validation: Learn (the untrustworthy), and verify. In 2018 IEEE International Test Conference (ITC), pages 1–10. IEEE, 2018.

[49] Faiq Khalid Lodhi, I. Abbasi, Faiq Khalid, Osman Hasan, F. Awwad, and Syed Rafay Hasan. A self-learning framework to detect the intruded integrated circuits. In 2016 IEEE International Symposium on Circuits and Systems (ISCAS), pages 1702–1705. IEEE, 2016.

[50] Faiq Khalid Lodhi, Syed Rafay Hasan, Osman Hasan, and Falah Awwadl. Power profiling of microcontroller's instruction set for runtime hardware trojans detection without golden circuit models. In Design, Automation & Test in Europe Conference & Exhibition (DATE), 2017, pages 294–297. IEEE, 2017.

[51] Halit Dogan, Domenic Forte, and Mark Mohammad Tehranipoor. Aging analysis for recycled fpga detection. In 2014 IEEE International Symposium on Defect and Fault Tolerance in VLSI and Nanotechnology Systems (DFT), pages 171–176. IEEE, 2014.

[52] Annelie Heuser and Michael Zohner. Intelligent machine homicide. In International Workshop on Constructive Side-Channel Analysis and Secure Design, pages 249–264. Springer, 2012.

[53] Stjepan Picek, Annelie Heuser, Alan Jovic, Simone A Ludwig, Sylvain Guilley, Domagoj Jakobovic, and Nele Mentens. Side-channel analysis and machine learning: A practical perspective. In 2017 International Joint Conference on Neural Networks (IJCNN), pages 4095–4102. IEEE, 2017.

[54] Juan Jose Rodriguez, Ludmila I. Kuncheva, and Carlos J. Alonso. Rotation forest: A new classifier ensemble method. *IEEE Transactions on Pattern Analysis and Machine Intelligence*, 28(10):1619–1630, 2006.

[55] Yoav Freund and Robert E. Schapire. A decision-theoretic generalization of on-line learning and an application to boosting. *Journal of Computer and System Sciences*, 55(1):119–139, 1997.

[56] Manuel Fernandez-Delgado, Eva Cernadas, Sen´en Barro, and Dinani Amorim. Do we need hundreds of classifiers to solve real world classification problems? *The Journal of Machine Learning Research*, 15(1):3133–3181, 2014.

5 Application-Driven Fault Identification in NoC Designs

Ankur Gogoi and Bibhas Ghoshal
Department of IT, IIIT Allahabad, Allahabad, Uttar Pradesh, India

CONTENTS

5.1 INTRODUCTION

To cope with rapidly growing requirements of computation and bandwidth, Network-on-Chip (NoC) has been accepted as a promising communication infrastructure for multicore architectures. NoC communication infrastructure facilitates fast and reliable transmission of data packets among the cores employing various routing algorithms. These algorithms ensure that the router and link-based architecture of NoC, as shown in Figure 5.1, deliver packets and satisfy the different design requirements such as low latency, high throughput, etc. The traffic fed to the NoC gets transferred as flits (small chunks of a packet) through these routers and links [1].

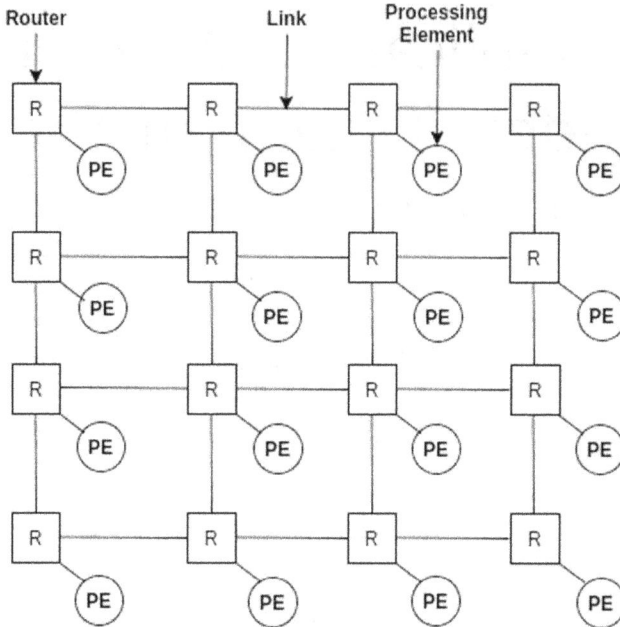

FIGURE 5.1 A typical 4 × 4 NoC architecture.

As found in literature, researchers have considered two techniques for routing packets in NoC – source and distributed routing [2]. In source routing, the packet header is encoded with the whole path from source to destination. In the case of distributed routing, the destination address is injected in the packet header, and the routing path is determined in each intermediate router using the routing function contained in the routers. NoC supports various applications, and each application gives rise to a unique traffic pattern while processing different applications. That is, the routing pattern of each application traffic corresponding to different applications is different from each other, and the number of flits processed by each router with respect to different application traffic is not fixed. For a router, the more flit it processes, it is more likely to become faulty as the health status of a router in NoC is directly related to the amount of flits it has processed. Therefore, if a set of application traffic with different traffic patterns is injected into an NoC, then for each traffic pattern, the health status of each router will vary. Thus, routers while handling various applications, have to deal with varied workloads. As a result, the lifetime (health status) of routers vary. This variation in health status of individual routers in turn affects the overall performance of the NoC. Thus, it becomes imperative to identify and isolate the faulty routers before they start malfunctioning. During designing a Multi-Processor SoC (MPSoC), identifying probable faulty routers is a difficult task since it is an application-dependent problem. That is, MPSoCs are used for applications that are not known during design time [3].

Thus, it becomes essential to have an application-traffic–driven approach for probable faulty router prediction so that necessary fault recovery mechanisms can be applied for the NoC routers. We propose the use of traffic classification to identify probable fault sites in NoC. A supervised machine learning (ML) approach is used to classify traffic and identify the fault sites or the vulnerable routers with respect to that traffic.

5.2   RELATED WORK

For a network of computers, working on various traffic patterns, traffic classification is utilized to inspect and achieve improvement in network performance. In contrast, to the best of our knowledge, only a limited number of approaches exist for application traffic classification for NoC architectures. Initially, fault-tolerant routing algorithms and structural redundancy methods were prime approaches of fault-detection and recovery mechanisms. Of late, researchers have started experiencing the need for the classification of NoC application traffics to achieve improvement in performance of a NoC-based system. Sooner or later, the NoC traffic classification will play a key role in examining various aspects of NoC, such as security, bandwidth, etc. Some notable works that have classified the NoC traffics are by Lin et al. [4], who proposed a method to achieve guaranteed latency or bandwidth considering three different types of traffic: GL (guaranteed latency), GB (guaranteed bandwidth), and BE (best effort). Bolotin et al. [5] classified the NoC traffics into four common classes: signaling, real-time, RD/WR and block transfer based on throughput, end-to-end delay, and relative priority to ensure the achievement of required Quality-of-Service (QoS). A traffic-aware crosstalk fault-detection method with minimum intrusions has been presented in Liu et al. [6].

Above-mentioned traffic classification of these approaches has been performed to evaluate NoC in terms of bandwidth and latency. Fault-tolerance, though a major concern, was not addressed. In this work, we present a supervised ML-based technique of predicting possible fault sites (vulnerable routers) in NoC for a certain set of real-world application traffics [7].

5.3   IDENTIFICATION OF VULNERABLE ROUTERS

In this section, we derive an approach that identifies the vulnerable routers for given application traffic. Our approach considers the reliability of a router in terms of traffic processed by the router [8] and calculates the workload of individual routers to find out the vulnerable routers under the influence of different application traffic. Both mathematical derivation and our experimental work for the identification of vulnerable routers are presented in this section. The mathematical reliability model provides an insight into how the workload affects the temperature generated in a router which in turn makes an impact on the reliability of the NoC system. A simulation-based experimental work is also presented, which provides the total number of flits processed by each router and later the vulnerable routers are identified by determining a threshold. Noxim [9] simulator has been used for the experimental purpose to acquire the necessary knowledge for both identifications of vulnerable routers and to classify the application traffics.

5.3.1 Proposed Mathematical Model for Router Reliability

We propose a reliability model for NoC routers that involves defining router reliability in terms of traffic workload defined in Rosing et al. [10], which can be expressed as:

$$R(t) = e^{-\int_0^1 \lambda(t)dt} \tag{5.1}$$

Where λ is the failure rate that varies with time t.

Traditionally, the failure rate of an NoC router is assumed to be constant with a value of 0.00315 (times/year) [11–13]. Though it is a suggested value, a deeper inquiry leads us to the fact that the actual value of failure rate is a complex function of not just time but also of other parameters, such as temperature, power states, and frequency of switching between power states [10]. Reliability is mainly affected by factors, including accelerated aging mechanisms (time-dependent dielectric breakdown, or TDDB, and electromigration, or EM), process variations, dynamic power, thermal management, and workload [8]. Modeling reliability in terms of these failure mechanisms (TDDB and EM) leads us to the fact that reliability varies strongly with respect to temperature. Reliability models at the device or transistor level are easily available and have been estimated in terms of the different failure mechanisms, such as TDDB, etc. However, the reliability estimation component or system level takes a more complex form, and several different ways of its estimation have been proposed [10,14].

An approximate mathematical equation between failure rate (λ) and the temperature with respect to electro-migration (EM), as derived in Lu et al. [15] is given as:

$$\lambda_{EM}(J, T) = Je^{\dfrac{-EaEM}{kT}} \tag{5.2}$$

Where E_a is the activation energy.

Similarly, for TDDB, the relationship between the failure rate and temperature can be expressed as [16],

$$\lambda(J, T)_{TDDB} \propto Je^{\dfrac{x + \frac{Y}{T} + ZT}{kT}} \tag{5.3}$$

Where X, Y and Z are fitting parameters and kT is the thermal energy.

For both EM and TDDB, the current density J is,

$$J = \left(\frac{CVdd}{WH}\right) * f * \sigma$$

Where σ is the switching activity or bit transitions in one clock cycle, Vdd is the voltage, and f is frequency. If the voltage and frequency are assumed to be constant, then the current density depends on the switching activity, which in turn depends on the incoming flit rate [15]. The failure rate can then be approximated as follows:

For EM,

$$\lambda_{EM}(J, T) \propto (n_i/\Delta t_s)\sigma_{avg}\, e^{-\dfrac{EaEM}{kT}} \tag{5.4}$$

For TDDB,

$$\lambda_{TDDB}(J, T) \propto (n_i/\Delta t_s)\sigma_{avg}\, e^{\dfrac{x + \frac{Y}{T} + ZT}{kT}} \tag{5.5}$$

Here, σ_{avg} is the average bit transitions per flit in a router, where n_i is the total number of incoming flits. The incoming flit rate is the number of flits passing through the router per unit of time. The total incoming flits are the total flits crossing the router during the traffic processing in Δt_s time and also stands for the workload of a router [12].
For EM,

$$\lambda_{EM}(J, T) \propto n_i.\; e^{-\dfrac{EaEM}{kT}} \tag{5.6}$$

For TDDB,

$$\lambda_{TDDB}(J, T) \propto n_i.\; e^{\dfrac{x + \frac{Y}{T} + ZT}{kT}} \tag{5.7}$$

From equations (5.6) and (5.7), it can be concluded that reliability is a function of the number of bit transitions taking place in an NoC router and consequently is a function of the total number of flits crossing through the router, i.e. their individual traffic load. Higher router workload contributes to an increased temperature, which in turn affects the reliability adversely and making a router more prone to failures.

5.3.2 Determination of the Vulnerable Routers Using Simulation

For fault-tolerant NoC design, the router plays the most important role and all other components (memory, I/O, PEs) are simply considered to be nodes. The traffic data for an application is taken as input and the traffic load of each router. In terms of the number of incoming flits, f_i is obtained from the simulator and summed up to find out the total number of flits F processed by the whole NoC during the processing of that traffic. A comparison of these results with the threshold flit value leads to the identification of the vulnerable routers that need redundancy. The threshold traffic (N_{th}) is obtained as,

$$N_{th} = \dfrac{F}{ActiveRouters}$$

Here, $F = \Sigma f_i$, $i = 0, 1, 2 \dots n$ where n is the total number of active routers

Where F is the total number of flits crossing through the routers of the NoC and *Active Routers* are the number of routers (participating routers) that have been processing the flits. We have considered a 4×4 2D mesh NoC architecture for our experiment to detect the vulnerable routers. Simulations were performed to estimate the workload of every node in the network under the influence of a particular traffic profile and XY routing algorithm. The simulations were run using Noxim simulator, and the outcome of the simulation results are shown below.

Figure 5.2 shows the router load in terms of total incoming flits for encoder/decoder traffic in a 4×4 architecture with threshold $N_{th} = 1883$ flits. The traffic load in routers 0, 2, 4, and 8 are seen to exceed the N_{Th} and fall under the category of vulnerable routers. The routing algorithm affects the total incoming flits by adding flits that are in transit, thereby causing a change in the traffic load of each router [15].

As seen in Figure 5.3, routers 4, 5, 6, and 8 exceed their limit of 2696 flits/node (N_{th}). As a result, the mesh-based NoC architecture when used for MPEG application the routers 4, 5, 6, and 8 are identified as vulnerable routers.

Similarly, for consumer application traffic, shown in Figure 5.4, router numbers 2, 3, 4, 8, 9, and 10 crossed the threshold limit and need fault-tolerant support for working properly. For MWD traffic shown in Figure 5.5, the locations 1, 2, 3, and 4 become vulnerable to fault. In the case of networking traffic shown in Figure 5.6, router positions 1, 2, 3, 4, and 8 are identified as faulty. From Figure 5.7, it can be observed that for office application traffic our experiment result identified router positions 1 and 2 as vulnerable routers.

5.3.3 LOOK-UP-TABLE (LUT) GENERATION FROM EXPERIMENTAL DATA

The vulnerable routers are identified using the approach mentioned in the previous section (Section 5.3.2) and stored in the form of a table for classification convenience. The structure of the table is shown below:

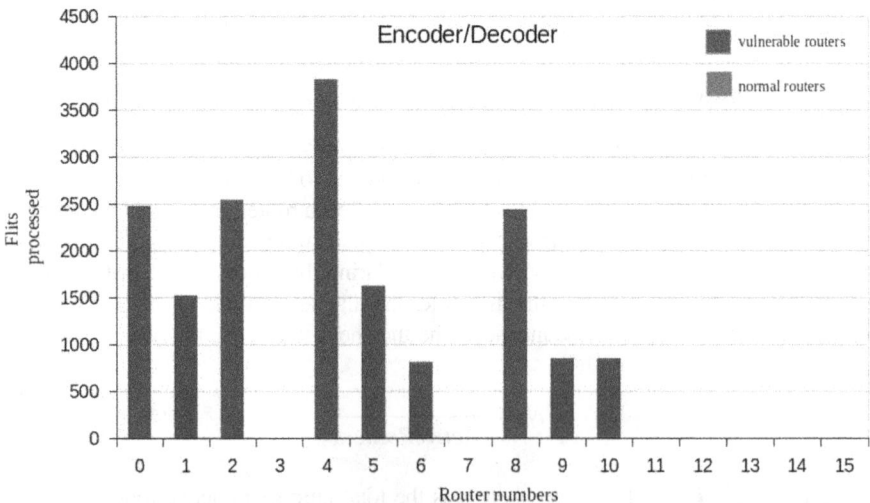

FIGURE 5.2 Vulnerable router identification for encoder/decoder traffic.

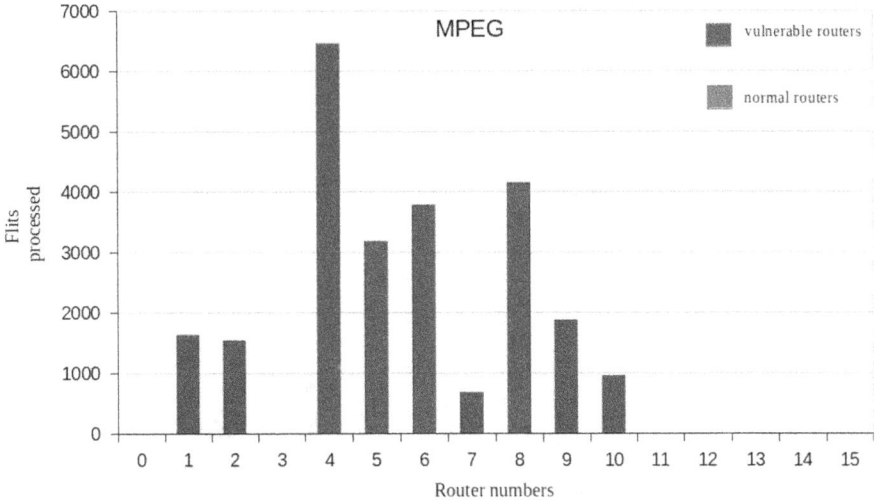

FIGURE 5.3   Vulnerable router identification for MPEG traffic.

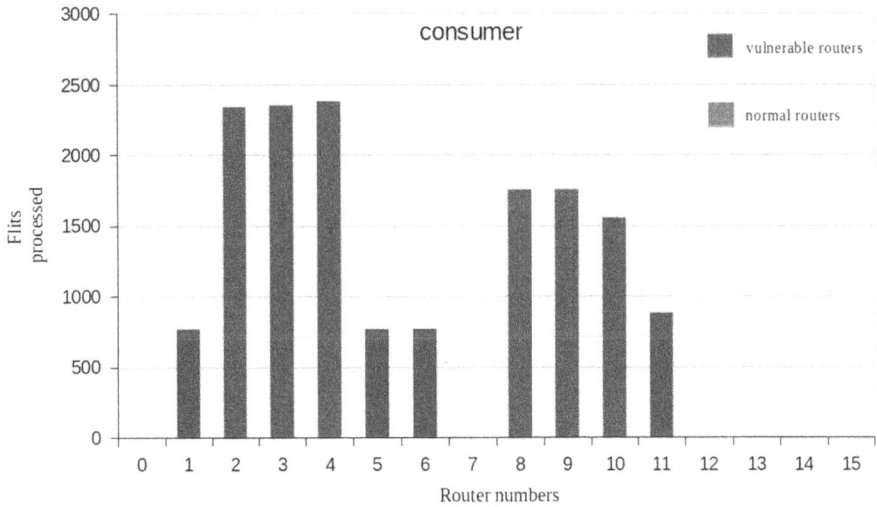

FIGURE 5.4   Vulnerable router identification for consumer traffic.

Traffic Label	**List of Vulnerable Routers**
t_i	$l_1, l_2, l_3 \ldots l_N$

Here, t_i is the processed traffic, and l_i is the list of vulnerable routers determined against the traffic t_i. The size of the LuT can be defined as $M*N$, where M is the number of traffic classes considered for classification and N is the number of routers

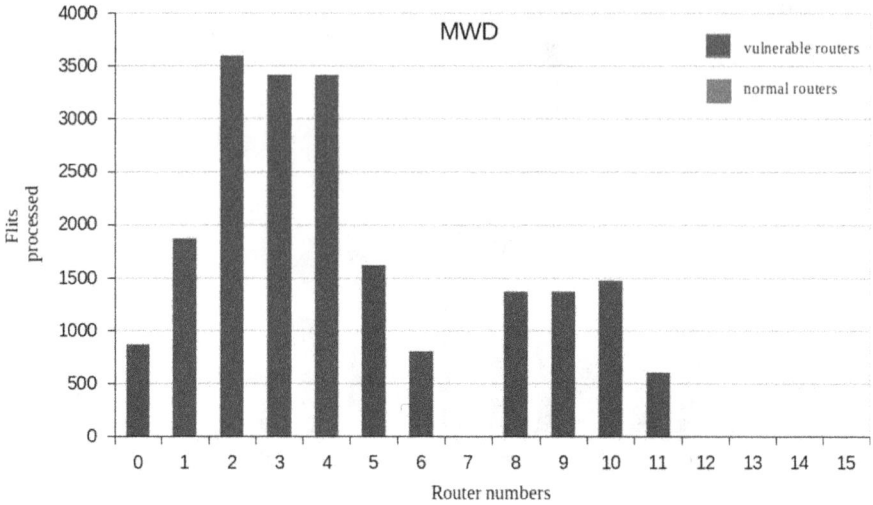

FIGURE 5.5   Vulnerable router identification for MWD traffic.

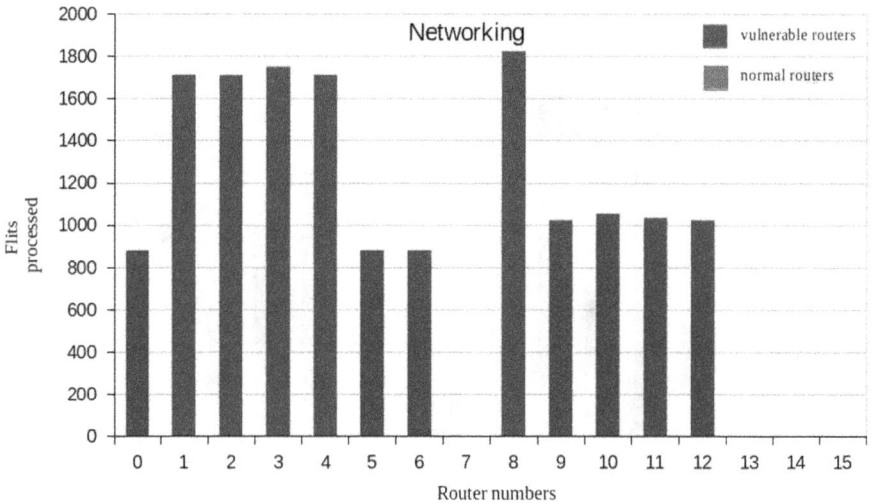

FIGURE 5.6   Vulnerable router identification for networking traffic.

in the NoC. For experimental purposes, we have considered M = 12 and N = 64. This LuT plays an important role in successful identification of vulnerable routers in NoC for a given traffic.

5.4   THE PROPOSED METHODOLOGY FOR THE IDENTIFICATION OF VULNERABLE ROUTERS

The distinguishing property of an NoC design is the ability to predict the vulnerable routers before the traffic is injected into the network. As shown earlier, the

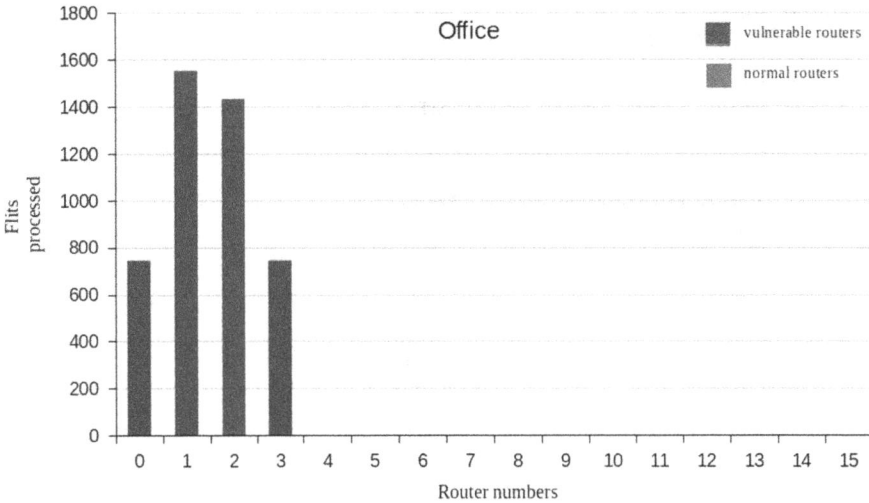

FIGURE 5.7   Vulnerable router identification for office traffic.

vulnerability of routers depends upon the different application patterns. Depending on different application traffic, network usage patterns are also changed, which leads to different locations of vulnerable routers in NoC for different application traffic. Therefore, the classification of different application traffic becomes necessary to identify incoming application traffic. The identified traffic class is further processed for determining vulnerable routers in the NoC and later different fault-tolerant techniques can be enforced to these vulnerable routers. The steps to identify the vulnerable routers for application traffic is shown in Figure 5.8. Our proposed methodology can be divided into two parts: Traffic classification using supervised ML techniques and using the classified traffic label to identify the vulnerable routers in NoC. These phases together result in a set of vulnerable routers. The algorithm for identification of vulnerable routers are shown below.

ALGORITHM 1   IDENTIFY THE VULNERABLE ROUTERS FOR APPLICATION TRAFFIC

Input: Application traffic   **Output:** Vulnerable router set

getVulnerableRouterSet (appTraffic)

1. Mesh-dimension: 8 × 8, flit_size: {16, 32, 64, 128}
2. buffer_depth :8, routing_algo= XY
3. for i = 1 to 1000 cycles
4. warmupNoxim()
5. end for

```
6. for j = 1 to 10000 cycles
7. noximResult = runNoxim(appTraffic)
8. end for
9. featureVector = extractFeatures(noximResult)
10. predictClass = classificationML(featureVector)
11. for k = 1 to 12
12. bool = exist(predictClass, LuT_class(k))
13. If bool is true
14. result = getVulnerableSet(LuT_class(k))
15. end if
16. end for
17. print result
```

The above algorithm takes application traffic as input and provides vulnerable router sets as output. The algorithm can be divided into two phases, as mentioned earlier. The first phase comprises Noxim setup and traffic classification, and the second phase takes care of the LuT-based identification of vulnerable routers. The setup parameters are made available and ready before warming up of the simulator and can be seen in lines 1 and 2 of the algorithm. From line 3 to line 5 the Noxim simulator is warmed up with the defined setup present in line 1 and line 2. After successful warm-up, the given application traffic will simulate for 10,000 cycles (from lines 6 to 8). The simulation outcome is stored in *noximResult* (line 7). The features are extracted from the stored result of Noxim using the *extractFeatures* function (line 9) to construct the feature vector (*featureVector*). Next, the feature vector is fed to the ML traffic classification model (*classificationML*) to predict the class of the given traffic (*appTraffic*) (line 10). The first phase of the algorithm comprises up to line 10, and after the successful classification of traffic, the predicted class is passed to the second phase of the algorithm to get the probable vulnerable set. First, the predicted class is validated or matched with the available traffic class present in the LuT (lines 11 and 12). If the predicted class is present in the LuT list (line 12), then its corresponding vulnerable router set is extracted from the LuT list using the function *LuT_class* (line 14).

5.4.1   CLASSIFICATION OF APPLICATION TRAFFIC USING MACHINE LEARNING

The traffic classification is performed on an 8 x 8 mesh-based NoC experimental data. Prior to training the ML model, required traffic data are generated from the rigorous simulation using Noxim, and the feature vector is generated from those extracted traffic data. All the necessary steps for training the ML model are presented below.

5.4.1.1   Dataset Generation

For proper traffic classification, we trained the ML classification model with the generated data, which were obtained from the Noxim simulator with the simulation setup shown in Table 5.1. To have detailed information of application traffic

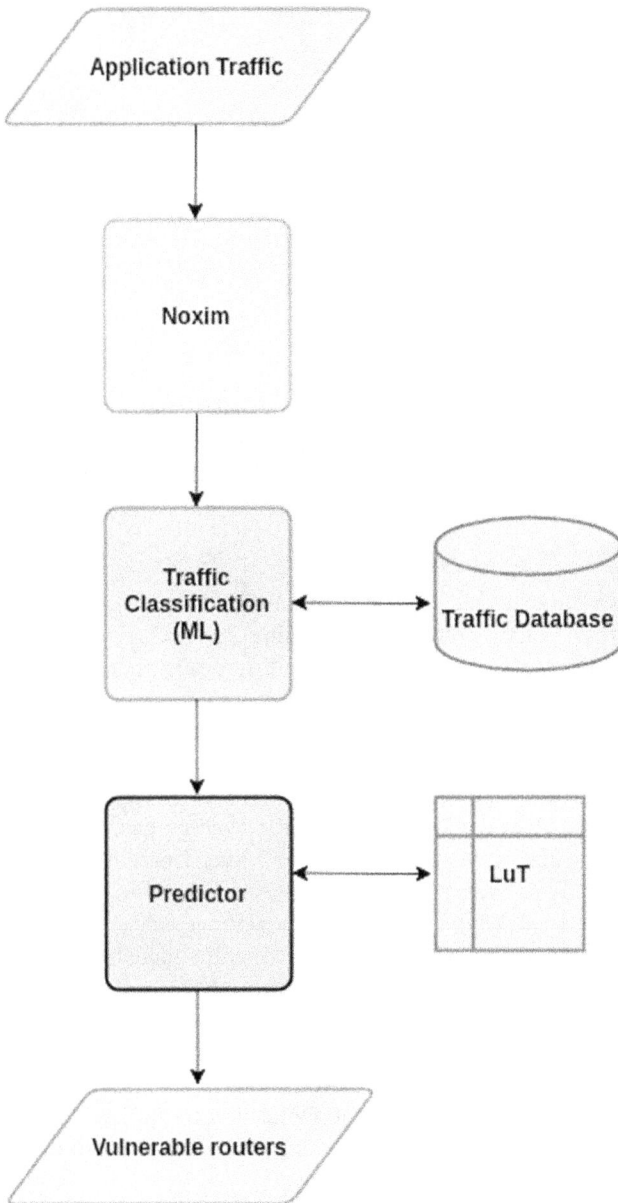

FIGURE 5.8 Steps for identification of vulnerable routers.

without any anomalies, flit-level traffic routing information has been considered. Twelve different types of application traffic are considered for the experiment, and the total number of application traffic can be further extended according to the interest of the individuals. Along with throughput and delay, the power consumption of application traffic has also been considered for a more accurate and specific

TABLE 5.1
Simulation Setup

Traffic	263decmp3dec, 263encmp3dec, auto indust, consumer, mp3encmp3dec, MPEG, networking, office, pip, telecom, VOPD
Mesh-Dimension	64 nodes
Warm-Up Cycle	1000
Simulation Cycle	10000
Flit Size	16 bit, 32 bit, 64 bit, 128 bit
Routing Algorithm	XY

classification. To enhance the classification model with different bit-level architecture data, the application traffics have been routed with different flit sizes (16 bit, 32 bit, 64 bit and 128 bit). Each application traffic is evaluated in Noxim by varying its packet-injection-rate (PIR) from 0.01 to 0.99.

5.4.1.2 Feature Vector Extraction

The data that were extracted from the output of Noxim have been studied to get a suitable optimal feature vector by combining six features that show justifiable variation for different traffic patterns. As different application traffics are considered for the experiments, the six factors that vary the overall performance of the traffics are taken as features. Those features are combined as the feature vector that can be represented as six tuples: Feature_Vector < Global Average Delay, Max Delay, Network throughput, Average IP Throughput, Dynamic Energy, Static Energy >. Global Average Delay implies the average delay incurred due to the processing of all flits of a specific traffic, and Max Delay presents the maximum delay experienced in a cycle processing all flits of that traffic. The throughput of the NoC can be obtained using the network throughput value, and this value varies depending on application traffic. Average IP throughput gives the average flit processing throughput of each individual router of the NoC. Dynamic Energy (in joules) indicates the fluctuations in energy consumption with respect to different operations. Static Energy (in joules) is fixed for all kinds of operations as the power consumption of the components of the NoC remains the same in all operations. All the above-mentioned features represent application traffic in a parameterized way and are suitable for traffic classification using ML. The sample of the feature vector of MPEG traffic is presented in Table 5.2.

5.4.1.3 Training of the ML Model

During the training phase, the feature vector that has been acquired during the profiling of application traffics is used to train the ML model, and all the trained knowledge is stored in the training database. The training database stores both feature vector data and the respective traffic label for all traffics. The supervised ML approach has been adopted for the classification. Among various supervised ML algorithms, K-NN algorithm is chosen as we assume that traffics having identical feature vectors belong to the same class. The ML operation is performed in the

TABLE 5.2

Feature Vector of MPEG Traffic with Respect to Different PIR

PIR	Max Delay	Global Average Delay	Network Throughput	Average IP Throughput	Dynamic Energy	Static Energy
0.01	3241	604.901	1.34344	0.0209913	1.9912e-07	7.901e-06
0.10	9339	3913.62	1.16622	0.0182222	1.89936e-07	7.901e-06
0.20	9604	4768.97	1.16611	0.0182361	1.95737e-07	7.901e-06
0.30	9736	4949.53	1.16711	0.0182361	1.95737e-07	7.901e-06
0.40	9755	5091.8	1.16667	0.0182292	1.89695e-07	7.901e-06
0.50	9780	5183.96	1.16633	0.018224	1.96968e-07	7.901e-06
0.60	9782	5220.72	1.167	0.0182344	1.97159e-07	7.901e-06
0.70	9780	5256.24	1.16711	0.0182361	1.96795e-07	7.901e-06
0.80	9785	5289.14	1.16656	0.0182274	1.96721e-07	7.901e-06
0.90	9780	5304.79	1.16796	0.0182270	1.96671e-07	7.901e-06

MATLAB environment. The training of the ML model for the classification of different applications is highlighted in Figure 5.9.

5.4.1.4 Working of the Trained Model

Application is fed as input to Noxim to extract the feature data of that particular traffic from the output of Noxim. Then, the application traffic's feature data are fed as input to the supervised ML algorithm (KNN), and with the help of the previously trained model, the algorithm classifies the given application traffic based on its training knowledge acquired from the training database.

Steps for predicting the class of application traffic are explained in Figure 5.10. An application traffic Tu is fed to the experimental NoC. Then, the features of the application traffic are extracted with the help of Noxim and are fed to the classifier, which has trained using the supervised ML algorithm. Based on the knowledge stored in the training database, the trained classifier or trained ML model predicts the best possible class.

5.4.2 VALIDATION OF THE ML MODEL FOR TRAFFIC CLASSIFICATION

As shown in Figure 5.11, the proposed ML classification results are compared with Manual Classification, and overall classification accuracy of approximately 89.23% is achieved. Nearest neighbor value (K value in MATLAB) is adjustable according to the requirement. The accuracy and capability of the ML model will depend upon the amount of training data and different types of application traffic that are used to train the ML model.

5.4.3 IDENTIFICATION OF VULNERABLE ROUTERS USING LOOK-UP-TABLE (LUT)

After successful classification of application traffic, the next step is to identify the vulnerable routers for that classified application traffic. As previously shown in

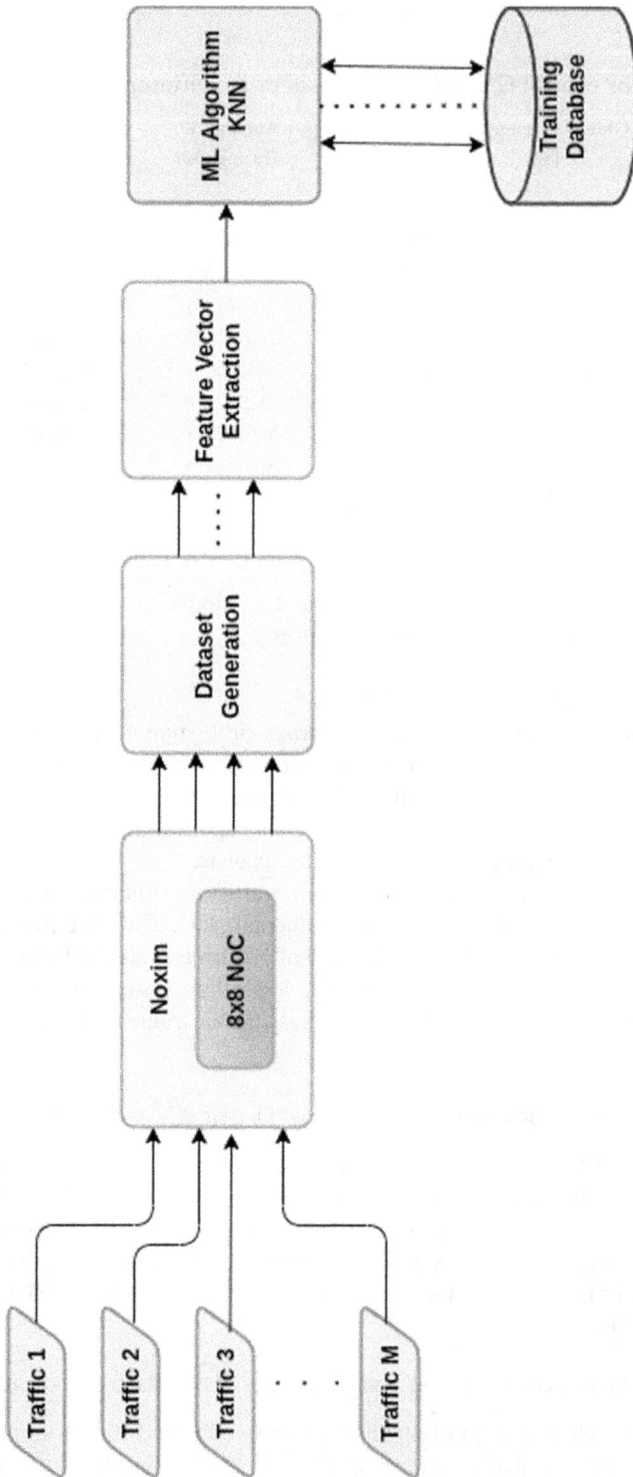

FIGURE 5.9   Training using ML algorithm.

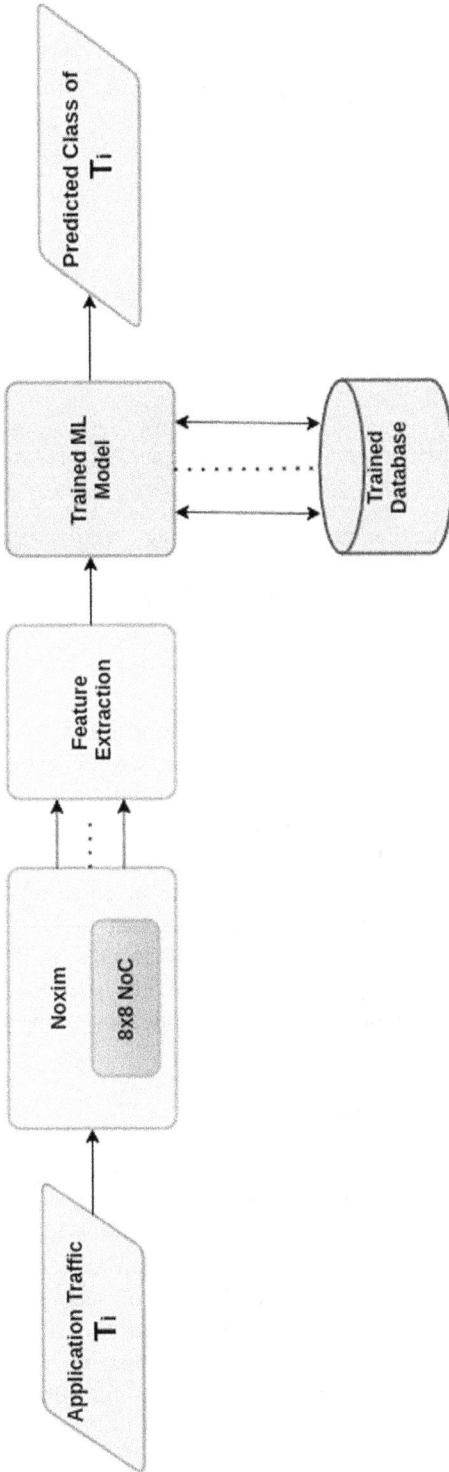

FIGURE 5.10   Traffic class prediction.

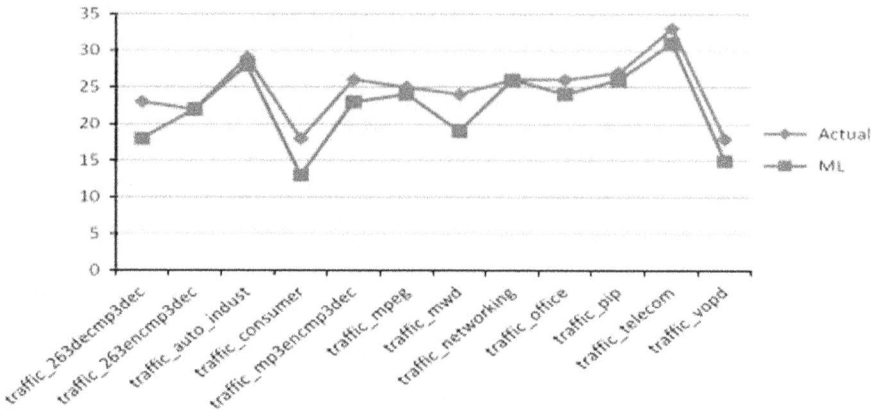

FIGURE 5.11 Comparison between actual (manual) and ML classification.

Figure 5.8, the identification of vulnerable routers is done using a LuT-based technique (*LuT_class* in algorithm 1). The LuT contains traffic labels and their respective set of vulnerable routers. The function *F* of LuT is expressed as

$$F(Ti) = \{l_1, l_2, l_3 \dots .. l_N\}$$

Here, l_i is the vulnerable router location in NoC for traffic *Ti*.

To get the glimpse of working of the algorithm, two application traffics – **MPEG** and **consumer** – are shown here. We have compared the output of our algorithm with manual outcome to detect vulnerable routers. From the experiment, we found that for MPEG traffic with PIR = 0.60, the *result* = {1,2,3,4,5} and *manual_approach* = {1,2,3,4,5}. Here, the *result* set is the outcome of our algorithm, and *manual_approach* is the formal calculated approach to identify the vulnerable routers. Similarly, for the consumer traffic, we found that at PIR = 0.60, *result* = {1,2,3,4,5,6}, and *manual_approach* = {1,2,3,4,5,6}. Hence, our approach successfully identifies the vulnerable routers for application traffic prior to processing the traffic.

5.5 FUTURE WORK AND SCOPE

Along with performances, the traffic classification helps the designers or developers to classify various types of application traffics, and based on traffic classification, the fault-locations or vulnerable routers in NoC can be identified. Hence, with the help of traffic classification, various needs-based fault-tolerant mechanisms can be applied to the NoC during design time. Preface of such possible fault-tolerant mechanism is presented below, and to the best of our knowledge, no literature has ever proposed this approach. The authors have no objection if the below-mentioned methodology is adopted by interested individuals for research or commercial purpose.

5.5.1 Pooling of Unused Routers: A Structural Redundancy Approach

During the execution of application traffic, all the routers of NoCs that are not utilized as different application traffic exhibit different traffic patterns, and different traffic patterns are mapped to different tiles or cores of NoC. After the classification of routing algorithms, the idea about the probable fault sites can be acquired, and the position of ideal cores and routers can be determined. Whenever any router of a NoC processes the flit's threshold amount, then the responsibility of the core attached to that router can be handed over to any other core whose router is idle at that time.

5.6 CONCLUSION

We have proposed a supervised ML-based approach to classify application traffic and identify the fault locations in NoC. With the help of traffic classification, further fault-tolerant mechanisms can be applied to make the NoC fault tolerant. We performed the experiment in real traffic and achieved classification accuracy of 89.23%. Further, our proposed algorithm also successfully identifies the fault locations or vulnerable routers for application traffic prior to the injection of that traffic to the NoC system. We presented a mathematical model for identifying faulty or vulnerable routers based on workload and also supported the presented mathematical model with experimental results. Based on this work, interested individuals may further improve and implement the fault-tolerant approach that has been mentioned in the previous section.

REFERENCES

[1] Sleeba, S. Z., Jose, J., Palesi, M., James, R. K., and Mini, M. G., 2018. Traffic aware deflection rerouting mechanism for mesh network on chip, in: 2018 IFIP/IEEE International Conference on Very Large Scale Integration (VLSI-SoC), Verona, Italy, pp. 25–30. doi:10.1109/VLSI-SoC.2018.8645011.

[2] John, M. R., James, R., Jose, J., Isaac, E., and Antony, J. K., 2014. A novel energy efficient source routing for mesh NoCs, in: 2014 Fourth International Conference on Advances in Computing and Communications, Cochin, pp. 125–129. doi:10.1109/ICACC.2014.36.

[3] Francis, Rosemary M., 2013. Exploring networks-on-chip for FPGAs. PhD. diss.

[4] Lin, S., Su, L., Su, H., Zhou, G., Jin, D., and Zeng, L., 2009. Design networks-on-chip with latency/bandwidth guarantees. *IET Computers Digital Techniques* 3, 184–194. doi:10.1049/iet-cdt:20080036.

[5] Bolotin, E., Cidon, I., Ginosar, R., and Kolodny, A., 2004. QNoC: QoS architecture and design process for network on chip. *Journal of Systems Architecture* 50, 2–3, 105–128. doi:10.1016/j.sysarc.2003.07.004.

[6] Liu, J., Harkin, J., Li, Y., and Maguire, L., 2014. Online traffic-aware fault detection for networks-on-chip. *Journal of Parallel and Distributed Computing* 74, 1, 1984–1993. doi:10.1016/j.jpdc.2013.09.001.

[7] Sahu, Pradip, Manna, Kanchan, Shah, Nisarg, and Chattopadhyay, Santanu, 2014. Extending Kernighan-Lin partitioning heuristic for application mapping onto network-on-chip. *Journal of Systems Architecture* 60. doi:10.1016/j.sysarc.2014.04.004.

[8] Sajjadi-Kia, H., and Ababei, C., 2013. A new reliability evaluation methodology with application to lifetime oriented circuit design. *IEEE Transactions on Device and Materials Reliability* 13, 192–202. doi:10.1109/TDMR.2012.2228862.

[9] Catania, V., Mineo, A., Monteleone, S., Palesi, M., and Patti, D., 2015. Noxim: An open, extensible and cycle-accurate network on chip simulator, in: 2015 IEEE 26th International Conference on Application-specific Systems, Architectures and Processors (ASAP), pp. 162–163. doi:10.1109/ASAP.2015.7245728.

[10] Rosing, T. S., Mihic, K., and De Micheli, G., 2007. Power and reliability management of socs. *IEEE Transactions on Very Large Scale Integration (VLSI) Systems* 15, 391–403. doi:10.1109/TVLSI.2007.895245.

[11] Chang, Y., Chiu, C., Lin, S., and Liu, C., 2011. On the design and analysis of fault tolerant noc architecture using spare routers, in: 16th Asia and South Pacific Design Automation Conference (ASP-DAC 2011), pp. 431–436. doi:10.1109/ASPDAC.2 011.5722228.

[12] Chatterjee, N., Chattopadhyay, S., and Manna, K., 2014a. A spare router based reliable network-on-chip design, in: 2014 IEEE International Symposium on Circuits and Systems (ISCAS), pp. 1957–1960. doi:10.1109/ISCAS.2014.6865545.

[13] Khalil, K., Eldash, O., and Bayoumi, M., 2017. Self-healing router architecture for reliable network-on-chips, in: 2017 24th IEEE International Conference on Electronics, Circuits and Systems (ICECS), pp.330–333. doi:10.1109/ICECS.2017.8292030.

[14] Xiang, Y., Chantem, T., Dick, R. P., Hu, X. S., and Shang, L., 2010. System-level reliability modeling for mpsocs, in: 2010 IEEE/ACM/IFIP International Conference on Hardware/Software Codesign and System Synthesis (CODES+ISSS), pp. 297–306.

[15] Lu, Z., Huang, W., Stan, M. R., Skadron, K., and Lach, J., 2007. Interconnect lifetime prediction for reliability-aware systems. *IEEE Transactions on Very Large Scale Integration (VLSI) Systems* 15, 159–172. doi:10.1109/TVLSI.2007.893578.

[16] Yamamoto, A. Y., and Ababei, C., 2014. Unified reliability estimation and management of noc based chip multiprocessors. Microprocessors and Microsystems 38, 53–63. URL: http://www.sciencedirect.com/science/article/pii/S014193311300195 6, doi:10.1016/j.micpro.2013.11.009.

6 Online Test Derived from Binary Neural Network for Critical Autonomous Automotive Hardware

Dr. Philemon Daniel
Assistant Professor, NIT Hamirpur, Hamirpur,
Himachal Pradesh, India

CONTENTS

DOI: 10.1201/9781003201038-6

6.1 AUTONOMOUS VEHICLES

There has been a huge revolution in the way we travel, especially in the past decade. Autonomous driving has come from just a conversation into almost a reality. In the future, we should see personal and shared vehicles, both for short distances and longer distances. It won't take long before delivery and logistics gain autonomy as in (Kapser and Abdelrahman 2020). Autonomous driving is going to revolutionize the way everything moves around. Even though there is enough technology to accomplish this, there is one thing that is stopping companies: the safety aspect of the vehicles (Hulse, Xie, and Galea 2018).

There is huge potential in autonomous vehicles. They can increase productivity, decrease accidents, minimize traffic efficiently, utilize energy, assist mobility stricken persons, and even enhance the travel experience.

The promises of an autonomous vehicle are plenty. Some of the major benefits are

1. Increased road safety: Better analysis of the environment, better decision making, and hence reduced accidents.
2. Decreased traffic and congestion: Since autonomous vehicles can plan their route and re-adjust dynamically according to the traffic conditions, they can self-manage traffic and avoid creating congestion.
3. Productive free time: Often people spend hours on the road, and hence their hands and brain are preoccupied with the road. These hours could be redeemed by introducing autonomous vehicles. Many commute hours each day can be better utilized.
4. Faster travel: Travel time is also expected to reduce because of efficient automatic planning and driving.
5. Dependence on others: People with disabilities, seniors, or people who have not learned to drive can now travel without depending on others.
6. Less expenditure: Because the vehicles are better managed, there are fewer accidents, reduced wear and tear, reduced medical expenditures, and possibly even reduced insurance.
7. Less pollution: The vehicles would save fuel and are expected to have reduced carbon emissions, and many of them are expected to be electric.
8. Drop off/pickup solution: A drop-off at the airport, railway station, or bus stations is now a possibility. Likewise, pickup from these places is possible as well. Even a drop-off and park solution is possible, in case parking is far away from the destination.
9. Parking solution: Since the vehicles are expected to locate a free slot and park themselves, that would solve a major problem in busy places.
10. Simple planning: Tedious and long route planning is no longer required as the vehicle would do the job for us.
11. Additional services: Pickup and delivery, refilling fuel, etc. can be automated as well.
12. Precise predictions: More accurate estimate of arrival timings will be a reality.
13. Improved lane capacity: As the lanes are better managed, more vehicles could utilize the free lanes, and there could be an increase in the vehicle count without any congestion.

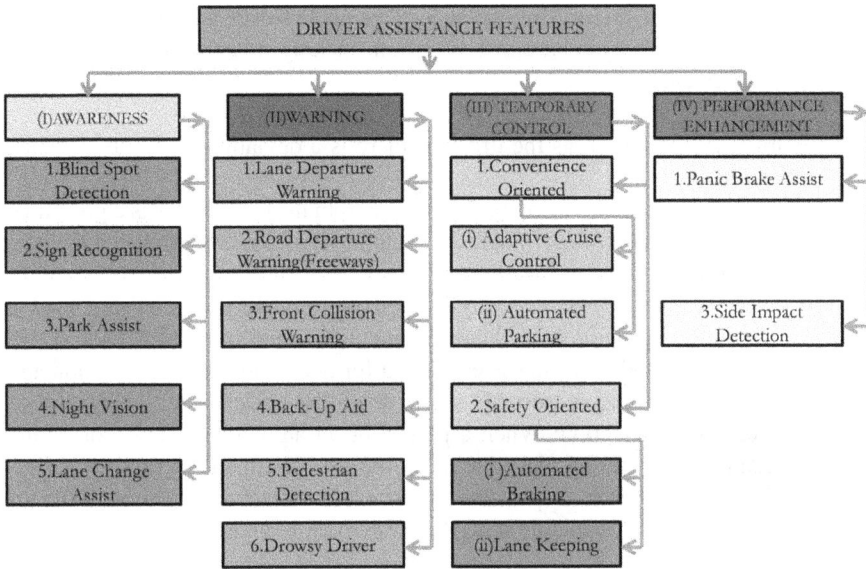

FIGURE 6.1 Classification of ADAS features.

14. Economical taxis: Since driverless taxis have reduced overall expenses, a cheaper commute is possible.

There is time before we fully realize the potential of autonomous vehicles (Hancock 2019), but already the driver assistance systems and partial autonomy have improved our experience in a big way. These systems use sophisticated hardware along with a combination of sensors, radar, lidars, cameras, and software to take decisions or make predictions based on the present and past data stream.

The following are some of the features already in place, as shown in Figure 6.1. These features can be categorized into four major sub-sections (Manoharan and Daniel 2018).

1. Awareness: This subsection briefs the implementations that assist a driver in being aware of one's surrounding.
   a. Blind spot detection: Some corners on either side of the car are not clearly visible to the driver. Various cameras and radar assistances provided to the driver can potentially minimize accidents.
   b. Sign recognition: Reading a road sign ahead of time and taking preventive or corrective measures is important. The driver could either miss the sign or read the sign incorrectly. An assistance of this sort through an image detection camera module could help in making the right decision.
   c. Park assist: A combination of cameras and ultrasonic or radar sensors is employed to guide the driver to park correctly. Assistance is also provided in locating and assessing if the parking space is sufficient for the vehicle.
   d. Night vision: IR cameras with recognition are helpful in seeing in the dark where there is not enough lighting.

e. Lane change assist: Multiple checks are made before a lane change can be initiated. Sensors can assist greatly in making a decision in this critical movement.

2. Warning: This set warns the driver if there is a deviation from the expected threshold.
   a. Lane departure: Unintentional departure from the current lane is a good time to receive a warning. Single or multiple cameras are employed for this purpose.
   b. Road departure: On freeways, when the vehicle drifts off the road a warning is generated by the edge detection algorithms.
   c. Collision: Radar or lidar is employed for detecting possible collision and issues a warning or possibly slows down the vehicle.
   d. Pedestrian detection: When a person or an animal suddenly crosses the road, radar can pick up more instantly and alert the driver.
   e. Drowsy driver: It is possible to doze off unknowingly, either fully or partially, and a camera facing the driver can recognize the threshold and warn the driver.

3. Temporary control: Although we are yet to reach full autonomy, there are times when vehicles can complete single tasks fully autonomously.

   a. Adaptive cruise control: Sometimes it is helpful to have speed controlled automatically based on the traffic or the slope of the road.
   b. Automated parking: Unlike assist, the entire parking process can be completely automated.

4. Performance enhancement: There are other assists that are more for comfort than for safety.

   a. Panic brake assist: At times the driver could unnecessarily brake but release the brake shortly. A hardware system can watch and smooth the brake pulse.
   b. Side impact detection: Although lane-assist systems should take care of drifting off, there could be occasions when another vehicle is too close to the side. Then, a gentle alert to stay clear could help.

6.1.1 LEVELS OF AUTONOMY

According to the NHTSA (Snyder 2016), there are six levels of autonomy, as shown in Figure 6.2. Most of the levels are initially experimented on cars and are extended to other vehicles later. Level 0 (No automation): No autonomy at all and the driver performs all the tasks. Level 1 (Driver assistance): The driver drives and controls the vehicle entirely but is assisted by sensors and minimal hardware. This segment of vehicles is the most common currently. Level 2 (Partial automation): The vehicle takes control from the driver at times and performs specific tasks, but driver is expected to continually stay in control of the events. Some cars are at this level

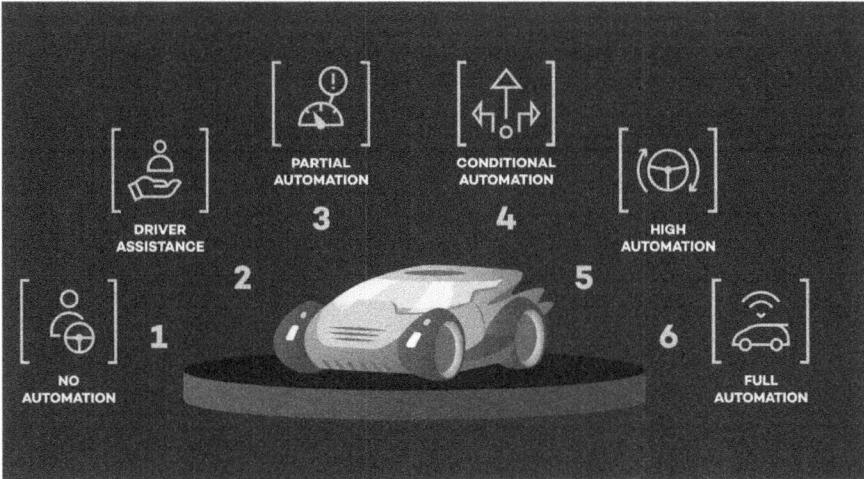

FIGURE 6.2   Level 1 to Level 6 of autonomous vehicles.

today. Level 3 (conditional automation): This is a big step as the driver is more passive in his role, and the car is expected to assume control in situations when the vehicle cannot decide or manage after a short warning. There are very few cars at this level. Level 4 (High automation): The car is able to manage most of the functions if the conditions are good and the driver is not expected to stay attentive. This is a much higher level of automation and requires a mountain of safety features, both at the hardware level and at the software level. This level and the next would lay the premise and be the major focus of the discussions further in this chapter. Level 5 (Full automation): This level would be complete autonomy at all conditions, which includes cars without a driver or even a steering wheel. No cars to date have achieved this level, and this level is still a far-fetched dream majorly because of the lack of sophisticated safety mechanisms required.

6.1.2   Safety Concerns

Autonomous driving is risky (Demmel et al. 2019). The risks are mainly due to the varying and unpredictable nature of the surroundings and because of the technological limitations, as Figure 6.3 shows. Most of the risks are directly attributed to the design and manufacturing process. The electronics, VLSI chip design and implementation play a major role in enhancing safety of the systems. Accidents happen because of reasons like malfunction of one or more components, operational errors, unanticipated situations, failures of parts or circuits or software, or a combination of these.

Although software errors do occur in real time, they are easier to manage and identify. Hardware design and reliability is paramount, and it is hard to predict and manage a breakdown. Hardware design errors are major and fall into the following types:

FIGURE 6.3 Risky navigation.

- Incorrectly assumed or underestimated environmental conditions such as weather, road conditions, an object at an awkward angle, sunlight directly opposite to the vehicle, etc.
- Unreliable design or process. Designs that are pushed to the limits or process with little or no buffer are prone to failure. Even good designs and processes could fail but more so when the design is not measured and calculated.
- Incorrect or erroneous design.

Prevention of error and risk management is an important part of any good VLSI implementation. Some of the errors creep in during the design and manufacturing process and can go undetected during the testing process. These ICs then reach the autonomous cars and can manifest anytime during the run time of the cars, which could become catastrophic.

The autonomous systems are becoming very complex, and varied systems integrated together compounds the complexity of error detection and management. Unfortunately, safety and management has become a specialized expertise of few learned on the job and is not taught in schools.

According to Thorn, Kimmel, and Chaka (2018), safety of autonomous vehicles should happen in four segments. The first one is the sensing part, which includes sensors like LiDAR, radar, camera, GPS, vehicle-to-vehicle, and vehicle-to-everything. There has to be hardware mechanisms in place when these sensors fail. It is usual to keep redundant sensors and either takes the majority opinion or switch over when a fault is detected.

The second one is the perception, where the data streamed from each of the sensors is assembled together to make sense of the environment and the position of the vehicle with respect to the surroundings. This is where the detection and understanding of both static and moving objects and the interpretation of each of them is accomplished. This is a key step where sophisticated hardware is used for the

FIGURE 6.4 Faults leading to accidents.

purpose of correct perception. Any error here could be disastrous. This is a key place where the controllers and hardware are expected to be perfect. Not much room for deviation or even an additional delay could be problematic. In real-time systems, correct decisions arriving late are wrong answers.

The next phase is the planning phase, where the vehicle's immediate future is planned and a route is prepared for the vehicle to maneuver. Perception and planning phases are the key phases where VLSI hardware faults are hard to detect and manage and shall be focus of further discussion. These two phases are the brain of the vehicle when in Level 5 and Level 6 autonomy.

The last phase is the control phase, where the execution of the plan takes place and the car is steered in the desired path and speed. These are mostly mechanical components.

There were a number of cases where partially autonomous vehicles were involved in accidents, and multiple times they were fatal because of multiple error factors, as in Figure 6.4. In one incident, a Tesla autopilot crashed fatally on May 2016 when it collided with a truck with a sun-lit background (BBCNews 2019). This time, the sensors failed to pick up the truck and hence failed to apply the brake. In the same year, a Google car crashed into a bus when shifting lanes (Associated Press 2016). This time, the perception and planning went wrong. More recently, in March 2018 according to Channel News Asia (2018), a self-driving Uber car killed a woman in Arizona while she was crossing a street.

6.2 TRADITIONAL VLSI TESTING

Traditional testing techniques, both offline and online testing, are predominantly done by logic BISTs, but logic BISTs are not very helpful for autonomous vehicles because of the complex safety critical devices and requirements in present and future vehicles (Haq et al. 2020). The current top models have in excess of 200 microcontrollers in a single car, like in Figure 6.5. The test techniques for such

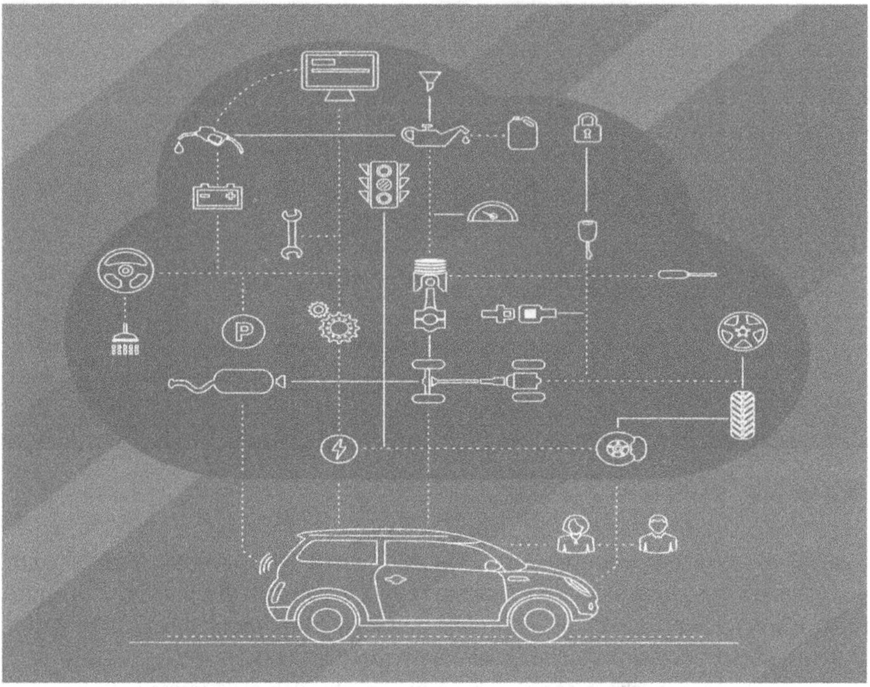

FIGURE 6.5 Electronics in cars.

systems have to be of very high quality and of long-term reliability. The test has to be performed continually during the entire lifetime of the vehicle. These techniques are usually employed during key-off, key-on, and periodic online tests, so ensuring functional safety using traditional test techniques is impossible.

The real- time faults in a system on the basis of their existence and effect can be classified as below:

a. Permanent faults: These faults exist indefinitely or for the entire lifetime in the system unless any corrective action is taken. Mostly these faults are caused due to manufacturing or design errors, shorts and opens in VLSI circuits, and a few of these faults are caused by catastrophic environmental disturbances or due to any physical damage to the chip. These defects may remain for entire lifetime and can cause long-term malfunctioning of components. Although these faults are easiest to detect compared to other faults, they could go undetected because of a certain gap in fault coverage.

b. Intermittent faults: These faults may appear regularly for a small time period. These faults disappear and reappear after a relatively longer time period. This entire process is repeated again and again. The cause of these faults is difficult to predict, because there are no such fixed parameters that causes these faults, but the effects are highly correlated. The system works well most of the time but fails under unusual environmental conditions. These faults are caused mainly due to marginal and unstable hardware. These faults are much harder to detect (Guilhemsang et al. 2011).

c. Transient faults: These faults appear for a much shorter time period and then disappear quickly and are uncorrelated with each other. These faults are hard to detect because of their short duration and are mainly caused due to random environmental disturbances. These are the hardest to detect and need high-quality live online test hardware to detect and manage them.

The goal of online testing is to detect faults in the system during the normal operation without disturbing its operation, and to take suitable corrective actions to mitigate its effect. For example, in some critical applications, after detecting a malfunction the system has to repair on the fly or configure a different hardware to take over or shut down its operation. Online testing can be categorized into two types: concurrent and non-concurrent testing.

- Non-concurrent testing: This testing is performed by temporarily suspending the system's normal operation or when the system is in an idle state like the key-off and key-on tests. It is not able to detect faults whose effects disappear quickly like transient or intermittent faults.
- Concurrent testing: This testing is performed concurrently without disturbing the normal operation of the system. This type of testing is required in critical applications where the system's normal operation cannot be suspended during testing. The tests have to be non-intrusive but at the same time watch all the lines and detect faults.

6.3   FUNCTIONAL SAFETY

ISO 26262 (The International Organization for Standardization 2011) is derived from IEC 61508 the International Electrotechnical Commission. The IEC 61508 is a general functional safety standard generally employed for electronic systems, whereas ISO 26262 is for the automotive industry. This addresses functional safety in the cars on the highway. ISO 26262 mainly focuses on the possible risks caused because of breakdowns or malfunctioning behavior of E/E systems from their expected work. The primary objective of ISO 26262 is to provide safety throughout the life cycle to all automotive electronic systems; it can be employed on system during the product development at the hardware or software level, and also can be applied during production, and operation of the system. Automotive vehicles must be able to operate safely, even if system fails work as intended. The purpose of functional safety for automotives is to identify the cause of the malfunction and focus on specific actions and techniques to be adopted to minimize the overall risk, when hardware or software of the system does not work as intended. Today, deep learning and artificial intelligence techniques implemented in hardware are playing a key role in the evolution of autonomous vehicles, and hence these vehicles are facing many new safety challenges. ISO/PAS 21448 focuses on avoiding undue risk and accidents caused because of unintended functionality or if the system is being misused by human. The intention behind ISO/PAS 21448 is to look into external causes and guaranteeing safety of the intended functionality (SOTIF), whereas the traditional functional safety standard ISO 26262 only concentrates on reducing risk caused because of malfunctioning of the system. Nardi (2021) proposed automotive functional safety using LBIST and other detection methods and compared all these based on area overhead and test time. They also considered the need of ISO 26262 in automotives and the failures that can be induced for malfunctioning of the system components.

The deficiencies in systems that lead to defective performance can be organized as below:

- Systematic failures: The failures that appear in the system in a deterministic manner due to human flaws, during the development and operation of the system, are known as systematic failures. These failures can be brought about in any stage of the system life cycle, such as specification, design, development, manufacturing, or maintenance. Some of these failures can be disposed of by modifying the design or manufacturing process.
- Random failures: These failures appear unpredictably during the lifetime of the system. Mainly these failures are produced from accidental defects to the process or could come from permanent or transient occurrences due to disturbed environment conditions to the system. These failures cannot be detected readily, so to distinguish and control such failures a special safety process is called for.

The automotive safety integrity level (ASIL) is a key segment of the ISO 26262 functional safety standard. It is established during the development stage and is determined by observing the hazard analysis and risk assessment (HARA) of the potential hazards based on feasibility and influence of the damage. The HARA process is performed to identify the malfunctions that could perhaps contribute to system risks and evaluate the compromise related with them. Khastgir et al. (2017) discussed the automotive HARA process and presented a different approach to overcome the security issues. They formed a ruleset for conducting automotive HARA to determine the ASIL by parameterizing the individual automotive segment based on harshness, exposure, and controllability of the risk.

There are four standards ASILs identified by the automotive safety standard ISO 26262: ASIL A, ASIL B, ASIL C, and ASIL D. ASIL D represents the largest level of the automotive risk and enforced strict steps to reduce it, whereas ASIL A represents the least level of automotive risk that can be warded off to some degree. ASIL are divided by checking out the safety objectives of the system. For each electronic segment of the automotive, the safety level is set up on the basis of three parameters: severity, exposure, and controllability of the hazard, and each of these parameters are further broken down into sub-classes. Severity measures the magnitude of damage to the driver, commuters, and other people on the road. Exposure measures the possibility of the occurrence of the hazardous conditions. Controllability measures the degree to which the vehicle can be dealt with by the driver when the hazardous condition takes place owing to malfunctioning of any system component.

The ASILs in the vehicle are where the key systems like electric steering, airbag deployment, and antilock braking systems are considered ASIL D, because the risks related with malfunctioning of these systems is greatest. On the other end of the safety spectrum, the systems which require smaller safety objectives are considered ASIL A, because the compromise of accident caused owing to failures of such systems is negligible and can be endured to some degree. Therefore, the process of deciding the ASIL category is a very demanding process for automotive functions that require significant functional safety and reliability.

With the infusion of safety techniques to identify or remedy the functional deficiencies in automotive, the undesirable risks can be played down to a considerable degree. The security mechanism is a technological solution achieved by electronic functions to identify the imperfections or control deficiencies to work out a safe state and, if possible, repair the deficiencies.

6.3.1 FAULT DETECTION TIME INTERVAL

For the functional safety of electronic systems in automotives, the following discussion is of utmost importance. The time available to observe that something has failed and take preventive steps to alleviate the threats caused due to this failure is of extreme importance. This time period is referred to in Denomme, Hooson, and Winkelman (2019) as Fault Tolerant Time Interval (FTTI). The functional safety system must be able to carry out a safe series of steps to take the system into a safe state in this period of time so that unwanted hazards can be mitigated. When an internal error takes place, it will take some time to transmit through the system before it produces an effect on the actual function of the vehicle. Safety-related issues must be dealt with by safety procedures, and the safety mechanism can be achieved with a combination of hardware and software form. Safety mechanism comprises of detection of fault and set of actions to mitigate its affect. The time interval between the occurrence of a fault to its detection is referred as FDTI (Fault Detection Time Interval). Thus, in plain words, FDTI gives us knowledge about the time being taken to find a fault, and it should be as minimal as possible. Once the presence of fault is established by detection, again the system has to react to that fault to bring about a safe state. Denomme, Hooson, and Winkelman (2019) discussed the FTTI development process and the security development life cycle for the recognition of timing constraints. They also came up with a system for determining FTTI for functional safety in automotive to avoid safety goal violations.

FTTI: According to automotive functional safety standard ISO 26262, FTTI is the minimal time span from fault occurrence in a system to a potential occurrence of an uncertain event, without any security mechanism being switched on, as shown in Figure 6.6. The automotive functional safety mechanism must bring about a healthy state within this time period for intended operation of the vehicle. It can be broken down into two parts:

 I. Malfunctioning Behavior Manifestation Time (MBMT): It is the minimum time period from the error occurrence in a system to the appearance of its malfunctioning behavior.
 II. Hazard Manifestation Time (HMT): It is the minimum time period from the beginning of the malfunctioning behavior in the system to the occurrence of a hazardous event or any safety goal violation.

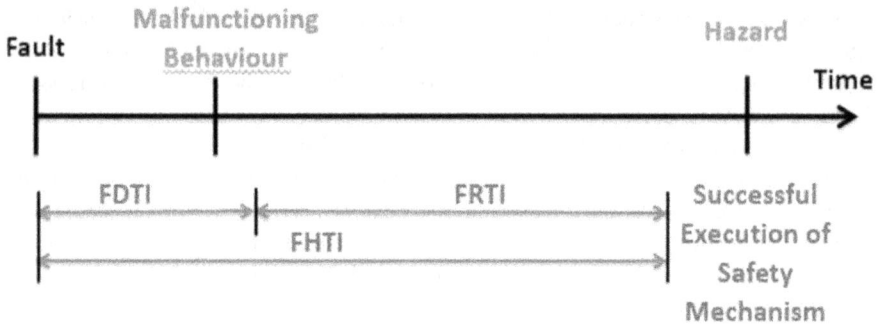

FIGURE 6.6 Fault detection time interval.

Fault Handling Time Interval (FHTI): It is the time span from the occurrence of the defect to the successful execution of the safety mechanism in the automotive system to mitigate the risk and to come to a safe state. In this time period, both detection and reaction processes are executed. The safety mechanism with low FHTI is needed. It is further broken down into two parts:

    I. Fault Detection Time Interval (FDTI): It is spelled out as the time period from the occurrence of a fault in the system to its detection, and it should be as small as possible.

    II. Fault Reaction Time Interval (FRTI): It is represented as the time period between the detection of faults in a system to achieve a safe state with the help of safety mechanism.

FDTI plays a significant part in functional safety development as it provides us knowledge about the time taken to observe a flaw, it should be as minimal as feasible. Later, the presence of fault is established by detection; thus, the system has to respond to that fault to bring about a safe state. The system has to proceed to achieve a safe state before manifestation of any other fault. The FDTI must be as short as possible to reduce the risks produced owing to deterioration in the performance of any automotive system.

    In order to achieve a very low FDTI through online concurrent test hardware for a Level 5 or Level 6 autonomous vehicle, deep learning is helpful in deriving suitable and very small test hardware. The consecutive sections demonstrate different techniques and the experimental results for deriving online test hardware from various binary neural networks.

6.4   DISCUSSION 1: BINARY CONVOLUTIONAL NEURAL NETWORK

One of the experiments is to derive a decompressor logic for online test from a binary neural network (Daniel et al. 2019). Let us discuss the design using a binary convolutional neural network (CNN) (Rastegari et al. 2016). It is a convolutional feedforward neural network with binary filters and activations that reconstruct the output the same as an input.

A deep neural network is stacked using multiple layers of binary convolutional network that compress the input bits into the lower dimension bits and then use an activation layer to skew the output decision to either a pass or a fail. Binary convolutional networks are trained to learn binary filters that are capable of extracting features present in the input bits and then combine them into multiple useful patterns that are rearranged to regenerate the output identical to its input.

Two different designs of CNNs are shown, one using Convolution 1D and the other with Convolution 2D. The first set of models use 1D binary filters while the later employs 2D filters. It includes an encoder section that compresses the input bits into lower dimensions and a decision-making activation section. The neural network is trained for minimal loss. To make the network synthesizable, instead of the high precision values, binarized activations, inputs, outputs, and filters are used that take on only two possible values +1 or −1.

The second sets of models are the 2D CNNs where the filters are 2D. They are also trained for binary filters and activations. By binarizing the weights and activation, complex matrix multiplications are replaced by a sequence of addition and subtraction operations. Further, the values are quantized to limit the bits from exploding. Binary filters give an efficient way of implementing convolutional operations. If all the parameters of convolutional networks are binary, then the convolutional operation can be approximated by a simple combinational circuit. Only binarization is used but if it is used with additional optimization techniques like Gordon et al. (2018), the resultant hardware will be small while at the same time maintain similar levels of accuracy.

6.4.1 ONE LAYER OF THE CONVOLUTIONAL NETWORK

The input is fed as a 1D or 2D matrix based on the dimensions of the filter. The input is convolved with a filter or a set of filters, and then added with an optional bias to obtain the outputs of one layer. This output is then taken through an activation function to limit the range of the outputs of that layer. As shown in the equation.

$$y^{[1]} = f^{[1]} * x^{[1]} + b^{[1]}$$

$$a^{[1]} = A(y^{[1]})$$

Where $y^{[1]}$ is output of first layer after convolving with filter and addition of bias term, $f^{[1]}$ is filter matrix, $x^{[1]}$ is input, $b^{[1]}$ is the bias matrix, A is activation function, $*$ indicates convolutional operation and $a^{[1]}$ is final output after applying activation function.

6.4.2 FORWARD PROPAGATION

There are three steps employed during the training process, forward propagation, backward propagation, and parameter update. While the focus is to train a binary

(a)

(b)

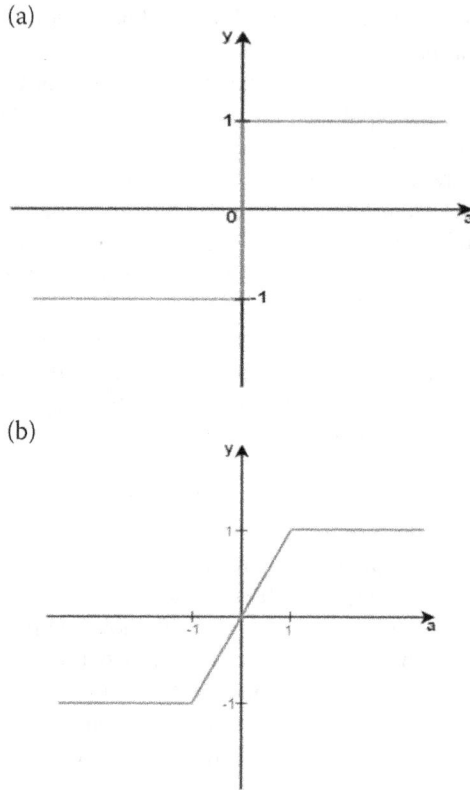

FIGURE 6.7 Sign function. Binary tanh activation function.

convolutional network, we convert the weights into binary values only during the forward. The weights are clipped during the backward pass. This is a requirement for stochastic gradient descent to be able to calculate the direction of minimum loss and be able to update weights. For binarizing floating weights and activation in binary convolutional networks, we used deterministic binarizes function, which is easy to implement on hardware, as described in the following equation and as shown in Figure 6.7.

$$a_b = sign(a) = \begin{cases} +1 & if a >= 0 \\ -1 & otherwise \end{cases}$$

If the value is greater than or equal to 0, the sign function represents it with a +1, otherwise by −1. Binary tanh is used as the activation function, as shown in the equation below and in Figure 6.7.

$$y = 2 * \sigma(a) - 1$$

where sigma is computed as shown.

$$\sigma(a) = clip(0.5a + 0.5, 0, 1) = max(0, min(0.5a + 0.5, 1))$$

6.4.3 BINARY NEURAL AUTOENCODER MODEL WITH CONVOLUTIONAL 1D

The forward propagation is illustrated for a 4-2-4 convolutional 1D autoencoder structure, having three layers as in Figure 6.8. The input layer, one hidden layer (encoding layer), and an output layer (decoding layer), as shown in the figure. An input matrix of size 4×1 is taken, convolve it with four filters of size 2×1, no padding with stride as 1. This gives an output of $3 \times 1 \times 4$. This output is then convolved with $2 \times 1 \times 4$ size filter to get an output of shape 2×1. These are the encoded bits of size 2 bits. This output is then convolved further with four filters of size 2×1 to get 4 bits output as the decompressed bits.

ALGORITHM 1   FORWARD PROPAGATION FOR AUTOENCODER

Input: x - input, f- filter, A - binary tanh activation function;
   Output: a - output
   Consider bias zero, f (2×1) filter

1. Output of first layer: we used four filter so we got four featured map

$$y^{[1]} = f^{[1]} * x^{[1]}$$

$$a^{[1]} = A(y^{[1]})$$

2. Output of Encoder : we used one filter and got encoded output of 1 bit.

$$y^{[2]} = f^{[2]} * x^{[2]}$$

$$a^{[2]} = A(y^{[2]})$$

3. Output of Decoder : we used four filter and got 4 bits output.

$$y^{[3]} = f^{[3]} * x^{[3]}$$

$$a^{[3]} = A(y^{[3]})$$

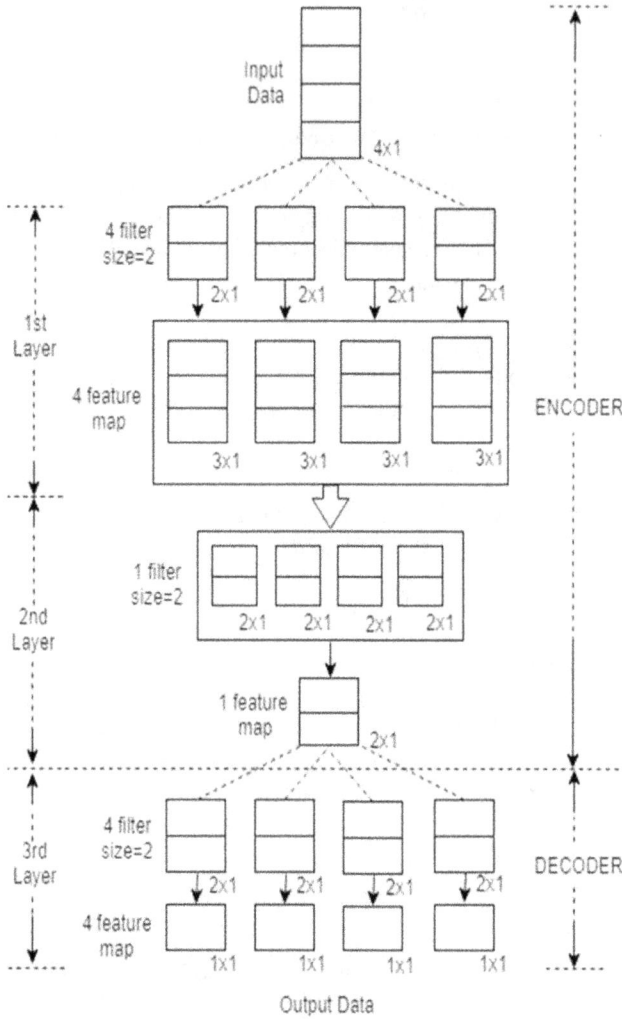

FIGURE 6.8   4-2-4 convolutional 1D autoencoder.

6.4.4   BINARY NEURAL NETWORK MODEL WITH CONVOLUTIONAL 2D

Figure 6.9 shows a 16-4-16 convolutional autoencoder model designed having three layers that is the input layer, one hidden layer (encoding layer), and output layer (decoding layer). The 16-bit input is fed in the shape of a 4×4 matrix, then convolved with four filters of size 2×2 without padding with stride as 1. This will reduce the size to $3 \times 3 \times 4$. This is further convolved with a filter of 2×2 size and gets a 2×2 encoded output. Then, convolved this output with 2×2 size of 16 filters to get 16 1-bit outputs because of decoded bits.

Input Data

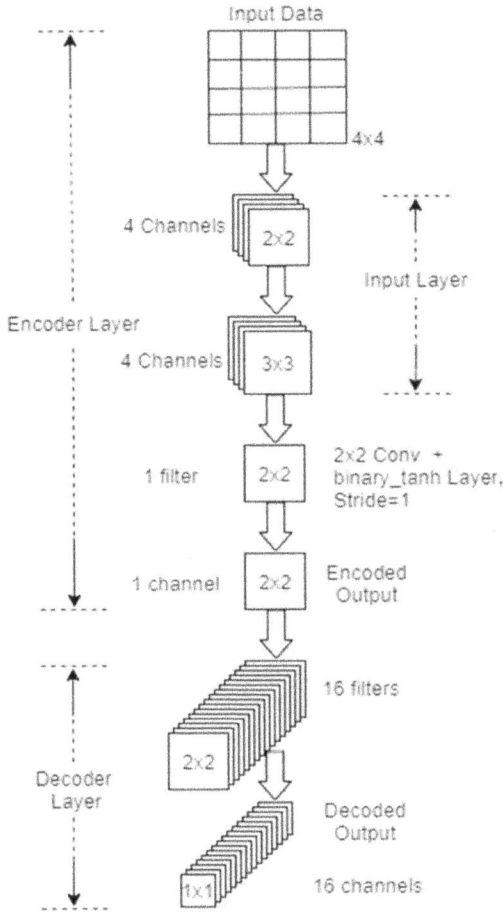

FIGURE 6.9 16-4-16 convolutional 2D autoencoder.

6.4.5 BACKWARD PROPAGATION

During backward propagation, the gradient is computed with respect to the real weights. We adopt squared hinged loss function to measure the inconsistency between the actual output and the predicted output. It measures the performance of the proposed neural network. The squared hinge loss is represented by "L", which is defined as in the equation below.

$$L = \frac{1}{n} \sum_{n=1}^{i} (max(0, 1 - y . \bar{y}))^2$$

Where y is the actual output and $\bar{y}$ is equal to w.x which is the predicted output. This indicates that when y and $\bar{y}$ have the same sign; the loss is zero otherwise we get some loss. Squared hinge loss is a modified version of hinge loss, as in Figure 6.9. It solves the problem of discontinuity that occurs in hinge loss at $y.\bar{y} = 1$.

The gradient for squared hinge loss is calculated with the help of chain rule.

$$\frac{\partial L(\bar{y})}{\partial x} = \frac{\partial L(w.\,x)}{\partial x} = \frac{\partial L}{\partial x} \cdot \frac{\partial (w.\,x)}{\partial w} = -2xy(1-y\bar{y})$$

$$\frac{\partial L(\bar{y})}{\partial x} = \begin{cases} -2xy(1-y.\,\bar{y})) & \text{if } xyw < 1 \\ 0 & \text{otherwise} \end{cases}$$

ALGORITHM 2   BACKWARD PROPAGATION FOR AUTOENCODER

Input: input x, weights/filter f, learning rate α

Output: the gradient of the input $\frac{\partial y}{\partial x}$, the gradient of the weight $\frac{\partial y}{\partial f}$, update the weights

new_weights = existing_weights learning_rate * gradient

$$Wnew = Wold - \alpha * \frac{\partial L}{\partial f}$$

6.5   DISCUSSION 2: ON-CHIP COMPACTION

An alternative to the traditional MISRs for data compaction is also presented for test vector outputs to generate a golden signature. Compaction is done to reduce the number of bits in the response of the circuit during testing. Multiple input signature register (MISR) compacts outputs over longer periods into a signature. An example 3-bit MISR, as in Figure 6.10, for a characteristic polynomial of x^3+x+1 is shown below:

$$\begin{bmatrix} X_0(t+1) \\ X_1(t+1) \\ X_2(t+1) \end{bmatrix} = \begin{bmatrix} 001 \\ 101 \\ 010 \end{bmatrix} \begin{bmatrix} X_0(t) \\ X_1(t) \\ X_2(t) \end{bmatrix} \oplus \begin{bmatrix} d_0(t) \\ d_1(t) \\ d_2(t) \end{bmatrix}$$

Where d_i denotes the circuit output at time t on PO i.

$X_i(t+1)$ denotes the future output of $X_i(t)$.

$X_i(t)$ denotes present input.

This MISR can be replaced by a more compact binary recurrent neural network (B-RNN) as an alternative for one or more MISR.

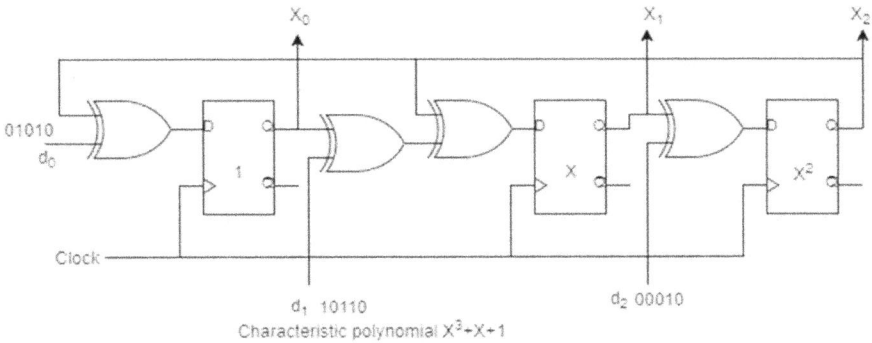

FIGURE 6.10   3-bit MISR.

6.5.1 BINARY RECURRENT NEURAL NETWORKS

Recurrent neural network (RNN) is a type of neural network where the output of the current time step is fed as input into the next time step, as in Figure 6.11. They are usually designed to implement time series problems. Their applications usually include speech and handwriting recognition, temporal pattern recognition, language understanding, and translation. But here, RNNs are used to train as data compaction. RNN behaves as if it has a memory that remembers the recent past. The parameters/weights are shared across time steps. The weights of RNN are also binarized to just -1, $+1$.

6.5.2 FORWARD PROPAGATION

As shown in Figure 6.12, x_t is supplied to the input.
   current state(h_t) using present input and previous state.
   current state (h_t) becomes previous state (h_{t-1}) for next time step.
   Combining all the information from the previous state, output state (y_t) is calculated.

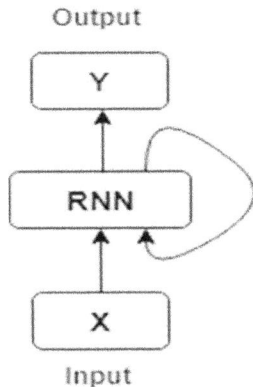

FIGURE 6.11   Basic RNN structure.

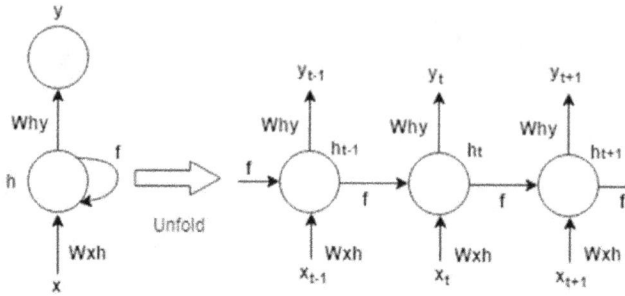

FIGURE 6.12 Unrolled RNN structure.

Current State:

$$h_{t=f}(h_{t-1}, x_t)$$

In RNN, activation function is tanh, the weight at recurring neuron is W_{hh}, and at input neuron is W_{xh}.

So, current state can be given as:

$$h_t = \tanh(W_{hh}h_{t-1} + W_{xh}x_t)$$

The weight at the output neuron is W_{hy}.

So output state is given by:

$$y_t = W_{hy}h_t$$

6.5.3   BACKPROPAGATION

In RNN, we go back in time to change the weights, so this algorithm is known as Backpropagation through time (BPTT). We adopt the Logcosh loss function to measure the inconsistency between the actual output and the predicted output. The Logcosh loss is represented by "L", which is defined as in the equation below.

$$L(y.\,\hat{y}) = \sum_{i=1}^{n} log\,(cosh\,(\hat{y}_i - y_i))$$

where, $\hat{y}$ denotes predicted value.

y denotes true value

For unrolled network, for each time step, gradient is calculated with respect to the weights as in Figure 6.13.

Consider for two-layer RNN BPTT, as shown in Figure 6.14:

Here,

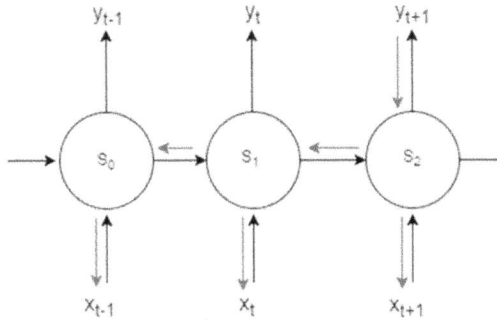

FIGURE 6.13 RNN backpropagation.

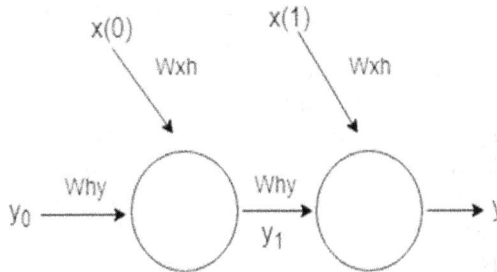

FIGURE 6.14 Backpropagation through time.

$$y_2 = f(W_{xh}x(1) + W_{hy}y(1))$$

To update W_{xh} and W_{hy}
   For second layer:

$$W_{xh}^2 = \mu\ \delta_2 x(1)\ \text{(standard delta rule)}$$
$$W_{hy}^2 = \mu\ \delta_2 y(1)$$

here, $\delta_2 = e_2 . \dfrac{dL(y.\hat{y})}{dt}$

$$e_2 = \hat{y}_2 - y_2$$

where μ denotes learning rate
   For first layer:

$$W_{xh}^1 = \mu\ \delta_1 x(0)$$

$$W_{hy}^1 = \mu\ \delta_1 y(0)$$

here, $\delta_1 = .\frac{dL(y.\hat{y})}{dt} (e_1 + \delta_2 W_{hy})$.

$$e_1 = \hat{y}_1 - y_1$$

hence, for ith layer

$$W^{new} = W^{old} + \Sigma (\Delta W)^i$$

For all time steps, gradients can be combined as the weights are same for each time step.

A 16-bit ALU with 36-bit input and 16-bit output performing 16 operations is chosen for this experiment. Using ATPG, 117 patterns were generated to get 100% stuck at fault coverage for 5786 faults. These test patterns were compressed and decompressed using the convolutional autoencoder network. Table 6.1 shows the performance of CNN architectures for decompression, including training statistics. The first column gives the number of layers in the network, followed by the input shape during training. The third column shows the network structure where the line separates the encoder from the decoder. The structure below the line is the decoder, which does the decompression. The next two columns display the training accuracy with float weights and binarized weights. Further two columns present the efficiency of the B-CNN decompressor with float weights and binarized weights in terms of fault coverage when trained for 2^N patterns where N is the input bit length for the decompressor. The next column gives the actual compression ratio obtained, and the last column confirms the small area overhead in terms of gates.

The hardware overhead can be calculated for each decompressor layer by this expression:

$$h[i] = \sum_{m=0}^{k-1} w[m] * x[m + i]$$

Hardware equation for 2×1 kernel size

$$h = h = [w_0 \odot x_0] + [w_1 \odot x_1]$$

Table 6.2 shows the results of compaction using B-RNN. Each RNN can generate a stable and unique signature. The table also shows the number of unique values each of them can generate. There are several improvements possible to the RNN structure, so they can be improved greatly.

6.5.4 ADVANTAGES AND LIMITATIONS

The following are the advantages of the binary neural network on-chip decompressor:

- The circuit is combinational and has no sequential elements.
- Hence, it has very little area and performance overhead.

TABLE 6.1
CNN as DECOMPRESSOR

Layers	Input shape	Network structure	Training accuracy %		Fault Coverage %		Compression ratio N/36	Gate count
			Foal tweights	Binary weights	Float weights	Binary weights		
3	4 × 9	3 × 322 × 6 1 × 36	76.07	47.86	99.03	97.08	1/3	108
	6 × 6	5 × 64 × 3 3 × 12	47.29	23.93	94.04	93.04	1/3	108
	4 × 9	3 × 92 × 9 1 × 36	70.09	44.44	99.40	99.34	1/2	108
	4 × 9	3 × 322 × 10 1 × 36	78.63	49.57	99.29	99.25	5/6	108
	12 × 3	11 × 12810 × 2 9 × 4	63.72	63.53	83.63	82.36	5/6	108
5	6 × 6	5 × 2564 × 1283 × 642 × 6 1 × 36	70.94	46.15	98.91	99.07	1/3	108
	9 × 4	8 × 1287 × 646 × 2 5 × 1284 × 9	65.38	47.44	97.36	97.81	1/3	2,028
	6 × 6	5 × 1284 × 643 × 3 2 × 1281 × 36	83.76	38.46	98.72	97.79	1/4	876
	6 × 6	5 × 1284 × 643 × 6 2 × 1281 × 36	99.15	44.44	99.43	99.33	1/2	876
	6 × 6	5 × 1284 × 643 × 322 × 10	90.60	46.15	99.33	99.34	5/6	108

(Continued)

TABLE 6.1 (Continued)
CNN as DECOMPRESSOR

Layers	Input shape	Network structure	Training accuracy %		Fault Coverage %		Compression ratio N/36	Gate count
			Foal tweights	Binary weights	Float weights	Binary weights		
		1 × 36						
	9 × 4	8 × 1287 × 646 × 325 × 4	66.67	42.09	99.05	99.42	5/6	108
		4 × 9						
6	12 × 3	11 × 12810 × 649 × 2	67.81	44.59	98.32	97.63	1/2	4,332
		8 × 647 × 1286 × 6						
	12 × 3	11 × 12810 × 649 × 1	58.98	42.92	86.95	88.54	1/4	4,332
		8 × 647 × 1286 × 6						

TABLE 6.2
RNN as MISR

No of Input Bits	Structure	Trained With	No of Unique Values
16	3-RNN + 1DNN	5-Stacked MISRs	10,606
	2-RNN + 1DNN		8,385
	1-RNN + 1DNN		3,047

- It is scalable and can be easily adopted for larger circuits.
- The input output relationship is non-linear, which is a major advantage of deep neural networks.
- Binary-encoded neural network autoencoders can be used to obtain very high fault coverage.
- Generic architecture for a fixed number of input channels and scan chains. Only the weights change.
- Hence, it has a smaller number of test cycles.
- The training method is reduced to a generic and simpler algorithm independent of the decompressor.
- Traditional scan ATPG can be used to generate test patterns.
- A one-bit change in one of the weights can generate a fresh set of outputs.
- Requirements of a sophisticated compression tool and multiple iterations like in EDT are eliminated.
- Overfitting is generally avoided in neural networks since they are not expected to memorize their inputs but rather generalize. In our case, overfitting is our objective.

The following are the limitations of the proposed on-chip decompressor:

- There is no deterministic way to know which network might perform well and which wouldn't for achieving certain fault coverage except to try various structures.
- Since there is no seed value, there is no pseudo randomness to the test patterns. They are always deterministic.

6.6 DISCUSSION 3: BINARY DEEP NEURAL NETWORK FOR CONTROLLER VARIANCE DETECTION

A controller has a certain invariance property that can be used for ensuring a certain level of safety with a very small hardware. Each controller has a certain set of valid and a certain set of invalid states. Neural networks are very good in differentiating between a valid state and an invalid state. After a neural network is trained, tiny hardware can be derived from the neural network to actually differential invalid

TABLE 6.3
Area for Controller Variance

Area Overhead	Binary Conv1D Layer	Binary Dense Layer
NO of 2-input XNOR gates required for multiplication	55	11
NO of 2-input AND gates required for addition	44	10

states in hardware. A special kind of neural network called the binary neural network is used for the purpose, so there is no additional requirement of quantization.

For demonstration purpose, an SDRAM controller is utilized. This 256 Mb (16 Mb × 16 data bit bus) SDRAM is a high-speed DRAM designed with CMOS to work in 3.3 V supply voltage with a synchronous data transfer. This SDRAM controller employs a pipelined design and therefore has a significant speed data transfer rate. All the inputs and the output signals are also full synchronized and are registered. The busy signal guides the interaction with the SDRAM. It is a quad bank SDRAM, and each bank has 8192 rows by 512 columns by 16 bits.

A dataset is prepared by exciting all possible controller states in the SDRAM and thereby capturing all the valid states. The valid states are then subtracted from the total possible states to obtain the invalid states. The controller has the following control signals – the read enable(rd_enable), write enable(wr_enable), reset(rst_n), clock enable(clk_enable), chip select(cs_n), Column address strobe (cas_n), Row address strobe(ras_n), busy, we_n, data mask low (data_mask_low), and data mask high (data_mask_high). A set of 2048 states are captured for these 11 signals, out of which 547 states are valid. After training, binary neural network coverage of 98.2% is achieved. The following is the hardware overhead required for the online controller test, which is approximately equivalent to 120 gates (Table 6.3).

6.7 CONCLUSION

A new and efficient DFT technique for on-chip test decompression, compaction, and controller variance check is presented. This method, although new, does not require any modifications in the existing test architectures nor in the design. They are non-intrusive since hardware overhead and FDTI are key factors to achieve safety in Level 5 and Level 6 autonomous vehicles. Binary convolutional auto-encoders, neural networks, and recurrent neural networks are promising for the future of several on-chip test strategies. From these experiments, I am convinced that these structures would outperform the existing on-chip test techniques. Also, the results of using binary RNN/LSTM as output data compactor for signature generation are very good. The controller variance detection hardware is small for watching control lines in real time and can perform quick detection of faults. The steps can be automized as the steps are a standard process.

ACKNOWLEDGMENT

We gratefully acknowledge the support of NVIDIA Corporation with the NVIDIA GPU Grant of Titan X Pascal GPU used for this research.

REFERENCES

Associated Press. 2016. "Google Self-Driving Car Caught on Video Colliding with Bus." *The Guardian*. https://www.theguardian.com/technology/2016/mar/09/google-self-driving-car-crash-video-accident-bus.

BBCNews. 2019. "Tesla Model 3: Autopilot Engaged during Fatal Crash." *BBCNews*. https://www.bbc.com/news/technology-48308852.

Channel News Asia. 2018. "Self-Driving Uber Car Kills Arizona Woman Crossing Street." *Channel News Asia, 20 March 2018*. https://www.reuters.com/article/us-autos-selfdriving-uber-idUSKBN1GV296.

Daniel, Philemon, Shaily Singh, Garima Gill, Anshu Gangwar, Bargaje Ganesh, and Kaushik Chakrabarti. 2019. "Demonstration of On-Chip Test Decompression for EDT Using Binary Encoded Neural Autoencoders." In *2019 IEEE International Test Conference India, ITC India 2019*. doi:10.1109/ITCIndia46717.2019.8979710.

Demmel, Sébastien, Dominique Gruyer, Jean Marie Burkhardt, Sébastien Glaser, Grégoire Larue, Olivier Orfila, and Andry Rakotonirainy. 2019. "Global Risk Assessment in an Autonomous Driving Context: Impact on Both the Car and the Driver." *IFAC-PapersOnLine*. doi:10.1016/j.ifacol.2019.01.009.

Denomme, Daniel, Sam Hooson, and James Winkelman. 2019. "A Fault Tolerant Time Interval Process for Functional Safety Development." In *SAE Technical Papers*. Vol. 2019-April. SAE International. doi:10.4271/2019-01-0110.

Gordon, Ariel, Elad Eban, Ofir Nachum, Bo Chen, Hao Wu, Tien Ju Yang, and Edward Choi. 2018. "MorphNet: Fast & Simple Resource-Constrained Structure Learning of Deep Networks." In *IEEE Computer Society Conference on Computer Vision and Pattern Recognition*, no. 1: 1586–1595. doi:10.1109/CVPR.2018.00171.

Guilhemsang, Julien, Olivier Heron, Nicolas Ventroux, Olivier Goncalves, and Alain Giulieri. 2011. "Impact of the Application Activity on Intermittent Faults in Embedded Systems." In *IEEE VLSI Test Symposium*. doi:10.1109/VTS.2011.5783782.

Hancock, P. A. 2019. "Some Pitfalls in the Promises of Automated and Autonomous Vehicles." *Ergonomics* 62 (4). doi:10.1080/00140139.2018.1498136.

Haq, Fitash U. L., Donghwan Shin, Shiva Nejati, and Lionel C. Briand. 2020. "Comparing Offline and Online Testing of Deep Neural Networks: An Autonomous Car Case Study." In *2020 IEEE 13th International Conference on Software Testing, Verification and Validation, ICST 2020*. doi:10.1109/ICST46399.2020.00019.

Hulse, Lynn M., Hui Xie, and Edwin R. Galea. 2018. "Perceptions of Autonomous Vehicles: Relationships with Road Users, Risk, Gender and Age." *Safety Science* 102. doi:10.1016/j.ssci.2017.10.001.

Kapser, Sebastian, and Mahmoud Abdelrahman. 2020. "Acceptance of Autonomous Delivery Vehicles for Last-Mile Delivery in Germany – Extending UTAUT2 with Risk Perceptions." *Transportation Research Part C: Emerging Technologies* 111. doi:10.1016/j.trc.2019.12.016.

Khastgir, Siddartha, Stewart Birrell, Gunwant Dhadyalla, Håkan Sivencrona, and Paul Jennings. 2017. "Towards Increased Reliability by Objectification of Hazard Analysis and Risk Assessment (HARA) of Automated Automotive Systems." *Safety Science* 113870731. doi:10.1016/j.ssci.2017.03.024.

Manoharan, K., and P. Daniel. 2018. "Survey on Various Lane and Driver Detection Techniques Based on Image Processing for Hilly Terrain." *IET Image Processing* 12 (9). doi:10.1049/iet-ipr.2017.0864.

Nardi, Alessandra. 2021. "Automotive Functional Safety Using LBIST and Other Detection Methods." *CADENCE*. Accessed January7. https://www.cadence.com.

Rastegari, Mohammad, Vicente Ordonez, Joseph Redmon, and Ali Farhadi. 2016. "XNOR-Net: Imagenet Classification Using Binary Convolutional Neural Networks." *Lecture Notes in Computer Science (Including Subseries Lecture Notes in Artificial Intelligence and Lecture Notes in Bioinformatics)* 9908 LNCS: 525–542. doi:10.1007/978-3-319-4 6493-0_32.

Snyder, Ryan. 2016. "Implications of Autonomous Vehicles: A Planner's Perspective." *Institute of Transportation Engineers. ITE Journal* 86 (25).

The International Organization for Standardization. 2011. "Road Vehicles — Functional Safety." *ISO 26262*.

Thorn, Eric, Shawn Kimmel, and Michelle Chaka. 2018. "A Framework for Automated Driving System Testable Cases and Scenarios." *Dot Hs 812 623*.

7 Applications of Machine Learning in VLSI Design

Sneh Saurabh, Pranav Jain, Madhvi Agarwal,
and OVS Shashank Ram
IIIT DelhiDelhi, India

CONTENTS

7.1   INTRODUCTION

With refinement and decreasing abstraction level, the number of components in a design increases. At lower levels of abstraction, electronic design automation (EDA) tools routinely are required to handle billions of entities. Therefore, these tools need to operate on voluminous data and complex models to accomplish various tasks that can be categorized as follows:

a. Estimate: Typically, we need to make estimates based on incomplete information or designer-provided hints.
b. Infer dependencies: The given design data can be voluminous, and finding dependencies among various input parameters can be challenging.
c. Transform: We need to refine design information based on some optimality criteria and constraints. It can also involve making transformations among behavioral, structural, and physical views of a design.

The quality of results (QoR) in accomplishing the above tasks can often be improved by statistical data analysis, learning from examples, and encapsulating the designer's intelligence into an EDA tool. These considerations motivate applying machine learning (ML) in VLSI design.

In recent times, there have been tremendous advancements in ML tools and technology. These advancements were facilitated by novel mathematical formulations

DOI: 10.1201/9781003201038-7

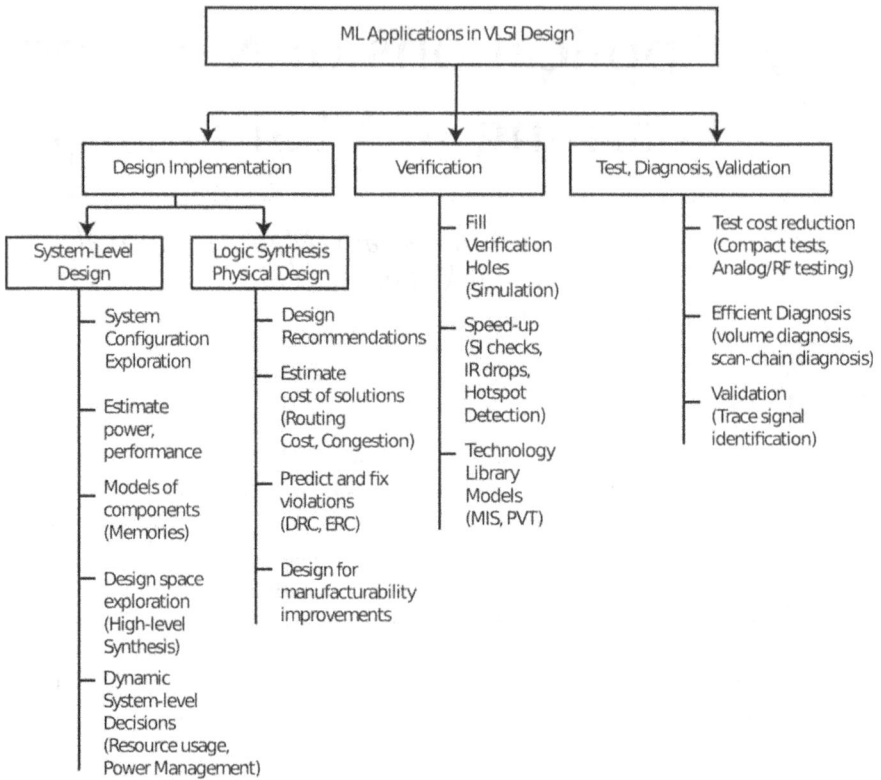

FIGURE 7.1   Applications of ML in VLSI Design covered in this chapter.

and powerful computing resources that allow massive data processing. Consequently, a plethora of freely-available tools to implement ML techniques and algorithms are available to us. Therefore, ML techniques are now widely employed in VLSI design. These techniques have improved the designer's productivity, tackled complex problems more efficiently, and provided alternative solutions.

VLSI design is a complex process. We decompose the design process into multiple steps, starting with system-level design and culminating in chip fabrication. We can apply ML and related techniques in design implementation, verification, testing, diagnosis, and validation stages. In this chapter, we review the applications of ML in all these stages. We have summarized the applications of ML in VLSI design covered in this chapter in Figure 7.1. However, note that this chapter does not exhaustively list relevant work in these areas. Instead, this chapter demonstrates the application of ML on a few examples taken from literature. It will help readers in appreciating the link between ML and VLSI design. Subsequently, readers can apply ML techniques to their specific problems by making appropriate modifications to the illustrated examples.

We can identify three main steps in the design implementation: a) system-level design, b) logic synthesis, and c) physical design. There are many opportunities to apply ML in design implementation. It can help in exploring the design space more efficiently. It can capture the designer's knowledge and reuse them by making recommendations to the designers. It can also make smart transformations based on past experiences and learning.

The design implementation is supported by the analysis and verification steps in design flows. The verification process is integral to VLSI design since it ensures that the design functionality matches the given specification. The verification process involves some analysis of the design data, and it can often be time-consuming. ML techniques can help in inferring complex dependencies and handling voluminous data during the analysis of design data. It can help in the more accurate and faster analysis and can ease or supplement the verification process. However, we should ensure that the errors obtained during ML-based analysis are acceptable for verification purposes.

Post-designing, we carry out fabrication in semiconductor foundries and test the fabricated die to rule out failures due to defects. If there are failures, we diagnose the root cause of failures and make a fix to avoid the problem. Additionally, we carry out pre-production validation of chips functionality based on the real usage scenario. In all these steps, we often encounter voluminous data. We can improve the conventional approaches of test and diagnosis by employing statistical ML and data mining techniques. We can also improve the fabrication and manufacturing of chips by ML-driven pattern matching and correcting the masks efficiently.

The rest of this chapter is organized as follows. We describe the basic concepts of ML that are relevant for this chapter in Section 7.2. We describe the application of ML in system-level design in Section 7.3 and logic synthesis and physical design in Section 7.4. We describe the application of ML in verification in Section 7.5 and the test, diagnosis, and validation in Section 7.6. We highlight key challenges in adopting ML-based solutions in VLSI design flow in Section 7.7. We conclude this chapter by outlining future directions in Section 7.8.

7.2 MACHINE LEARNING PRELIMINARIES

Before describing the applications of ML, we explain some of the basic concepts related to ML.

ML creates a model based on given examples. Subsequently, it produces output (s) for an unseen input. Typically, we divide the known examples into a training set and a test set. We use the training set for creating the ML model. We use the test set to assess the quality of learning. We can measure the learning quality by examining whether the ML model can generalize or produce the desired output for unseen inputs. Based on the given examples, we can classify ML techniques into three main categories: supervised learning, unsupervised learning, and reinforcement learning.

In supervised learning, we provide example inputs and their desired outputs. The goal of supervised learning is to adjust the set of parameters that map inputs to outputs. Some examples of supervised learning are:

  a. Regression: It involves mapping the inputs onto real-valued outputs.
  b. Classification: It requires mapping the inputs onto one of possibly several groups or classes.

In unsupervised learning, we provide example inputs with no desired outputs. The goal of unsupervised learning is to find some structure in the given input. Some examples of unsupervised learning are:

a. Clustering: It involves separating inputs into several groups or clusters based on some similarity measure.
b. Generating probability density function (PDF) for the inputs.
c. Dimensionality reduction: It involves mapping the input space onto a lower-dimensional space.

In reinforcement learning, training is done in a dynamic environment using a reward function. The reward function is designed to provide feedback. The goal is to produce a response that maximizes the reward value.

For the above problems, we can build various types of ML models. Some popular ML algorithms and models are as follows:

1. Linear regression: The model consists of a vector of coefficients. The coefficients are determined to minimize the error on the training set. The output is produced by taking the linear combination of input features multiplied by corresponding coefficients.
2. Decision Trees: The model produces output based on decision rules. The decision rules are inferred from the features present in the given training data. These models are easy to visualize and understand the factors that impacted the result.
3. k-Nearest Neighbors (kNN): The output is produced by computing the average output value of k-most similar training samples.
4. Naive Bayes: It is a classification technique. It employs Bayes' theorem and assumes that the presence of each feature in a class is independent.
5. Support Vector Machine (SVM): The model consists of a hyperplane that distinctly classifies the data points. The hyperplane is in an N-dimensional space, where N is the number of features. The data points on different sides of the hyperplane belong to distinct classes.
6. Artificial Neural Network (ANN): The model consists of layers of neurons interconnected through weights. While training, these weights are adjusted to alter the importance of some inputs over others. Each neuron includes an activation function that produces an output based on input value and the corresponding weight.
7. Principal Component Analysis (PCA): It analyzes given data and determines a new coordinate system that captures the maximum variance in the data. The axes of the new coordinate system are called principal components. Each component is orthogonal to one another.

In VLSI design, we often employ a combination of the above models and learning strategies. The choice of the ML model strongly depends on the available training data. Moreover, the model complexity and whether it fits in the existing design flows are critical considerations.

7.3   SYSTEM-LEVEL DESIGN

We face many challenges in designing state-of-the-art systems consisting of heterogeneous components such as CPUs, GPUs, and hardware accelerators. The difficulty arises due to the following reasons:

1. At the system-level, we do not have full implementation detail of many components. Therefore, based on some abstract models, we need to estimate the performance, power, and other attributes. It is difficult and error-prone.
2. We need to determine near-optimal system parameters for multiple objectives and multiple constraints. Moreover, these constraints and objectives often depend on the application and change dynamically.
3. Due to an increase in the number and multitude of heterogeneous components, the design space or the solution space becomes big.
4. Traditional methods such as simulations are often too slow and expensive in exploring the design space for modern heterogeneous systems.

Some of these challenges can be efficiently tackled by ML, as explained in the following paragraphs.

In designing a new system, we often need to choose a particular system configuration by estimating the system performance. For example, for a laptop, we need to choose the CPU type, CPU frequency, memory type, memory size, bus type, and motherboard. However, to make a choice, we need to predict the real system performance. It has been shown that we can estimate the system performance by developing a predictive model based on neural networks and linear regression [1]. These models can accurately predict system performance by using a small fraction of the overall design space and employing data of already built systems.

We can estimate the performance and power of heterogeneous systems with more detail also by applying ML techniques. In particular, we can measure these attributes with many test applications running on real hardware with various configurations [2]. These measurements can train an ML model to produce the expected performance and power attributes for a given system design parameter. Subsequently, we can predict them for a new application running on a range of system configurations. We measure these values for a single configuration and feed it to the ML model. Using these inputs, ML can quickly produce the estimate for various configurations and with accuracy comparable to cycle-level simulators [2].

At the component-level also, we need to quickly estimate performance during design space exploration. For example, in the scaled heterogeneous systems there are advantages of implementing the hybrid memory architectures. We can combine non-volatile memories (NVM) with the traditional DRAM to improve performance [3]. To implement the hybrid memory architecture, we need to analyze the performance of various configurations. Traditionally, architectural-level memory simulators are used in these analyses. However, this approach suffers from long simulation time and inadequate design space exploration [3]. A better technique is to build an ML model that predicts various figures of merit for a memory. We can train the ML model by a small set of simulated memory configurations. These predictive models can quickly report latency, bandwidth, power consumption, and other attributes for a given memory configuration. In Sen and Imam [3], neural network, SVM, random forest (RF), and gradient boosting (GB) are tried for the ML model. It is reported that the SVM and RF methods yielded better accuracies compared with other models [3].

At the system-level, we often use high-level synthesis (HLS) for design space exploration. The exploration goal is to finally settle on a near-optimal solution that meets the given constraints. We can use ML techniques to determine the Pareto-optimal designs efficiently. We can build an ML model by training with samples obtained by synthesizing a fraction of the design space [4]. The ML model can quickly predict attributes such as area and throughput for a design. We can also refine the model by smartly selecting the next synthesized sample [4]. We can employ a regression model based on Gaussian processes, as reported in Zuluaga et al. [4]. Alternatively, we can use tree-based RF, as reported in Liu et al. [5]. During design space exploration, we often need to make binary decisions. For example, whether to inline a given function. It is easy to model binary decisions in a tree-based RF using two branches [5]. Note that RF consists of multiple regression trees. It produces the final result by a collective vote of these trees. Therefore, we can minimize both the generalization error and prediction variance using RF [5].

The design space exploration during HLS can be made more efficient by using ML techniques. For example, conventional search strategies such as simulated annealing can generate a training set for implementing a decision tree [6]. Subsequently, the trained decision tree prunes the search space. Thus, design space exploration becomes more efficient. We can improve local search algorithms such as simulated annealing by choosing the starting point smartly [7]. For example, first, we perform a conventional local search. Using these searches, we train an eva-luation function to predict the outcome of a given starting point. Subsequently, the evaluation function guides design space exploration and filters out non-promising start points. Thus, the local search becomes more efficient [7].

Another challenge in a heterogeneous system is allocating resources and managing power during runtime. Traditionally, we depend on a priori information about workload and thermal model of chips while designing. However, a priori information is sometimes unavailable. Moreover, a priori information cannot ade-quately model the temporal and spatial uncertainties. These variations can be due to the variations in the workloads, devices, and environment. Therefore, we need to make appropriate decisions during runtime. We can employ ML techniques, such as reinforcement learning, that adapt to the varying workload and environment [8]. They can handle scenarios for which we have not trained the system. It has been shown that we can employ Q-learning to find an optimal policy in dynamic resource allocation and improve the system performance [9].

7.4 LOGIC SYNTHESIS AND PHYSICAL DESIGN

Logic synthesis and physical design are the main tasks in the VLSI design flow. Due to computational complexity, we divide them into multiple steps. Some key steps are multi-level logic synthesis, budgeting, technology mapping, timing optimization, chip planning, placement, clock network synthesis, and routing. Nevertheless, each of these steps is still computationally difficult. Typically, EDA tools employ several heuristics to obtain a solution. The heuristics are guided by tool options and user-specified settings. The QoR strongly depends on them. Note that these steps are sequential. Therefore, the solution produced by a step impacts all subsequent tasks. A designer often adjusts the tool settings and inputs based on experience and intuition to achieve the desired QoR. We can reduce design effort in these tasks and improve the QoR by employing ML tools and techniques, as explained in the following paragraphs.

One of the earliest attempts of reducing design effort using ML was Learning Apprentice for VLSI design (LEAP) [10]. LEAP acquires knowledge and learns rules by observing a designer and analyzing the problem-solving steps during their activities. Subsequently, it provides advice to the designer on design refinement and optimization. A designer can accept LEAP's advice or can ignore it and manually carry out transformations. When a designer ignores the advice, LEAP considers it as a training sample and updates its rule.

Recently, there is a renewed interest in ML-based design adviser tool. In Beerel and Massoud [11], the authors reported developing DesignAdvisor. It will monitor and record example design problems and the corresponding action taken by a designer when using standard EDA tools. For example, the design problem can be multi-level Boolean network optimization using logic restructuring and Boolean simplification. The corresponding designer action can be setting tool options or directives for necessary logic transformation and optimization. Subsequently, DesignAdvisor will learn to produce the best solution for a given problem. It can make recommendations to designers and help to tune EDA tools and their optimization algorithms.

An ML-based tool for logic synthesis and physical design, such as DesignAdvisor, need to consider to implement following tasks [11]:

1. Developing a training set: A training set consists of data points with a design problem and its corresponding solution. For example, a data point can be an initial netlist, constraints, cost function, optimization settings, and the final netlist. We need to generate these data points for training or can acquire them from designers.
2. Reduced representations of the training set: The training data points typically contain many features. However, for efficient learning, we can reduce the dimensionality of the training set. For example, we can perform PCA and retain the most relevant input features.
3. Learning to produce the optimum output: The training data points that we collect from the existing EDA tools are, typically, not the mathematical optimum. These tools give the best possible solution that could be acceptable to the designers. Therefore, training data does not represent the ground truth of the problem. Moreover, data can be sparse and biased because some specific tools generate those results. We can employ statistical models such as Bayesian neural networks (BNNs) to tackle this problem [11]. BNNs have weights and biases specified as distributions instead of scalar values. Therefore, it can tackle disturbances due to noisy or incomplete training set.

4. Dynamic updates: We expect the ML-based design adviser to continue learning from the new design problems. We can use reinforcement learning to adjust the model dynamically.

In addition to making recommendations for the designers, ML techniques can be employed directly in EDA algorithms to obtain better QoR and make efficient computations. In physical design, we often encounter problems in later design steps due to decisions made in earlier design steps. Traditionally, the strategy is to detect design problems and fix them at the appropriate design step. However, it creates loops or iteration in the design flow, and the design effort increases. Recently, the strategies of EDA tools and designers have shifted towards predict and avoid paradigm. However, predicting the downstream problems, such as the problems encountered during placement and routing, is challenging. Conventional approaches, such as carrying out light-weight placement or routing during early design stages, suffer from high runtime. The accuracies of these estimates are also questionable. To tackle these problems, we can employ ML techniques for prediction. We can train ML models using decisions taken later in design flows. Furthermore, we can update these models with more training samples and reinforcement learning techniques.

It is often easier to apply ML techniques for estimation than making or suggesting changes in a circuit. ML techniques have been reported to build an accurate and fast routing-cost model [12]. It can quickly predict the total wire length associated with a given power distribution network (PDN). Thus, we can search for the optimal PDN configurations efficiently. As a result, we can reduce the total wire length of a PDN. The training data consists of several basic predictor features such as attributes of PDN, block design, and placement. An ML model based on Gaussian process regression (GPR) is employed in Chang et al. [12] because a routing-cost model involves a complex high dimensional function.

An attempt has been made in Kirby et al. [13] to predict routing congestion by employing only the logical structure and the cell attributes such as the number of pins, size, and function. The authors use Graph Attention Networks (GAT) in their work [13]. A GAT employs one or more graph convolution layers. It is suitable for netlist problems that we encounter in EDA frequently. It produces output for each vertex (cell) in the graph (netlist) using neighborhood-based aggregated features. It can perhaps capture the essence of logic structure and predict congestion.

We can use ML techniques to predict detailed routing short violations from a placed netlist [14]. Factors contributing to routing violations, such as the location of routing tiles, routing accessibility, and track availability are given as features to the ML model. A neural network is trained to detect these violations. However, in this case, the training samples are skewed, with less than 1% of examples having routing violations. Therefore, greater weights are assigned to violating instances in the calculation of the loss function. In Tabrizi et al. [14], the authors penalized the wrong classification 20 times more for the violating instances than the non-violating instances. The model is reported to predict on an average 90% of the shorts and 7% false alarms.

Similarly, we can employ ML techniques to predict detailed-route DRC violations [15]. The global routing cells are classified into two classes based on whether it contains any DRC violation. Post-detailed routing samples have been used for training the classifier. The classifier detects DRC violations based on features such as local pin density, local overflow, pin proximity, and connectivity parameters [15]. An SVM was found to be the most suitable model for this classifier in Chan et al. [15]. Furthermore, using this predictor, we can avoid DRC problems by spreading cells appropriately after placement [15].

Another area in which ML can be effective is design for manufacturability (DFM) [16]. We can carry out VLSI mask optimization using ML techniques to reduce cost and computational effort. The capabilities of ML in efficiently handling big data are useful in these applications. It has been reported that efficient mask layout optimization can be done in the optical proximity correction (OPC) framework [17]. We can train a deterministic classification ML model to identify the best OPC engine for a given design [17].

7.5   VERIFICATION

We can employ ML techniques to improve and augment traditional verification methodologies in the following ways:

1. Traditional verification often makes certain assumptions to simplify its implementation. Consequently, it can leave some verification holes, or we sacrifice the QoR of a design. ML-based verifications can consider a larger verification domain. Thus, they can fill these holes and make verification more robust.
2. Traditional verification can take more runtime in finding patterns in a design. ML-based verification can search efficiently and produce results faster.
3. Traditional verification employs some abstract models for a circuit. ML can augment such models or replace them.

We will discuss some of the above applications of ML in the following paragraphs.

In simulation-based logic verification, we can employ ML to quickly fill the coverage holes or reduce the runtime [18,19]. Traditionally, we apply randomly generated test stimuli and observe their response. To improve coverage within a time limit, we often incorporate coverage-directed test generation (CDG). However, it is challenging to predict the constraints that can produce test stimuli with high coverage. ML techniques can generate the bias that directs CDG towards improved coverage. ML techniques are added in the feedback loop of CDG to produce new directives or bias towards obtaining stimuli that fill coverage holes [18]. The ML model can also learn dynamically and screen stimuli before sending them for verification. For example, ANN can extract features of stimuli and select only critical ones for verification [19]. Thus, we can filter out many stimuli and accelerate the overall verification process.

Some verification steps, such as signal integrity (SI) checks, take large runtime. At advanced process nodes, SI-effects are critical. It changes the delay and slew of signals due to the coupling capacitance and switching activity in the neighboring wires. We can employ ML techniques to estimate the SI-effects quickly [20]. First, we identify parameters on which SI-effects depend. Some of the parameters that impact SI-effects are: nominal (without considering SI) delay and slew, clock period, resistance, coupling capacitance, toggle rate, the logical effort of the driver, and temporal alignment of victim and aggressor signals. Using these parameters, we can train an ML model such as ANN or SVM to predict SI-induced changes on delay and slew [20]. Since ML models can capture dependency in a high-dimensional space, we can utilize them for easy verification. However, we should ensure that the errors produced by ML models are tolerable for our verification purpose.

Another approach to estimate SI-effects is by using anomaly detection (AD) techniques [21]. AD techniques are popularly employed to detect anomalies in financial transactions. However, we can train an ML model, such as a contractive autoencoder (CA), with the features of SI-free time-domain waveforms. Subsequently, we use the trained model to identify anomalies due to SI-effects. We can use both unsupervised and semi-supervised AD techniques for this [21].

We can employ ML techniques to efficiently fix IR drop problems in an integrated circuit [22]. Traditionally, we carry out dynamic IR drop analysis at the end of design flows. Any IR drop problem is corrected by Engineering Change Order (ECO) based on the designer's experience. Typically, we cannot identify and fix all the IR drop problems together. Consequently, we need to carry out dynamic IR drop analysis and ECO iteratively, until we have corrected all the IR drop issues. However, IR drop analysis takes significant runtime and designer's effort. We can reduce the iterations in IR drop signoff by employing ML to predict all the potential IR drop issues and fix them together [22]. Firstly, by ML-based clustering techniques, we identify high IR drop regions. Subsequently, small regional ML-based models are built on local features. Using these regional models IR drop problems are identified and fixed. After we have corrected all the violations, a dynamic IR drop check is finally done for signoff. If some violations still exist, we repeat the process till all the IR drop issues are corrected.

We can use ML techniques in physical verification for problems such as lithographic hotspot detection [23]. By defining signatures of hotspots and a hierarchically refined detection flow consisting of ML kernels, ANN, and SVM, we can efficiently detect lithographic hotspots. We can also employ a dictionary learning approach with an online learning model to extract features from the layout [24].

Another area in which we can apply ML techniques is the technology library models. Technology libraries form the bedrock of digital VLSI design. Traditionally, timing and other attributes of standard cells are modeled in technology libraries as look-up tables. However, these attributes can be conveniently derived and compactly represented by using ML techniques. The ML-models models can efficiently exploit the intrinsic degrees of variation in the data.

In Shashank Ram and Saurabh [25], we demonstrate this by modeling multi-input switching (MIS) effects using ML techniques. Traditionally, we ignore MIS effects in timing analysis. We employ a delay model that assumes only a single input switching (SIS) for a gate during a transition. For SIS, the side inputs are held constant to non-controlling values. However, ignoring MIS effects can lead to either an overestimation or an under-estimation of a gate delay. We have examined the impact of MIS on the delay of different types of gates under varying conditions. We can model the MIS-effect by deriving a corrective quantity called MIS-SIS difference (MSD) [25]. We obtain MIS delay by adding MSD to the conventional SIS delay under varying conditions.

There are several benefits of adopting ML-based techniques for modeling MIS effects. We can represent multi-dimensional data using a learning-based model compactly. It can capture the dependency of MIS effects on multiple input parameters and efficiently exploit them in compact representation. In contrast, traditional interpolation-based models have large disk-size and loading time, especially at advanced process nodes. Moreover, incorporating MIS effects in advanced delay models will require a drastic change in the delay calculator and is challenging. Therefore, we have modeled the MIS effect as an incremental corrective quantity over SIS delay. It fits easily with the existing design flows and delay calculators. Additionally, the approach proposed in Shashank Ram and Saurabh [25] is generic. Therefore, the ANN-based model can be employed to capture other non-ideal effects at advanced process nodes.

We have employed the ML-based MIS model to carry out MIS-aware timing analysis [25]. It involves reading MIS-aware timing libraries and reconstructing the original ANN. Since the ANNs are compact, the time consumed in the reconstruction of ANNs is insignificant. Subsequently, using the circuit conditions, we compute the MSD for each relevant timing arc. Using MSD, we adjust the SIS delay and generate the MIS-annotated timing reports. It is demonstrated that the ML-based MIS modeling can improve the accuracy of timing analysis [25]. For example, for some benchmark circuits, traditional SIS-based delay differs from the corresponding SPICE-computed delay by 120%. However, the ML-based model produces delays with errors less than 3%. The runtime overhead of MIS-aware timing analysis is also negligible. The methodology of Shashank Ram and Saurabh [25] can be extended to create a single composite MIS model for different process voltage temperature (PVT) conditions. In the future, we expect that we can efficiently represent other complicated circuit and transistor-level empirical models using ML models.

7.6   TEST, DIAGNOSIS, AND VALIDATION

We can employ ML techniques in post-fabrication testing, diagnosis of failures, and validation of functionality. The ability of ML models to work efficiently on large data sets can be useful in these applications.

We can reduce the test cost by removing redundant tests. We can use ML to mine the test set and eliminate the redundant tests. We can consider a test as redundant if we can predict its output using some other tests that we are not removing [26]. For example, we can use statistical learning methodology based on decision trees for eliminating redundant tests [26]. Note that while removing tests, we should ensure maintaining product quality and limiting yield loss.

We can apply ML-based techniques in testing analog/RF devices also to reduce cost. However, it is challenging to maintain test errors in ML-based analog/RF device testing to an acceptable level. To solve this problem, we can adopt a two two-tier test approach [27]. A neural system-based framework is developed in Stratigopoulos and

Makris [27] that produces both the pass/fail labels and the confidence level in its prediction. If the confidence is low for a prediction, traditional and more expensive specification testing is employed to reach a final test decision. Thus, the cost advantage of ML-based analog/RF testing is leveraged. Note that the test quality is not sacrificed in the two-tier test approach [27]. We can employ a similar strategy for other verification problems where ML-induced errors are critical.

We can use ML-based strategies for the diagnosis of manufacturing defects. They can provide alternatives to the traditional techniques of exploring the causal relationship. We can formulate a diagnosis problem as an evaluation of several decision functions [28]. It can reduce the run time complexity of the traditional diagnosis methods, especially for volume diagnosis. It has been reported in Wang and Wei [28] that even with highly compressed output responses, we can find defect locations for most defective chips. We can employ ML techniques in the diagnosis of failures in a scan chain also [29]. Note that the scan chain patterns are not sufficient to determine the failing flip-flop in a scan chain. Therefore, we need chain failure diagnosis methodologies to identify defective scan cell(s) on a faulty scan chain. We can employ unsupervised ML techniques based on the Bayes theorem that are more tolerant to noises for this purpose [29].

Another problem that can utilize the capabilities of ML is the post-silicon validation. Before production, we carry out post-silicon validation to ensure that the silicon functions as expected, under on-field operating conditions. For this purpose, we need to identify a small set of traceable signals for debugging and state restoration. Traditional techniques such as simulation take high runtime in identifying traceable signals. Alternatively, we can employ ML-based techniques for efficient signal selection [30]. We can train an ML with a few simulation runs. Subsequently, we can use this model to identify beneficial trace signals instead of employing time-consuming simulations [30].

7.7 CHALLENGES

In the previous sections, we discussed various applications of ML techniques in VLSI design. Nevertheless, there are some challenges involved in adopting ML techniques in conventional design flows. The effectiveness of ML techniques in VLSI design is dependent on complex design data. Therefore, producing competitive results repeatedly on varying design data is a challenge for many applications. Moreover, training an ML model requires extracting voluminous data from a traditional or a detailed model. Sometimes it is challenging to generate such a training data set. Sometimes these training data are far away from the ground truth or contains a lot of noises. Handling such a training set is challenging.

ML-based design flows can disrupt the traditional design flows and be expensive to deploy. Moreover, applying ML-based EDA tools may not produce expected results immediately. There is some non-determinism associated with the ML-based applications. In the initial stages, there are not enough training data. Consequently, an ML-based EDA tool cannot guarantee accurate results. Therefore, adopting ML-based solutions in design flows is challenging for VLSI designers. Nevertheless, in the long run, ML-based techniques could deliver rich dividends.

7.8 CONCLUSIONS

In summary, ML offers efficient solutions for many VLSI design problems [31–35]. It is particularly suitable for complex problems for which we have readily available data to learn from and predict. With the advancement in technology, we expect that such design problems will increase. The advances in EDA tools will also help develop more efficient ML-specific hardware. The ML-specific hardware can accelerate the growth in ML technology. The advancement in ML technologies can further boost their applications in developing complex EDA tools. Thus, there is a synergic relationship between these two technologies. In the long run, both these technologies together can deliver benefits to many other domains and applications.

REFERENCES

[1] Ozisikyilmaz, Berkin, Gokhan Memik, and Alok Choudhary. "Efficient system design space exploration using machine learning techniques." In *Proceedings of the 2008 45th ACM/IEEE Design Automation Conference*, pp. 966–969. IEEE, 2008.

[2] Greathouse, Joseph L., and Gabriel H. Loh. "Machine learning for performance and power modeling of heterogeneous systems." In *Proceedings of the 2018 IEEE/ACM International Conference on Computer-Aided Design (ICCAD)*, pp. 1–6. IEEE, 2018.

[3] Sen, Satyabrata, and Neena Imam. "Machine learning based design space exploration for hybrid main-memory design." In *Proceedings of the International Symposium on Memory Systems*, pp. 480–489. 2019.

[4] Zuluaga, Marcela, Andreas Krause, Peter Milder, and Markus Püschel. ""Smart" design space sampling to predict Pareto-optimal solutions." In *Proceedings of the 13th ACM SIGPLAN/SIGBED International Conference on Languages, Compilers, Tools and Theory for Embedded Systems*, pp. 119–128. 2012.

[5] Liu, Hung-Yi, and Luca P. Carloni. "On learning-based methods for design-space exploration with high-level synthesis." In *Proceedings of the 50th Annual Design Automation Conference*, pp. 1–7. 2013.

[6] Mahapatra, Anushree, and Benjamin Carrion Schafer. "Machine-learning based simulated annealer method for high level synthesis design space exploration." In *Proceedings of the 2014 Electronic System Level Synthesis Conference (ESLsyn)*, pp. 1–6. IEEE, 2014.

[7] Kim, Ryan Gary, Janardhan Rao Doppa, and Partha Pratim Pande. "Machine learning for design space exploration and optimization of manycore systems." In *Proceedings of the 2018 IEEE/ACM International Conference on Computer-Aided Design (ICCAD)*, pp. 1–6. IEEE, 2018.

[8] Pagani, Santiago, P.D. Sai Manoj, Axel Jantsch, and Jörg Henkel. "Machine learning for power, energy, and thermal management on multicore processors: A survey." *IEEE Transactions on Computer-Aided Design of Integrated Circuits and Systems* 39, no. 1 (2018): 101–116.

[9] Xiao, Yao, Shahin Nazarian, and Paul Bogdan. "Self-optimizing and self-programming computing systems: A combined compiler, complex networks, and machine learning approach." *IEEE Transactions on Very Large Scale Integration (VLSI) Systems* 27, no. 6 (2019): 1416–1427.

[10] Mitchell, Tom M., Sridbar Mabadevan, and Louis I. Steinberg. "LEAP: A learning apprentice for VLSI design." In *Machine Learning*, pp. 271–289. Morgan Kaufmann, 1990.

[11] Beerel, Peter A., and Massoud Pedram. "Opportunities for machine learning in electronic design automation." In *Proceedings of the 2018 IEEE International Symposium on Circuits and Systems (ISCAS)*, pp. 1–5. IEEE, 2018.

[12] Chang, Wen-Hsiang, Chien-Hsueh Lin, Szu-Pang Mu, Li-De Chen, Cheng-Hong Tsai, Yen-Chih Chiu, and Mango C-T. Chao. "Generating routing-driven power distribution networks with machine-learning technique." *IEEE Transactions on Computer-Aided Design of Integrated Circuits and Systems* 36, no. 8 (2017): 1237–1250.

[13] Kirby, Robert, Saad Godil, Rajarshi Roy, and Bryan Catanzaro. "CongestionNet: Routing congestion prediction using deep graph neural networks." In *Proceedings of the 2019 IFIP/IEEE 27th International Conference on Very Large Scale Integration (VLSI-SoC)*, pp. 217–222. IEEE, 2019.

[14] Tabrizi, Aysa Fakheri, Logan Rakai, Nima Karimpour Darav, Ismail Bustany, Laleh Behjat, Shuchang Xu, and Andrew Kennings. "A machine learning framework to identify detailed routing short violations from a placed netlist." In *Proceedings of the 2018 55th ACM/ESDA/IEEE Design Automation Conference (DAC)*, pp. 1–6. IEEE, 2018.

[15] Chan, Wei-Ting J., Pei-Hsin Ho, Andrew B. Kahng, and Prashant Saxena. "Routability optimization for industrial designs at sub-14nm process nodes using machine learning." In *Proceedings of the 2017 ACM on International Symposium on Physical Design*, pp. 15–21. 2017.

[16] Baker Alawieh, Mohamed, Yibo Lin, Wei Ye, and David Z Pan. "Generative learning in VLSI design for manufacturability: Current status and future directions." *Journal of Microelectronic Manufacturing* 2, no. 4 (2019).

[17] Yang, Haoyu, Wei Zhong, Yuzhe Ma, Hao Geng, Ran Chen, Wanli Chen, and Bei Yu. "VLSI mask optimization: From shallow to deep learning." In *Proceedings of the 2020 25th Asia and South Pacific Design Automation Conference (ASP-DAC)*, pp. 434–439. IEEE, 2020.

[18] Ioannides, Charalambos, and Kerstin I. Eder. "Coverage-directed test generation automated by machine learning--a review." *ACM Transactions on Design Automation of Electronic Systems (TODAES)* 17, no. 1 (2012): 1–21.

[19] Wang, Fanchao, Hanbin Zhu, Pranjay Popli, Yao Xiao, Paul Bodgan, and Shahin Nazarian. "Accelerating coverage directed test generation for functional verification: A neural network-based framework." In *Proceedings of the 2018 on Great Lakes Symposium on VLSI*, pp. 207–212. 2018.

[20] Kahng, Andrew B., Mulong Luo, and Siddhartha Nath. "SI for free: Machine learning of interconnect coupling delay and transition effects." In *Proceedings of the 2015 ACM/IEEE International Workshop on System Level Interconnect Prediction (SLIP)*, pp. 1–8. IEEE, 2015.

[21] Medico, Roberto, Domenico Spina, Dries Vande Ginste, Dirk Deschrijver, and Tom Dhaene. "Machine-learning-based error detection and design optimization in signal integrity applications." *IEEE Transactions on Components, Packaging and Manufacturing Technology* 9, no. 9 (2019): 1712–1720.

[22] Fang, Yen-Chun, Heng-Yi Lin, Min-Yan Su, Chien-Mo Li, and Eric Jia-Wei Fang. "Machine-learning-based dynamic IR drop prediction for ECO." In *Proceedings of the International Conference on Computer-Aided Design*, pp. 1–7. 2018.

[23] Ding, Duo, Andres J. Torres, Fedor G. Pikus, and David Z. Pan. "High performance lithographic hotspot detection using hierarchically refined machine learning." In *Proceedings of the 16th Asia and South Pacific Design Automation Conference (ASP-DAC 2011)*, pp. 775–780. IEEE, 2011.

[24] Geng, Hao, Haoyu Yang, Bei Yu, Xingquan Li, and Xuan Zeng. "Sparse VLSI layout feature extraction: A dictionary learning approach." In *Proceedings of the*

2018 IEEE Computer Society Annual Symposium on VLSI (ISVLSI), pp. 488–493. IEEE, 2018.

[25] Shashank Ram, O.V.S., and Sneh Saurabh. "Modeling multiple input switching in timing analysis using machine learning." *IEEE Transactions on Computer-Aided Design of Integrated Circuits and Systems* 40, no. 4 (2020).

[26] Biswas, Sounil, and Ronald D. Blanton. "Statistical test compaction using binary decision trees." *IEEE Design & Test of Computers* 23, no. 6 (2006): 452–462.

[27] Stratigopoulos, Haralampos-G., and Yiorgos Makris. "Error moderation in low-cost machine-learning-based analog/RF testing." *IEEE Transactions on Computer-Aided Design of Integrated Circuits and Systems* 27, no. 2 (2008): 339–351.

[28] Wang, Seongmoon, and Wenlong Wei. "Machine learning-based volume diagnosis." In *Proceedings of the 2009 Design, Automation & Test in Europe Conference & Exhibition*, pp. 902–905. IEEE, 2009.

[29] Huang, Yu, Brady Benware, Randy Klingenberg, Huaxing Tang, Jayant Dsouza, and Wu-Tung Cheng. "Scan chain diagnosis based on unsupervised machine learning." In *Proceedings of the 2017 IEEE 26th Asian Test Symposium (ATS)*, pp. 225–230. IEEE, 2017.

[30] Rahmani, Kamran, Sandip Ray, and Prabhat Mishra. "Postsilicon trace signal selection using machine learning techniques." *IEEE Transactions on Very Large Scale Integration (VLSI) Systems* 25, no. 2 (2016): 570–580.

[31] Wang, Li-C., and Magdy S. Abadir. "Data mining in EDA-basic principles, promises, and constraints." In *Proceedings of the 2014 51st ACM/EDAC/IEEE Design Automation Conference (DAC)*, pp. 1–6. IEEE, 2014.

[32] Capodieci, Luigi. "Data analytics and machine learning for continued semiconductor scaling." *SPIE News* (2016).

[33] Wang, Li-C. "Experience of data analytics in EDA and test—principles, promises, and challenges." *IEEE Transactions on Computer-Aided Design of Integrated Circuits and Systems* 36, no. 6 (2016): 885–898.

[34] Pandey, Manish. "Machine learning and systems for building the next generation of EDA tools." In *Proceedings of the 2018 23rd Asia and South Pacific Design Automation Conference (ASP-DAC)*, pp. 411–415. IEEE, 2018.

[35] Kahng, Andrew B. "Machine learning applications in physical design: Recent results and directions." In *Proceedings of the 2018 International Symposium on Physical Design*, pp. 68–73. 2018.

8 An Overview of High-Performance Computing Techniques Applied to Image Processing

Giulliano Paes Carnielli[1], Rangel Arthur[1],
Ana Carolina Borges Monteiro[2],
Reinaldo Padilha Franca[2], and Yuzo Iano[2]
[1]Faculty of Technology (FT), State University of Campinas
(UNICAMP), Limeira, São Paulo, Brazil
[2]School of Electrical Engineering and Computing (FEEC),
State University of Campinas (UNICAMP), Campinas,
São Paulo, Brazil

CONTENTS

8.1 INTRODUCTION

8.1.1 Context

Digital image processing (DIP) is a dynamic and expanding area, comprising technologies widely used in the most diverse areas and applications. For example, construction of geographic and atmospheric models over a time series of satellite images, detection of anomalies in manufactured products, support for diagnostics based on medical images, and security based on biometric recognition, among others [1,2].

Many of the techniques used in image processing were developed back in the 1960s, as shown by the work of Azriel Rosenfeld [3], one of the first to address the issue of computer image processing. It is worth noting that, in the following decade, several studies were developed with the objective of optimizing hardware architectures and algorithms to make parallel image processing feasible. For example, in a work dated 1975 [4], Stamatopoulos shows the efficiency of parallel digital processors in relation to serial digital computers for image processing. In a 1977 paper [5], Meilander discusses the evolution of the parallel processor architecture for image processing.

Since its early days, DIP has been a challenge to the processing capacity of hardware and software. Computer vision (CV) problems are computationally intensive [6,7]. The explanation for this is that images concentrate a large amount of data and are usually treated in large sets by applications that demand results in a short time. For these characteristics, image processing can be included in the context of Big Data. Traditional general-objective machines cannot drive the distinct I/O requisite of most image processing assignments, nor do it take benefit of the parallel computing scope present in most vision-associated applications [8,9].

Therefore, as Jon A. Webb [10] states, DIP and CV are natural applications for high-performance computing (HPC), which boost research, for example, in the areas of graphics hardware, standards specification, modeling, efficient algorithms, simulation, animation, and virtual reality [11].

The purpose of this study is to present an overview of four high-performance processing techniques, used in the area of image processing and CV, as well as to briefly address the use of artificial neural networks (ANNs) in the same context. DIP has several applications that, admittedly, demand great computational power due to either the amount of data usually addressed or, in certain contexts, the need to generate results in a very short time. Therefore, an overview of these HPC techniques, commonly applied to such research areas, becomes relevant.

8.1.2 Concepts

1. Digital Image: An analogue image can be determined, according to Monteiro, Gonzales and Woods, and Tyagi [7,12,13], as a two-dimensional function in which x and y are respective to spatial coordinates, and the amplitude f is known as the intensity or gray level of the digital image at the point.

So, a digital image is a matrix representation (two-dimensional) created through a process of sampling and quantization (digitization) of a real image (analogue). In

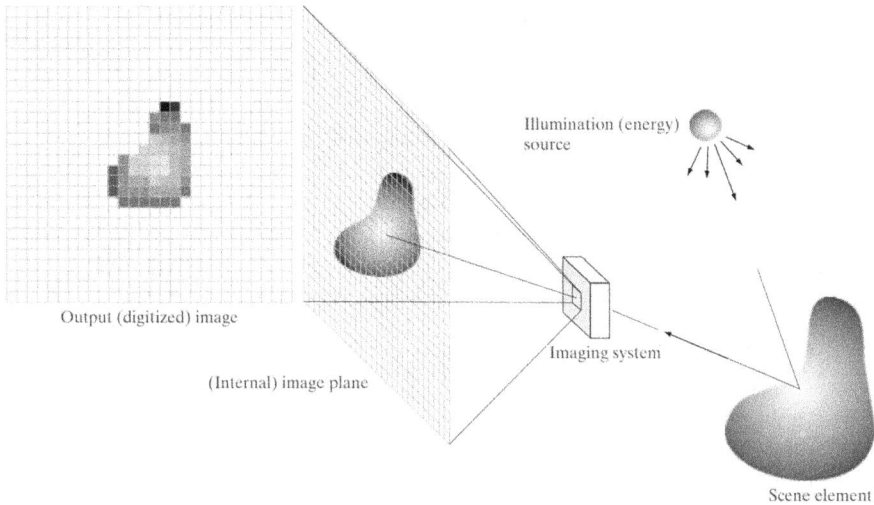

FIGURE 8.1 Scanning process [12].

such a process, the continuous coordinates of the original image are converted to discrete values, resulting in an array of values, usually real numbers.

Each sampled point is called a pixel or pel (picture element) and stores information, as a numerical value or a small set of numbers, of the properties recorded at the sampled point, such as color or brightness. The sampling intervals determine the spatial resolution of the image, that is, the number of points that form the matrix (Figure 8.1).

2. Image Processing and CV: According to Gonzales and Woods, and Nixon and Aguado [12,14], image processing and CV are two areas comprising a wide variety of applications, with no clear and defined border between them. Image processing is often defined as the activity of transforming one image into another, or a set of information. On the other hand, CV is the use of digital computers to emulate human vision, comprising being able to generate inferences and take actions based on visual inputs. Image analysis (understanding of images) is a discipline that stands between these two definitions.

Still, according to Monteiro, and Gonzales and Woods [7,12], it is useful to classify the spectrum of applications on images in three categories of computerized digital processes:

- Low Level: primitive digital operations such as contrast equalization, noise reduction, and sharpness. Processes whose inputs and outputs are images.
- Medium Level: segmentation, description, classification, and recognition of objects. Processes whose inputs are digital images, but the outputs are attributes (characteristics) extracted from the digital images.
- High Level: making sense of a set of objects extracted from the image, performing cognitive functions associated with vision.

8.2   HPC TECHNIQUES APPLIED TO IMAGE TREATMENT

Image processing and CV are areas of extreme interest due to their applicability in the most diverse contexts such as medicine, engineering, industrial production, and agriculture, among others. However, applications involving image processing and analysis require a lot of computational power. This differentiated processing capacity can be obtained through different techniques and approaches. The objective of this work is to present four HPC techniques, appropriate to different applications related to image processing, exemplifying each one with case studies taken from the literature of the area. In addition, briefly discussing the use of ANNs for image processing [15,16].

8.2.1   CLOUD-BASED DISTRIBUTED COMPUTING

1. Description: The Cloud Computing paradigm originated mainly from research on virtualization and distributed computing, and it is based on techniques developed and principles on the scope of these areas [17]. Although
   there is no consensus on the definition of Cloud Computing, a well-accepted
   statement is given by França et al. [18]:

   *Cloud Computing is the provision of computing services, including servers, storage,
   databases, network, software, analysis, and intelligence enabling convenient, on-demand
   network access to a shared pool of configurable computing resources. This offers faster
   innovations, flexible resources (eg, networks, servers, storage, applications, and services), and economies of scale, helping to reduce operating costs, run infrastructure
   more efficiently, with minimal management effort or service provider interaction, and
   even scale as needs arise.*

Platforms such as Amazon Web Services (AWS) or Google Cloud Platform (GCP) offer highly scalable, on-demand, accessible, and easily configurable environments without requiring service provider interference. Such configuration features allow for multiple processing units to be allocated in the same environment, or an en-vironment to be instantiated multiple times, providing an appropriate platform to perform tasks in parallel. In addition, data can be kept in the virtual environment, avoiding large transfers during processing (Figure 8.2) [19].

The aforementioned characteristics meet application scenarios in which the de-mand for HPC resources varies both over time and in the required specifications.

1. Case Study: The work presented by Shunxing Bao [21] discusses the processing of large quantities of medical images (Big Data Medical Image),
   acquired by heterogeneous magnetic resonance (MRI) methods, with analysis
   in multiple stages (multi-stage analysis), whose characteristics include the
   frequent use of high-performance clusters, incurring large operational costs,
   in addition to variability in processing time, sequential execution (pipeline)
   with errors identified only in later stages, wasting time and computational
   resources.

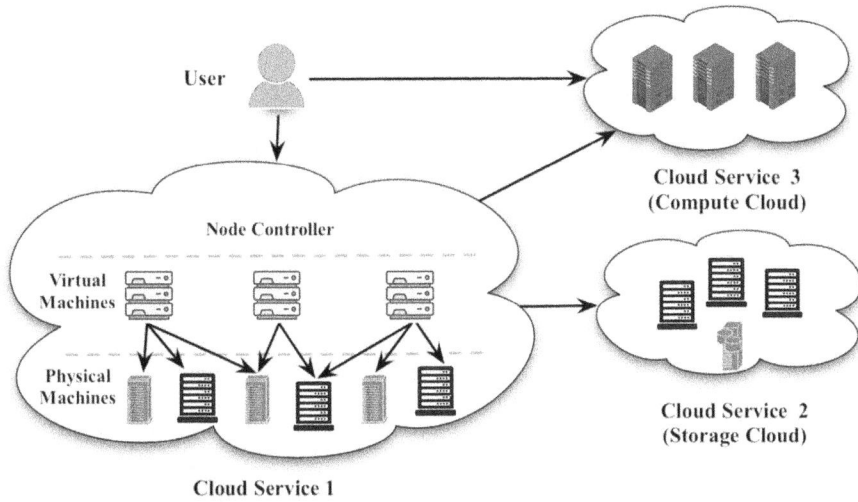

FIGURE 8.2 Cloud-based distributed computing is an architecture in which task-parallelism is achieved by allocating resources and services available in the cloud [20].

The article proposes an alternative approach, developing a concurrent framework for the processing pipeline, in which the constituting stages are executed concurrently (parallelism with dependencies) in an Apache Hadoop ecosystem, with tasks submitted to a MapReduce module and outputs stored in HBase, reducing the needs of data transfer during the processing. Additionally, the proposed framework relies on a monitoring mechanism that parses the quality of intermediate results, to identify outliers, and to detect problems in the early stages.

The new architecture was empirically assessed in a private research cloud, consisting of a Gigabit network and 188 CPUs. Results showed a reduction in the required execution time (wall time) of 76.75%, when compared with the traditional approach, executed in a cluster from Vanderbilt University (TN), which count with a 10 Gigabit network, GPFS (IBM's General Parallel Filesystem) and more than 6000 CPUs.

8.2.2 GPU-ACCELERATED PARALLELIZATION

1. Description: This technique relies on the combination of a graphics processing unit (GPU) and a conventional CPU, to speed up tasks that require extensive computing with both matrix and scalar characteristics [22].

The GPU benefits parallel applications, by offering the performance of hundreds of CPUs with a SIMD (single instruction multiple data) architectures, which is suitable for certain types of load. On the other hand, processes that are inherently sequential would achieve better performance taking advantage of the processing power and capacities of a CPU (Figure 8.3). Therefore, using a combination of CPUs and GPUs can extract the best of both technologies [23,24].

Application Code

Rest of sequential
CPU code

Intensive computation

CPU

GPU

CPU

GPU

+

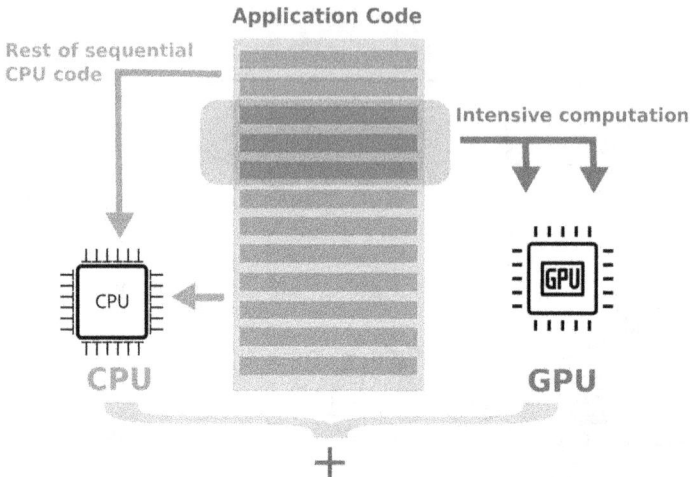

FIGURE 8.3 Conceptual illustration of a GPU-accelerated architecture [11].

GPU was initially developed for graphic processing and image rendering, but it has become a highly parallel programmable processor, used for more generic purposes. This change of application context was promoted by the publication of parallel programming APIs, such as Open Computing Language (OpenCL – Apple) and compute unified device architecture (CUDA - NVIDIA), which allocate the computing power of a GPU, for a variety of applications [25].

2. Case Study 1: An interesting application of the parallelization technique, with GPU acceleration, can be found at Tsai et al. [26]. In this work, the authors come up with a new paradigm for applying GPU acceleration to parallelize the extraction of characteristics, based on the gray-level co-occurrence matrix (GLCM[1]), in magnetic resonance images (MRI) of the brain, with the purpose of identifying types of tissue damage

A typical pipeline for texture analysis involves region recognition and segmentation, feature extraction (characteristics), and even classification. The performance of each of these steps depends on the accuracy of the previous step. In this context, Tsai et al. [26] focuses on accelerating the extraction of features from MRI images of the brain, using GPU.

The most widely used method for extracting statistical characteristics from textures is the GLCM. However, this approach implies a high computational cost, when performing several calculations on a Region of Interest (ROI) that slides over a high-resolution image. This task is not suitable for processing on a simple CPU, given the increase in the quantity and quality of data, and the need for a prompt response [27].

The purpose of this work is to use the power of GPU parallelism, to generate the GLCM and extract its characteristics, simultaneously, for many small overlapping ROI, covering the entire image area. The processing in each ROI is independent and very similar, with great potential for parallelism (Figure 8.4) [28].

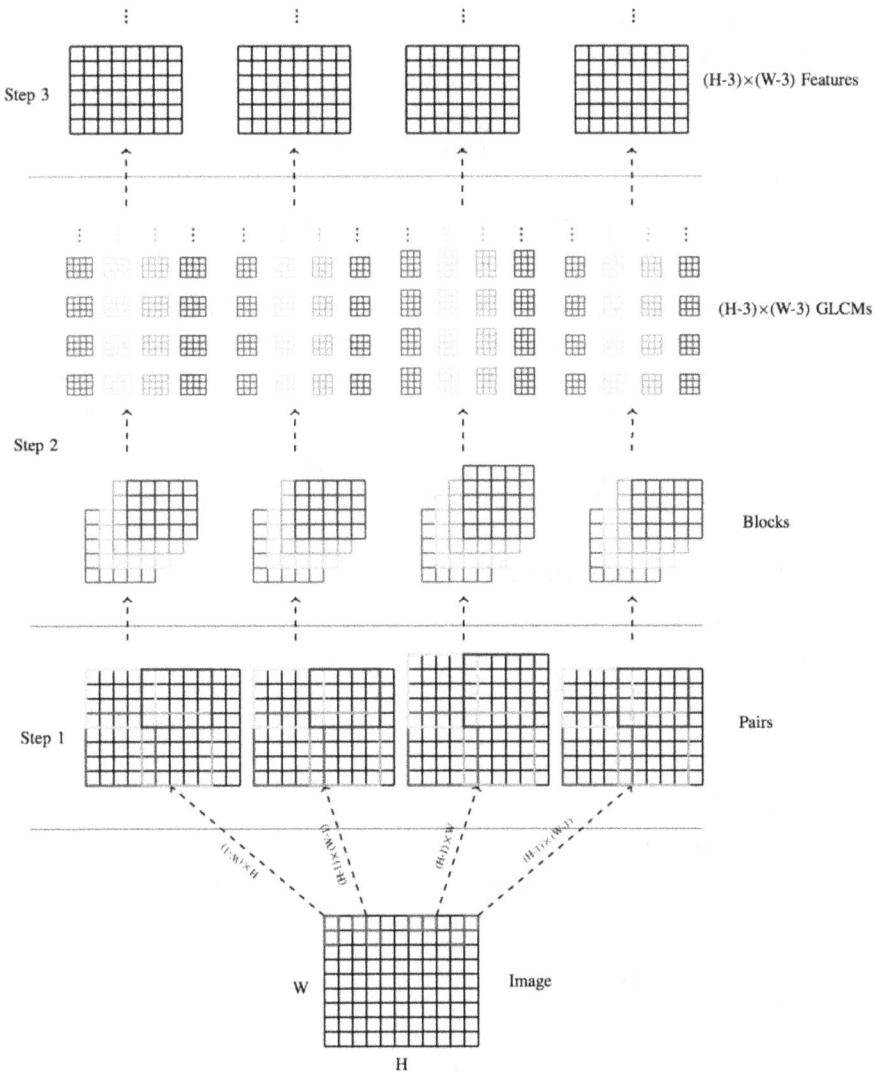

FIGURE 8.4 Proposed parallel processing scheme [26].

The experiments were performed on a GeForce GTX 1080 graphics device, applied to a computer system consisting of an Intel E3–1231 CPU with eight 3.40 GHz cores, 8GB RAM (1333 Mhz), and Ubuntu operating system (kernel 4.4.0-66 - generic) [26–28].

The experiment showed some interesting aspects, such as the fact that the method of extracting characteristics is very efficient when the size of the ROI is relatively small. The speedup in the total processing time (for computing in simple precision) varied between 26 and 103 times. The work also concluded that this acceleration was limited, in large part by operations of memory allocation and data transfer between CPU and GPU [26–28].

3. Case Study 2: Still related to the use of GPUs, to accelerate image processing techniques, Saxena et al. [29] propose a parallel implementation for the morphological algorithm vHGW (van Herk/Gill-Werman), considered one of the fastest for serial processing in CPU. Morphological operators (e.g. erosion, dilation, opening) are used in the extraction of components from images, used for representation and description of regions and shapes, through the application of a structuring element on the image.

This work presents a parallel strategy, based on the use of the NVIDIA GTX860M GPU, with implementation over the CUDA 5.0 API, with 640 CUDA cores. The computer system featured a 2.50GHz Intel i7 CPU, and 8GB of RAM.

The result was obtained from experiments with expansion and erosion operators, implemented by the vHGW algorithm in CUDA 5.0. There was an effective gain with the use of the GPU, but the speedup varied considerably with the size of the image, and with the size and shape of the structuring element. The work reported that the biggest gains were obtained with large images, and with the increase of CUDA cores.

8.2.3 PARALLELIZATION USING GPU CLUSTER

1. Description: Efforts to incorporate HPC technologies in several areas generally focus on three categories: data processing based on specialized hardware, such as GPUs; data processing in clusters; and large-scale distributed computing infrastructure, such as grid computing and Cloud Computing.

In modern times, GPUs have progressed into highly parallel, multithreaded processors with high memory bandwidth [30]. Therefore, it makes sense for multiple GPUs to be integrated in order to create an even more parallel and powerful architecture.

A GPU cluster is a cluster of computers in which each node is provided with a GPU, suitable for the GPGPU (General-Purpose Graphics Processing Unit) model, in which such devices are typically aimed at graphics processing, are used for processing general purpose.

In a GPU cluster, the nodes are connected to each other by some high-performance network technology (e.g. InfiniBand). An Ethernet switch connects the parallel infrastructure to a Head Node, responsible for the cluster interface, which receives and processes external requests and assigns work to the nodes (Figure 8.5).

2. Case Study: An example of using a GPU cluster in the image processing area can be found at Lui et al. [30]. This work presents a high-performance framework to handle time-consuming applications, and time-series quantitative retrieval of satellite images, in a cluster of GPUs.

Such quantitative sensing models are used to estimate diverse geophysical parameters such as Aerosol Optical Depth (AOD) and even Normalized Difference Vegetation Index (NDVI). That project investigates an efficient solution to derive a decade AOD dataset on 1 km resolution images over Asia, related to the SRAP-MODIS

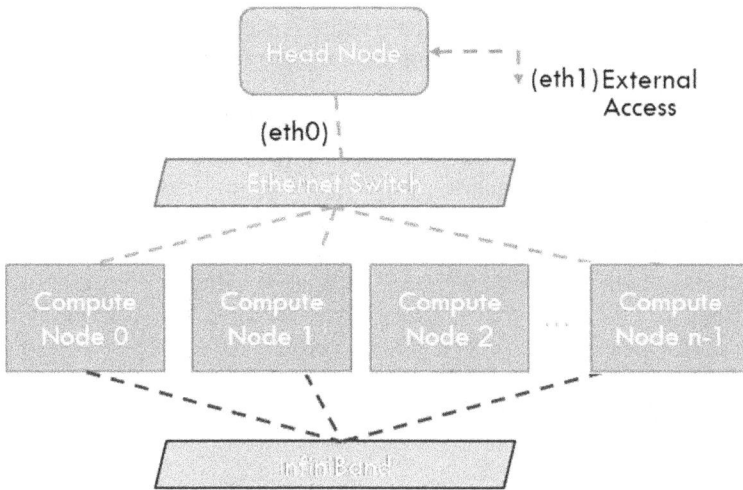

FIGURE 8.5 GPU cluster architecture, as proposed by NVIDIA.

	Master node	Slave node
CPU configuration	Intel Xeon E5-2660 server CPUs running at 2.2 GHz 8 physical cores 32 GB memory memory bandwidth of 76.8 GB/s	Intel Xeon E5-1650 v2 CPU running at 3.50 GHz 6 physical cores 64 GB memory memory bandwidth of 51.2 GB/s
GPU configuration	/ / / / / / /	NVIDIA Tesla K40 2880 cores 12 GB memory memory bandwidth of 288 GB/s

FIGURE 8.6 Experimental system configuration [14].

(Synergetic Retrieval of Aerosol Properties Model from MODIS[2]). For performance, the time series recovery task is distributed over the GPU cluster.

The experiments were carried out over a cluster with 14 nodes and one master node, whose configurations are detailed in Figure 8.6. The resulting speedup for the proposed complete processing, when compared with a parallel GPU solution, was 1.61. The overall performance improvement was 28.6%, using a hybrid MPI-CUDA parallel solution. However, it was observed that there is still room for improvement, since the CPU cores of multicore processors, installed in each node of the cluster, have not been completely utilized

8.2.4 MULTICORE ARCHITECTURE

1. Description: Architecture in which two or more processing units (cores) are mounted in a single integrated circuit matrix, or chip (CMP – Chip

multiprocessor), or even in multiple chips, composing a single package (chip package). Currently, most microprocessors are of the multicore type [31].

Cores can be tightly or loosely coupled, depending on how they share resources (e.g. cache, memory) and how they communicate (e.g. messages or memory sharing). In addition, multicore systems are also classified as homogeneous (Figure 8.7a), when the cores are identical, or support the same set of instructions, or are heterogeneous (Figure 8.7b), otherwise [32].

In general, multicore architectures provide more efficiency in resource management and data transfer due to the proximity of the cores. However, the Operating System must be able to use the resources offered by architectures like this [31,32].

2. Case Study: An interesting study on the application of multicore architectures in image processing applications can be found at Osorio et al. [33]. This work, whose main focus is hardware design, proposes the use of a Massively Parallel Processor Array (MPPA), available in an Ambric Am2045 system[3].

The Am2045 system, MPPA with distributed memory, had 336 processing cores with 300MHz and a memory bank with 4 blocks of 2KB. Each core had two SR units, a simple processor, with support for basic arithmetic and logic instructions, and two SRD units, similar to a DSP (digital signal processor), supporting more complex instructions (Figure 8.8).

The purpose of this work is to implement two image processing applications on the Am2045 architecture: begin itemize item algorithm for JPEG encoding and, item binary DIP (edge detection and hole filling) end itemize.

Regarding the JPEG encoding algorithm, the objective was to compare the results with those of 6 other platforms. In the case of the binary image processing, the result was compared with just one other approach based on dedicated hardware (FPGA – field-programmable gate array).

The results indicated that, for the JPEG encoding algorithm, the solution offered by Osorio et al. and Gu et al. [33,35] was surpassed by only one platform (Table 8.1), which indicates that hardware-based solutions (FPGAs) still have some advantage over software-based implementations. However, the second half of the table, shown in Table 8.1, which is composed of approaches based on FPGAs, with the exception of Altera Stratix, all the others had lower performance than the technique with MPPA.

In the case of binary image processing, the method was compared with a single work based on FPGA (Table 8.2). The technique proposed by Osorio et al. [33] had a better performance than the competitor, after normalization based on clock speed, since the metric was the count of machine cycles required for each processing.

8.3 NEURAL NETWORKS

Nowadays, it is impossible to deal with the subject of image processing without mentioning the growing use of neural networks as an important set of tools for the

(a)

(b)

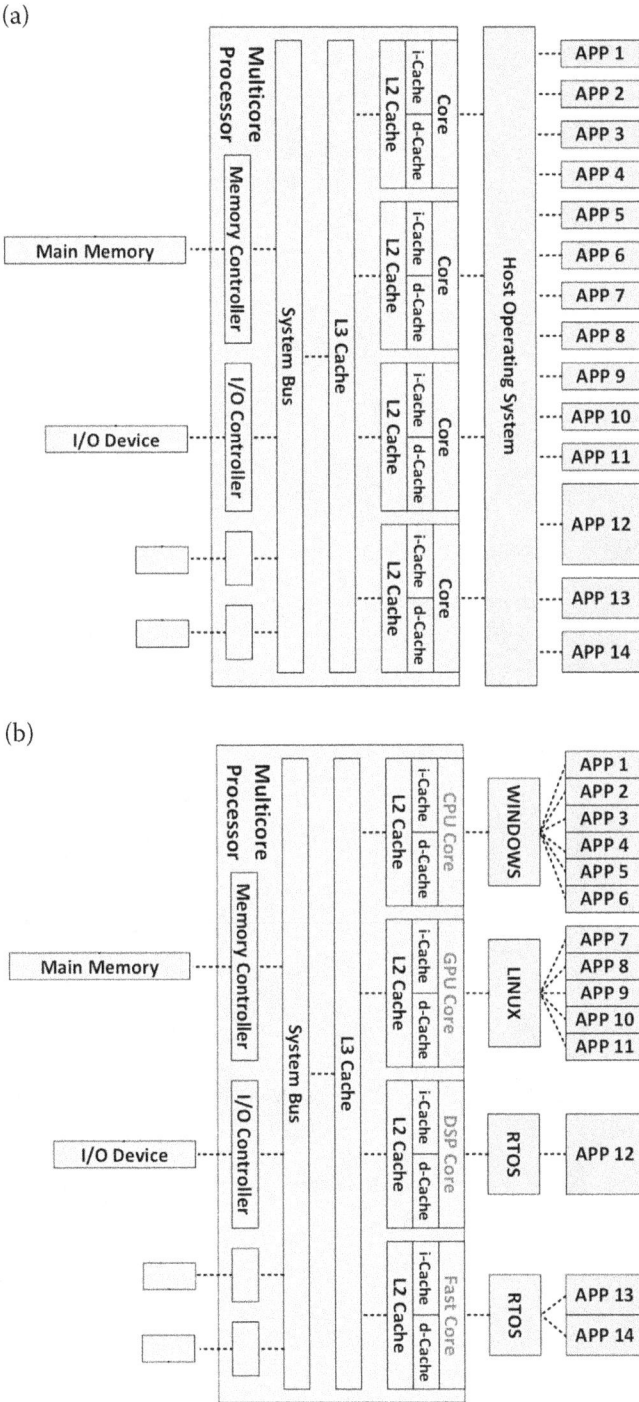

FIGURE 8.7 Multicore system: (a) homogeneous and (b) heterogeneous [15].

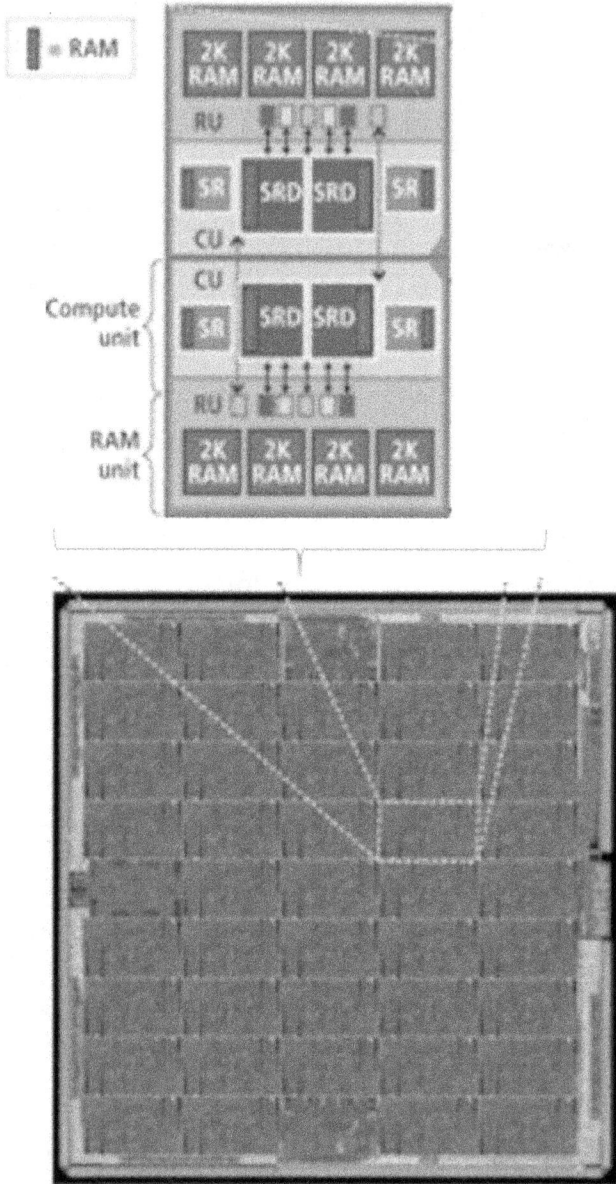

FIGURE 8.8 Am2045 architecture [34].

treatment of images. There are several types of neural networks applied to image processing, but two of these networks stand out for the vast amount of applications in which they are used: convolutional neural networks (CNNs) and generative adversarial networks (GANs) [15,35].

TABLE 8.1
JPEG Encoding Results [30]

Platform	Speed (MHz)	MPixels/se
[16]	300	31
BlackFin [18]	750	2
TI C5410 [18]	160	0.4
Xilinx Virtex II-Pro 20 [18]	50	2.6
Xilinx Spartan 3 200 [19]	49	0.75
Altera Flex 10KE [20]	40	13.2
Altera Stratix II [20]	161	54

TABLE 8.2
Results for Binary Image Processing [33]

	[16]	[21]	Normalized [21]
Edge detection	17	6	26.7
Hole filling	$17 \times n$	$7 \times n$	$31.2 \times n$

8.3.1 CONVOLUTIONAL NEURAL NETWORK (CNN)

This category of deep neural network (MLP – multilayer perceptron) is commonly applied in image analysis. The rudiments of this type of neural network date back to the early 1980s, in a work that developed a neural network known as neocognitron [35,36].

Its hidden layers comprise a series (set) of convolutional layers, in which a scalar product operation is applied between the image and a sliding kernel (a process similar to filtering in the spatial domain). This operation is able to identify patterns in the image, extracting them as attributes for further classification. The kernel applied to each layer is also selected as part of the learning process (Figure 8.9) [38].

The first step ends when an attribute vector is generated, using the process described above. This vector of attributes (characteristics) is then passed to the input layer of an MLP (fully connected layer), which performs the image classification. Such neural networks are widely used for image classification, element identification, and segmentation [39,40].

8.3.2 GENERATIVE ADVERSARIAL NETWORK (GAN)

Machine learning system developed in 2014, in Goodfellow et al. [41]. Like CNN, it is a technique widely applied to image analysis. This type of network learns, based on a training set, to generate new data (images) with the same statistics as the data used in the training.

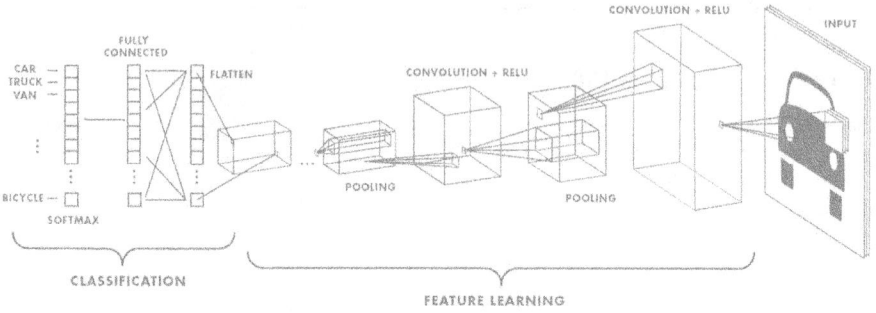

FIGURE 8.9 CNN structure [37].

The architecture of this elaborated system has two communicating networks, the generator network and the discriminator network (Figure 8.10). The generator takes samples of the input images (training data) and synthesizes new images, randomly changing their characteristics, adding noise. During training, the discriminator receives a random sampling of real and false images. The discriminator must then determine which images are real and which are false [42,43].

The learning process takes place based on the discriminator's output, which is validated with the data provided, and used both to adjust (optimize) the generator and the discriminator itself. The potential for application of these networks is huge, ranging from the generation of a face of a "virtual" model, to be used in advertising, to the identification of anomalies in products, inside a factory [42–44].

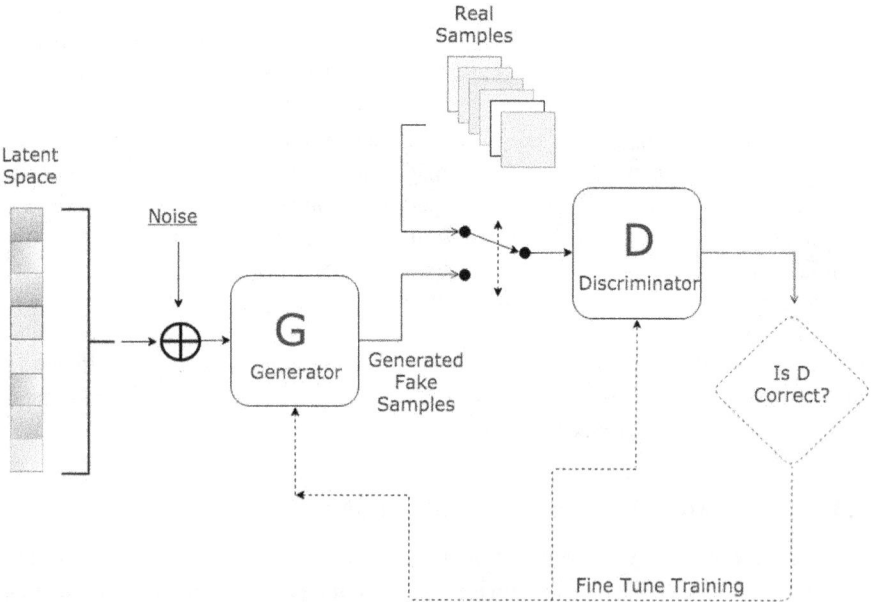

FIGURE 8.10 GAN structure [44].

8.3.3   HPC Techniques Applied to Neural Networks

Neural networks form another area extremely dependent on PAD techniques. Generally, the application of neural networks is not an extremely expensive activity, from a computational point of view. However, the process of training these networks, especially when it involves image processing, usually requires a lot of computational power [45,46].

Considering the type of data treated and the size of the training set, the process of generating a trained model can take days. Therefore, there is a need to use data and process parallelization techniques. The training of a network is a highly parallelizable task, as it replicates the same type of processing, repeatedly, in a large set of data, so it can benefit from several HPC techniques, including those that were discussed in this work [45,46].

However, due to the characteristics of this type of task, which involves large masses of data, being subjected to similar operations, data parallelism in GPUs is perhaps the most popularly used technique [45,46].

8.4   MACHINE LEARNING APPLICATIONS HARDWARE DESIGN

8.4.1   FPGA

FPGA is an implementation of interconnected digital circuits that perform common functions and, at the same time, have a high level of flexibility, with a general structure consisting of three basic components such as IOB (input and output blocks), a CLB (configurable logic blocks), and a Switch Matrix (interconnect switches) [47].

Currently, memory blocks, DSP (digital signal processing), and even an ARM processor are common in FPGA architecture. The logic blocks are organized in a two-dimensional matrix, still considering that the interconnection wires are arranged as vertical and horizontal routing channels, containing programmable wires and switches allowing logic blocks to connect in several different ways, between the rows and columns of the block [48,49].

Still considering that each logic block generally has a small quantity of inputs and outputs, as Look Up Table (LUT) containing storage cells capable of holding a logical value, 0 or 1, employed to effectuate a logic function. This value produces an output from the storage cell. It is possible to create LUTs of various sizes; this is defined by the number of entries [50].

In short, there are several logic functions, memory, DSP, input and output blocks, which can connect in several different ways to represent hardware described in a hardware description language (HDL) or in a block diagram. Emphasizing that the input file, be it HDL or block diagram is synthesized by a tool normally provided by the manufacturer of the FPGA, and a binary file is generated with the hardware configuration that FPGA must adopt [51].

FPGA applications are basically aimed at high processing, parallelism, and real-time are necessary, since the use of an FPGA must be considered. It is currently present in several sectors of the industry, where performance, parallelism, and real time are crucial. Because it is hardware, it is possible to perform an instruction per clock cycle, calculations running in parallel and delivering the result at the same clock pulse, given that this context is something completely impossible for software to perform [52].

Exemplifying FPGA applications, it is possible to stand out in the electric sector, for digital signal processing in real time; multimedia sector, for image processing in real time and high performance; telecom sector, in high-performance switches and routers; since FPGA chips are used for numerous applications, ranging from video games to areas such as aerospace, prototyping, HPC, medical, among others. It is possible to find the use of FPGA in applications for audio in circuits of DACs (digital to analog converter) where several aspects are modified to achieve a better reproduction of the sound, or even by alternative consoles using FPGA chips to reproduce via hardware the same quality as the original versions of video games, or even FPGA chips still allow the player the flexibility to play games from different consoles using a single device [52].

Still emphasizing that modern technology companies include FPGAs in data centers in order to speed up search engines, applying machine learning algorithms, such as support vector machines (SVMs). Highlighting the main objective in the parallel implementation of FPGAs both in the feedforward phase of a SVM and in its training phase [53].

In this context, it is possible to employ machine learning algorithms written in FPGAs, making them quite efficient and easily reprogrammable, evaluating that this specialization specifically converts parallel computing, synthesizing deep learning processor unit (DPU) or deep neural network (DNN) processing units in FPGAs [54].

In this sense, FPGAs applied to different types of machine learning models bring flexibility, making it easier to accelerate applications based on the most ideal numerical precision and the memory model being used. Considering the possibility of parallelizing the pre-training of DNNs in FPGAs to scale a given service horizontally. Still considering that DNNs can be pre-trained, as a deep resource transformer for transferring learning, or adjusted with updated weights [52,54].

8.4.2 SVM

SVMs represent an machine learning technique for modeling classifiers and regressors, seeking to minimize the empirical risk simultaneously with the ability to generalize. This is widely used mainly due to its mathematical properties that include good generalization capacity and robustness. SVM training models obtain good accuracy and low complexity, making it necessary to define the kernel and its parameters, since the kernel and training model parameters together are called hyperparameters of the SVM parameter selection problem [55,56].

SVM is a supervised learning algorithm whose objective is to classify a set of data points mapped to a multidimensional feature space (characteristics) employing a kernel function. In this supervised learning algorithm, the decision limit in the input space is represented by a hyperplane in a higher dimension in space. This, in general, is obtained from a finite subset of known data and their respective labels (training set), seeking to classify data that are not part of the training set [55,56].

SVMs realize the separation of a set of objects with distinct classes, that is, they utilize the logic of decision plans to have a wide range of applications in different areas, due to their application advantages related to good generalization performance, mathematical treatability, geometric interpretation, and use for the exploration of unlabeled data [56].

Emphasizing that the efficiency of a classifying function is measured by both complexity and generalizability. The generalization capacity is the function's ability to correctly classify data that does not belong to the training set. The complexity refers to the number of elements that are necessary to compose the function, such as, for example, the points of the SVMs training set, the centers of the radial basis function networks, or even relating the neurons in the layer hidden in the RNAs (ANN) [57].

In this sense, combining the importance of using FPGAs as computational accelerators in recent years, it is possible to specifically accelerate algorithms in search engines (machine learning algorithms) through SVM, through the efficient use of FPGA resources. Considering the parallel hardware implementation, allowing a larger number of kernels to be implemented on a chip of the same size, accelerating the training, both from the feedforward (inference) phase implemented using the SVM polynomial kernel and its training phase, possible to obtain the maximum possible acceleration at the cost of greater use of the available area associated with the use of the FPGA area [58].

Still reflecting that through FPGAs it is possible to meet the demand for performance of artificial intelligence and even Big Data, i.e., a high volume of processing in the face of a huge volume of data. FPGAs still tend to enhance processing velocity and decrease hardware costs by implementing a massive volume of processes concurrently and directing the flow of that data, removing cost and usability barriers, employing implementation accessibility for any, and all projects that require processing low latency of vast volume of data. Since FPGAs are programmable just like CPUs or even GPUs, but it aims for parallel, low latency, and high-speed issues, such as inferences and DNN [52,59].

8.5 CONCLUSIONS

Still emphasizing the FPGA technologies relating to machine learning and technologies derived from artificial intelligence highlighting benefits such as speed, since this performs at a remarkable clock speed in relation to actual CPUs. FPGAs are simultaneous, i.e., instead of executing sequential instruction streams, it operates in an optimal data stream between these parallel operations, following in an enhance in performance, running applications n times faster on the same code compared to CPUs/GPUs traditional.

Still reflecting that FPGAs technology is capable of containing millions of CLB that can be employed to execute numerous actions at the same time, providing a predominance of parallelism and competition, taking advantage of parallel architecture solving problems in well-structured and independent processes that can be carried out concurrently. As in the case of an image taken into processing non-simultaneously, a certain part would process the entire image pixel by pixel. However, when this digital image is processed simultaneously, it is divided into pieces processed at the same time by distinct parts and then reassembled together. This feature renders the process more complex, but much faster, considering that the data received must be optimally divided and efficiently distributed to the parties, after which the collected data are processed and reassembled, usually without blocking the work pipeline.

Parallel computing has an important meaning in various image processing techniques, such as image segmentation, edge detection, noise removal, histogram equalization, image registration, resource extraction, and distinct optimization techniques, among others.

The present study provided a brief introduction of high-performance processing techniques for DIP, with their respective resources and limitations, presenting case studies in the areas of medical images, time series analysis in satellite images, hardware architecture. However, the applications for these techniques are not limited to these cases.

Each HPC technique is best applied to different image processing scenarios. In addition, these techniques can be combined and overlapped, creating more possibilities for solutions.

It is a complex task to try to label and categorize these techniques, but an attempt in this direction, with the purpose of organizing knowledge, can be useful as a reference for future applications.

NOTES

1 Gray-Level Co-Occurrence Matrix: tabulation of occurrences of combined gray levels. Used in texture analysis
2 Moderate Resolution Imaging Spectroradiometer data
3 It was a company of parallel processors that developed the Am2045, used primarily in high performance embedded systems

REFERENCES

[1] Monteiro, A. C. B., França, R. P., Estrela, V. V., Razmjooy, N., Iano, Y., & Negrete, P. D. M. (2020). Metaheuristics applied to blood image analysis. In Metaheuristics and optimization in computer and electrical engineering (pp. 117–135). Springer, Cham.
[2] Monteiro, A. C. B., Iano, Y., França, R. P., & Arthur, R. (2020). Development of a laboratory medical algorithm for simultaneous detection and counting of erythrocytes and leukocytes in digital images of a blood smear. In Deep learning techniques for biomedical and health informatics (pp. 165–186). Academic Press, Cambridge, Massachusetts, United States.
[3] Rosenfeld, A. (1969). Picture processing by computer. ACM Computing Surveys (CSUR), 1(3), 147–176.
[4] Yang, Z., Zhu, Y., & Pu, Y. (2008, December). Parallel image processing based on CUDA. In 2008 International Conference on Computer Science and Software Engineering (Vol. 3, pp. 198–201). IEEE.
[5] Meilander, W. C. (1977, January). The evolution of parallel processor architecture for image processing. In COMPCON'77 (pp. 52–53). IEEE Computer Society.
[6] Choudhary, A., & Ranka, S. (1992). Guest editor's introduction: parallel processing for computer vision and image understanding. Computer, 25(2), 7–10.

[7] Monteiro, A. C. B. (2019). Proposta de uma metodologia de segmentação de imagens para detecção e contagem de hemácias e leucócitos através do algoritmo WT-MO. Master's thesis.

[8] Franca, R. P., Iano, Y., Monteiro, A. C. B., & Arthur, R. (2020). Big data and cloud computing: A technological and literary background. In Advanced deep learning applications in big data analytics (pp. 29–50). IGI Global, Pennsylvania, United States.

[9] Athanas, P. M., & Abbott, A. L. (1995). Real-time image processing on a custom computing platform. Computer, 28(2), 16–25.

[10] Webb, J. A. (1994, October). High-performance computing in image processing and computer vision. In 12th IAPR International Conference on Pattern Recognition, Vol. 2-Conference B: Computer Vision & Image Processing. (Cat. No. 94CH3440-5) (pp. 218–222). IEEE.

[11] Dougherty, E. R. (2020). Digital image processing methods. CRC Press, Boca Raton, Florida, United States.

[12] Gonzales, R. C., & Woods, R. E. (2002). Digital image processing, 2nd Edition. Prentice-Hall, Hoboken, New Jersey, United States.

[13] Tyagi, V. (2018). Understanding digital image processing. CRC Press, Boca Raton, Florida, United States.

[14] Nixon, M., & Aguado, A. (2019). Feature extraction and image processing for computer vision. Academic Press, Cambridge, Massachusetts, United States.

[15] Monteiro, A. C. B., Iano, Y., França, R. P., & Arthur, R. (2020). Deep learning methodology proposal for the classification of erythrocytes and leukocytes. In Trends in deep learning methodologies: algorithms, applications, and systems (p. 129). Elsevier.

[16] Walczak, S. (2019). Artificial neural networks. In Advanced methodologies and technologies in artificial intelligence, computer simulation, and human-computer interaction (pp. 40–53). IGI Global, Pennsylvania, United States.

[17] Panzieri, F., Babaoglu, O., Ferretti, S., Ghini, V., & Marzolla, M. (2011). Distributed computing in the 21st century: Some aspects of cloud computing. In Dependable and historic computing (pp. 393–412). Springer, Berlin, Heidelberg.

[18] França, R. P., Iano, Y., Borges, A. C., Monteiro, R. A., & Estrela, V. V. (2020). A proposal based on discrete events for improvement of the transmission channels in cloud environments and big data. In Big data, IoT, and machine learning: Tools and applications (p. 185). CRC Press.

[19] Dutta, P., & Dutta, P. (2019). Comparative study of cloud services offered by Amazon, Microsoft & Google. International Journal of Trend in Scientific Research and Development (IJTSRD), 3, 981–985.

[20] Jamkhedkar, P. A., Lamb, C. C., & Heileman, G. L. (2011, July). Usage management in cloud computing. In 2011 IEEE 4th International Conference on Cloud Computing (pp. 525–532). IEEE.

[21] Bao, S., Parvarthaneni, P., Huo, Y., Barve, Y., Plassard, A. J., Yao, Y., ... & Gokhale, A. (2018, December). Technology enablers for big data, multi-stage analysis in medical image processing. In 2018 IEEE International Conference on Big Data (Big Data) (pp. 1337–1346). IEEE.

[22] Kasap, B., & van Opstal, A. J. (2018). Dynamic parallelism for synaptic updating in GPU-accelerated spiking neural network simulations. Neurocomputing, 302, 55–65.

[23] Baji, T. (2018, March). Evolution of the GPU device widely used in AI and massive parallel processing. In 2018 IEEE 2nd Electron Devices Technology and Manufacturing Conference (EDTM) (pp. 7–9). IEEE.

[24] Nvidia.Com. (2017). NVIDIA on GPU computing and the difference between GPUs and CPUs. Available http://www.nvidia.com/object/what-isgpucomputing.html.

[25] Kalaiselvi, T., Sriramakrishnan, P., & Somasundaram, K. (2017). Survey of using GPU CUDA programming model in medical image analysis. Informatics in Medicine Unlocked, 9, 133–144.

[26] Tsai, H. Y., Zhang, H., Hung, C. L., & Min, G. (2017). GPU-accelerated features extraction from magnetic resonance images. IEEE Access, 5, 22634–22646.

[27] Xing, Z., & Jia, H. (2019). Multilevel color image segmentation based on GLCM and improved salp swarm algorithm. IEEE Access, 7, 37672–37690.

[28] Siahaan, R., Pardede, C., & Gurning, W. P. (2020). Another Parallelism Technique of GLCM Implementation with CUDA Programming. In 2020 4th International Conference on Advances in Image Processing (143–151).

[29] Saxena, S., Sharma, S., & Sharma, N. (2017). Study of parallel image processing with the implementation of vHGW algorithm using CUDA on NVIDIA'S GPU framework. In World Congress on Engineering (Vol. 1).

[30] Liu, J., Xue, Y., Ren, K., Song, J., Windmill, C., & Merritt, P. (2019). High-performance time-series quantitative retrieval from satellite images on a GPU cluster. IEEE Journal of Selected Topics in Applied Earth Observations and Remote Sensing, 12(8), 2810–2821.

[31] Jain, P. N., & Surve, S. K. (2020). A review on shared resource contention in multicores and its mitigating techniques. International Journal of High Performance Systems Architecture, 9(1), 20–48.

[32] Li, Y., & Zhang, Z. (2018, July). Parallel computing: review and perspective. In 2018 5th International Conference on Information Science and Control Engineering (ICISCE) (pp. 365–369). IEEE.

[33] Osorio, R. R., Diaz-Resco, C., & Bruguera, J. D. (2009, August). High-performance image processing on a massively parallel processor array. In 2009 12th Euromicro Conference on Digital System Design, Architectures, Methods, and Tools (pp. 233–236). IEEE.

[34] Imaging Boards and Software. (2008). Processor targets medical, video applications. Available https://www.vision-systems.com/boards-software/article/1673927 1/processor-targets-medical-video-applications.

[35] Gu, J., Wang, Z., Kuen, J., Ma, L., Shahroudy, A., Shuai, B., ... & Chen, T. (2018). Recent advances in convolutional neural networks. Pattern Recognition, 77, 354–377.

[36] Fukushima, K., & Miyake, S. (1982). Neocognitron: A self-organizing neural network model for a mechanism of visual pattern recognition. In Competition and cooperation in neural nets (pp. 267–285). Springer, Berlin, Heidelberg.

[37] Saha, S. (2018). A comprehensive guide to convolutional neural networks — the ELI5 way. Available https://towardsdatascience.com/a-comprehensive-guide-to-convolutional-neural-networks-the-eli5-way-3bd2b1164a53.

[38] França, R. P., Peluso, M., Monteiro, A. C. B., Iano, Y., Arthur, R., & Estrela, V. V. (2018, October). Development of a Kernel: A deeper look at the architecture of an operating system. In Brazilian technology symposium (pp. 103–114). Springer, Cham.

[39] Zeiler, M. D., & Fergus, R. (2014, September). Visualizing and understanding convolutional networks. In European Conference on Computer Vision (pp. 818–833). Springer, Cham.

[40] Shan, K., Guo, J., You, W., Lu, D., & Bie, R. (2017, June). Automatic facial expression recognition based on a deep convolutional-neural-network structure. In 2017 IEEE 15th International Conference on Software Engineering Research, Management and Applications (SERA) (pp. 123–128). IEEE.

[41] Goodfellow, I., Pouget-Abadie, J., Mirza, M., Xu, B., Warde-Farley, D., Ozair, S., ... & Bengio, Y. (2014). Generative adversarial nets. In Advances in neural information processing systems (pp. 2672–2680). NIPS, Montreal, Canada.

[42] Yi, X., Walia, E., & Babyn, P. (2019). Generative adversarial network in medical imaging: A review. Medical Image Analysis, 58, 101552.

[43] Nazeri, K., Ng, E., & Ebrahimi, M. (2018, July). Image colorization using generative adversarial networks. In International conference on articulated motion and deformable objects (pp. 85–94). Springer, Cham.

[44] Hakobyan, H. (2018). How GANs can turn AI into a massive force. Available https://www.techinasia.com/talk/gan-turn-ai-into-massive-force.

[45] Barnell, M., Raymond, C., Capraro, C., Isereau, D., Cicotta, C., & Stokes, N. (2018, September). High-performance computing (HPC) and machine learning demonstrated in flight using Agile Condor®. In 2018 IEEE High Performance Extreme Computing Conference (HPEC) (pp. 1–4). IEEE.

[46] Lynn, T., Liang, X., Gourinovitch, A., Morrison, J. P., Fox, G., & Rosati, P. (2018, January). Understanding the determinants of cloud computing adoption for high performance computing. In 51st Hawaii International Conference on System Sciences (HICSS-51) (pp. 3894–3903). University of Hawai'i at Manoa, Hawai'i.

[47] Amano, H. (Ed.). (2018). Principles and structures of FPGAs. Springer, Heidelberg, Germany.

[48] Dekoulis, G. (Ed.). (2017). Field: Programmable gate array. BoD–Books on Demand, London, UK.

[49] Dekoulis, G. (2020). Field programmable gate arrays (FPGAs) II. InTechOpen, London, UK.

[50] Kumar, T. N., Almurib, H. A., & Lombardi, F. (2014, November). A novel design of a memristor-based look-up table (LUT) for FPGA. In 2014 IEEE Asia Pacific Conference on Circuits and Systems (APCCAS) (pp. 703–706). IEEE.

[51] Shokrolah-Shirazi, M., & Miremadi, S. G. (2008, July). FPGA-based fault injection into synthesizable verilog HDL models. In 2008 Second International Conference on Secure System Integration and Reliability Improvement (pp. 143–149). IEEE.

[52] Romoth, J., Porrmann, M., & Rückert, U. (2017). Survey of FPGA applications in the period 2000–2015. Technical Report.

[53] Lopes, F. F., Ferreira, J. C., & Fernandes, M. A. (2019). Parallel implementation on FPGA of support vector machines using stochastic gradient descent. Electronics, 8(6), 631.

[54] Duarte, J., Harris, P., Hauck, S., Holzman, B., Hsu, S. C., Jindariani, S., … & Lončar, V. (2019). FPGA-accelerated machine learning inference as a service for particle physics computing. Computing and Software for Big Science, 3(1), 13.

[55] Müller, K. R. Supervised Machine Learning: Learning SVMs and Deep Learning. http://helper.ipam.ucla.edu/publications/mpstut/mpstut_13971.pdfIf.

[56] Cöltekin, C., & Rama, T. (2016, December). Discriminating similar languages with linear SVMs and neural networks. In Third Workshop on NLP for Similar Languages, Varieties and Dialects (VarDial3) (pp. 15–24).

[57] Pouriyeh, S., Vahid, S., Sannino, G., De Pietro, G., Arabnia, H., & Gutierrez, J. (2017, July). A comprehensive investigation and comparison of machine learning techniques in the domain of heart disease. In 2017 IEEE Symposium on Computers and Communications (ISCC) (pp. 204–207). IEEE.

[58] Padierna, L. C., Carpio, M., Rojas-Domínguez, A., Puga, H., & Fraire, H. (2018). A novel formulation of orthogonal polynomial kernel functions for SVM classifiers: The Gegenbauer family. Pattern Recognition, 84, 211–225.

[59] Huanrui, H. (2016). New mixed kernel functions of SVM used in pattern recognition. Cybernetics and Information Technologies, 16(5), 5–14.

9 Machine Learning Algorithms for Semiconductor Device Modeling

Yogendra Gupta[1], Niketa Sharma[1],
Ashish Sharma[2], and Harish Sharma[3]
[1]Swami Keshwanand Institute of Technology Jaipur, Jaipur, Rajasthan, India
[2]Indian Institute of Information Technology Kota, Kota, Rajasthan, India
[3]Rajasthan Technical University Kota, Kota, Rajasthan, India

CONTENTS

9.1 INTRODUCTION

This chapter describes machine learning algorithms from the viewpoint of semiconductor device modeling. Machine learning has become one of the most favorable modeling techniques for device, circuit, and system-level modeling. Electrical characteristics of the device such as I-V and C-V characteristics represent an input-output relationship, also temperature, process variation, and power variations can be modeled by machine learning algorithms. Semiconductor device modeling is an area of extensive

DOI: 10.1201/9781003201038-9

research. Technology computer aided design (TCAD) models are used to design and analyze semiconductor devices. They are also used as a first step in compact model development. Compact models are used for circuit design and functional verification. Accuracy of these models directly translates to design accuracy, high yield, and more profit. However, development of TCAD and compact models is a huge undertaking. It requires concrete understanding of underlying physics. It also requires a lot of simulation resources (computation time and memory), making it very complex and expensive. Additionally, models developed for one device cannot be ported, i.e. used for another device, if governing physical principles of two devices are different. This chapter is motivated by the need of acceleration of model development for semiconductor devices. Research efforts have been directed in semiconductor device modeling and machine learning fields separately; however, the use of machine learning in device modeling is yet to be explored by the research community. In this chapter, we describe various machine learning algorithms for semiconductor device modeling, challenges, and trade-offs associated with them. As a case study, we have proposed electrothermal modeling of GaN-based high electron mobility transistor (HEMT) devices. A data-driven approach has been implemented for a temperature range varying from 300 K to 600 K, based on one of the core methods of machine learning techniques, i.e. decision trees (DTs). The performance of the proposed models was validated through the simulated test examples. The attained outcomes show that the developed models predict the HEMT device characteristics accurately depending on the determined mean squared error between the actual and anticipated characteristics.

This chapter studies machine learning algorithms from the perspective of semiconductor device modeling and manufacturing. It aims to discuss and review major machine learning algorithms for semiconductor device modeling and manufacturing. The hypothesis behind this study is that semiconductor device modeling problems can be treated like data analysis and mining problems. We can model electrical characteristics of the device such as I-V, and C-V characteristics represent an input-output relationship by mathematical functions [1]. This study is based on the hypothesis that supervised machine learning algorithms can be used for development of such complex model equations.

Machine learning has also become popular for applications in semiconductor manufacturing, such as the etch anomaly analysis [2], lithographic hotspot detection [3], and optical proximity correction [4]. Today's analysis practices heavily rely on extensive device and material characterizations. An machine-learning–assisted variation analysis based on device electrical characteristics is highly desired to allow for more efficient material and device experimentation. Semiconductor manufacturing is among the most complex processes ever devised. It is also among the most data-rich, with extensive records of just about anything and everything that can be measured or observed. Much of this data goes unused, at least in part because of its huge volume. Unused data is sometimes called "dark" data. Artificial intelligence's (AI's) ability to mine these troves of dark data for relevant relationships is one area of great promise.

Semiconductor device modeling is an area of extensive research. TCAD models are used to design and analyze semiconductor devices. They are also used as a first step in compact model development. Compact models are used for circuit design and functional verification. Accuracy of these models directly translates to design accuracy, high yield, and more profit. However, development of TCAD and compact models is a huge undertaking. It requires concrete understanding of underlying physics. It also requires a lot of simulation resources (computation time and memory), making it very complex and expensive. Additionally, models developed for one device cannot be ported, i.e. used for other device, if governing physical principles of the two devices are different. This research is motivated by the need of acceleration of model development for semiconductor devices. Research efforts have been directed in semiconductor device modeling and machine learning fields separately; however, the use of machine learning in device modeling is yet to be explored by the research community. Earlier work [5] uses an artificial neural network (ANN) approach for region-specific FET modeling. This research is inspired by a similar philosophy; however, it uses different building blocks.

9.2 SEMICONDUCTOR DEVICE MODELING

With vast amount of data avialable as technology advances, an effective data analysis technique can contribute significantly to emerging technologies. Machine learning is the worldly technique for data analysis and modeling. According to Murphy [6], machine learning can be defined as "A set of methods that can automatically detect patterns in data and then use uncovered patterns to predict future data". Semiconductor device models are used to predict or reproduce the device characteristics using Silvaco-ATLAS. The modeling of semiconductor devices essentially needs a mapping algorithm, which enables the estimation of device current as a function of node voltages and device parameters. Generally, in semiconductor device modeling approaches attempt to understand the physical phenomena of the device. Modeling by this approach generally covers one region of device operation, which creates the problem modeling the transitions between the regions. In opposition to this, black-box modeling techniques produce models for all components. The model is often defined in terms of a mathematical equation that relates terminal voltages and node current. The model captures elecro-physical behavior of the device as a function of device characteristics like geometry, biasing, temperature variations, and process variations. Schneider and Galup-Montoro define semiconductor device model as a crucial link between the designers and the manufacturing foundries. The device model makes it possible to simulate the device circuits prior to their fabrication. We can perform all the parametric design space exploration and analysis before the fabrication. Therefore, we need an accurate device model for the computer-aided simulation of the circuit. The accurate device model results in better accuracy. This improves productivity as well as time to market. Figure 9.1 depicts classification of a semiconductor device model. The semiconductor model is broadly classified into two types. One models device design and analysis, and the second models circuit simulations (compact models). The device design and analysis models are classified into three types: Monte Carlo models, moment method models, and energy balance models. The circuit simulation models (compact models) are classified into four types: physical compact model, empirical model, black-box model, and look-up table model. The physical compact model is divided into three types: threshold voltage-based model, inversion

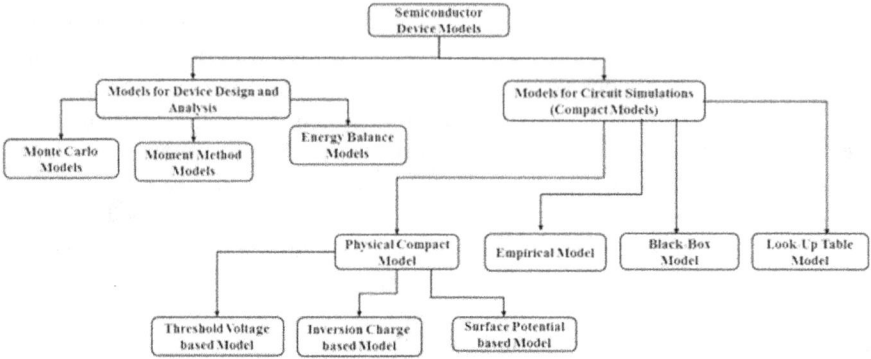

FIGURE 9.1   Semiconductor device model classification.

charge-based model, and surface potential-based model. The semiconductor device models, which are used for design and analysis of the devices, are called TCAD models. These models can be developed for different levels of abstraction. Semiconductor device models used for design and analysis of devices are called TCAD models. These models refer to the fabrication process and hence are most accurate. These models can provide insight into details of device operation and hence are very useful for device optimization for performance. However, large memory usage and long computation time prohibit their usage in circuit design flow. For use in circuits with sea-of-devices, simulation models are required, which are accurate, yet faster than TCAD models. These models are called compact models, used extensively for circuit design and functional verification. Tsividis and Andrews [7] have given guidelines pertaining to attributes of good compact models. For the sake of brevity, key attributes are mentioned below:

1. The semiconductor models should meet the characteristics of semiconductor devices such as current-voltage (I-V), transconductance, electron mobility, electric field, electron concentration, leakage currents, intrinsic and extrinsic capacitances, charges, etc.
2. All the models like charges and currents and their derivatives should be continuous with respect to each terminal voltage. A commercial circuit si-mulator often uses nonlinear KCL and the Newton-Raphson method based on linearization, which requires continuous derivatives for effective operations.
3. The outcome of the model should make physical sense and replicate real devices. The outcome should not be unphysical like negative transconduc-tance or capacitances.

A model can be thought of as a mathematical equation that governs behavior of a particular system. Model equations can be of various types, including but not re-stricted to dynamical systems, statistical models, and differential equations. It is also possible that these given types of models can overlap, depending on abstractions in context. Use of models spans across various disciplines, such as natural sciences (physics, chemistry, biology, astronomy, etc.), social sciences (psychology, sociology, economics, etc.), and importantly, engineering disciplines (artificial intelligence, electronics engineering, acoustic engineering, etc.).

Models are used to capture effects that contribute to behavior of the system under investigation. Many times, they are used to predict system response for unknown input stimuli. Stochastic (one or more input variables are of random nature) models are capable of modeling effects of random input variables on system behavior. Some of these models may be purely stochastic or hybrid (some of the input variables are deterministic, and some are random in nature). It is expected that models should be generalizable; that is, models should consider as many effects as possible so that they can be used accurately over broad input space. Models exhibit trade-off between accuracy and simplicity. Complex models have more predictive power; however, they are difficult to understand and analyze. Based on availability of information while developing models, they are broadly classified as black-box, white-box, and grey-box models. A white-box model is a model where all necessary system information was available at its inception. On the other hand, if there is no significant prior information available that can be used while developing the model, the model is then termed a black-box model. There are no models that are purely theoretical (white-box) or without prior information (black-box). They are called grey-box models, which are based on theoretical structures and use test data for completing development. It is imperative to assess model quality prior to deployment. There are multiple ways in which models can be evaluated.

9.3 RELATED WORK

Litovski et al. [5] proposed an ANN-based black-box modeling technique of semiconductor devices. They realized a mapping algorithm that is based on ANN for MOS transistor modeling. They used a unique continuous function to cover all regions of transistor operation.

Ahmad Khusro et al. [8] presented a support vector regression (SVR)-based modeling approach for HEMT devices. They used the nonlinear Gaussian or radial basis kernel function. The method has been illustrated by modeling the small signal behavior of a 4 × 100 μm GaN HEMT over a broad frequency range of 1–18 GHz with multi-biasing sets and measured sparameters.

N. Hari et al. [9] proposed an approach to use deep learning techniques to build a GaN-based regression model using a stochastic gradient algorithm by backpropagation. Among the different neural network architectures trained and tested, a deep feedforward neural network with 5 hidden layers and 30 neurons was found to be the best for prediction and optimization. The possibility of employing machine learning techniques for GaN can help open doors for faster commercialization of GaN power electronics.

Cai et al. [10] proposed a Pulsed IV (PIV) behavioral model for Gallium Nitride (GaN) HEMT. They used Bayesian inference, which is one of the primary methods of machine learning, to build this new model. They validated the performance of the proposed model through experimental test examples. The model can very accurately predict the PIV curves of a 10 W GaN HEMT at different pulse widths and duty cycles from isothermal up to the safe-operating area limit, with high voltage drain-source pulses, indicating that the proposed model can include both the thermal and dispersion effect in it precisely and effectively.

9.4 CHALLENGES

To explain common challenges faced during the process of semiconductor device modeling, the widely used HEMT technology is used as an example. GaN semiconductor devices have momentous properties contrasted with its silicon (Si) and silicon carbide (SiC) equivalents and have the possibility of replacing the ubiquitous Si power devices. The wide-bandgap semiconductor-based sensors have a number of advantages over other sensor technologies, such as operating at high temperatures, with high chemical stability, or under ionizing radiation. In any case, there are significant technical obstructions preventing their dynamic utilization by the biosensor community. This is performing as a limitation to the commercialization of GaN-based biosensors [11,12].

Considering the biosensing functioning of variety of GaN devices (which are based on different structures, design, and packaging) is the primary step in acquiring an indulgence into the operation of these devices. Nevertheless, since the modeling of these novel devices depends on the conventional semiconductor device physics methodology, there are practical difficulties for the sensor community to cop-up with them. Due to the complexity of the device structure, time, and the analytical procedures involved, the existing models are not suitable for validating all applications and cannot serve as a unified model for all existing GaN biosensing devices [13–17]. To address this issue, GaN simulation models, which are an accurate replica of the actual device, were designed using machine learning techniques.

Additional TCAD simulations are required to capture these effects with accuracy. As stated in the previous section, integrated circuit design and verification relies heavily on availability of accurate device models. To capture higher order effects, scaling of devices requires a complex model to accommodate as many scaling-imposed effects as possible. This requires fundamental understanding of device physics and the origin of such effects. This reinstates the requirement of physics-based modeling, due to its high accuracy compared to other methods. Similar endeavors are required to model a new semiconductor device, since the physical base of the new device may be different from existing devices. It may be possible to use some of the existing intellectual property such as model source codes, test benches, etc. However, it still takes a significant amount of time to make new models available to the designer community (Figure 9.2).

9.5 MACHINE LEARNING FUNDAMENTALS

Learning is the continuous process of benefiting from previous experiences and making decisions based on them. Learning and adapting is often regarded as the most basic form of animal intelligence. Throughout our lifespan, the ability to learn gives us the power to adjust and adapt to new situations. The decomposition of learning process into its constituents is somewhat difficult, mainly because of variety of processes that contribute to learning. Here are some formal definitions of learning. According to

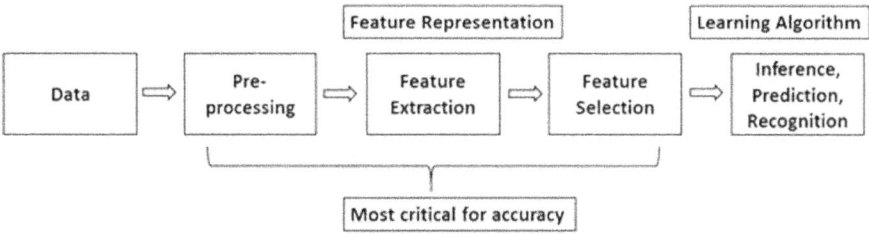

FIGURE 9.2   Standard flow of machine learning.

Merriam-Webster "Learning is an activity or a process of gaining knowledge or skill by studying, practicing, being taught or experiencing something". Another commonly accepted definition of learning is given with respect to living organisms as "modification of behavioral tendency by experience". Important steps in learning are remembering, adapting, and generalizing. Animals recollect being in this situation, take certain actions, and determine whether the actions worked or not. Should the actions work, they will be used again; otherwise, some different actions will be considered. Generalizing is the crucial part of animal intelligence. Owing to a variety of situations that arise, it takes a great deal of intelligence to recognize similarity in different situations so that previous experiences can be put in correct context.

9.5.1   Supervised Machine Learning Algorithms

Supervised learning is a task of inferring a function from training data. Supervised learning algorithms make use of available known data points and predict outcomes for unknown inputs. They analyze training data, generate an inferred function (also called hypothesis function), and use the same function for mapping unknown points. The generalization aspect of learning is very important in supervised learning, as these algorithms are required to "generalize" their experience to unknown inputs.

9.5.2   Unsupervised Machine Learning Algorithms

Unsupervised learning algorithms are applied primarily to problems of classification nature. It is analogous to certain learning experiences, where there is no right answer. Training data do not have "correct" responses; therefore, there is no supervision on algorithms. These algorithms try to identify similarities between input data points and categorize them together. The statistical approach to unsupervised learning is known as density estimation.

9.5.3   Deep Learning Algorithms

Deep learning algorithms are a subset of machine learning algorithms that use multiple transformations on training data. Deep learning algorithms resemble the human brain, where each section is responsible for a dedicated task and a final decision taken is an integration of responses of individual sections. Such sections are called "neurons" in both the cases. ANNs are nowadays referred as deep neural networks or deep learning algorithms. Continuing the neuron analogy, deep learning algorithms can be thought of as a multi-layer neuron network where neurons at each stage are performing a dedicated task on incoming data. All neurons in a network can implement unsupervised learning tasks such as classification, or all of them can implement supervised learning tasks such as classification and regression. In some cases, a neural network can have neurons performing unsupervised and supervised tasks independently and simultaneously. Such networks are called semi-supervised learning networks.

9.6 CASE STUDY: THERMAL MODELING OF THE GAN HEMT DEVICE

The gallium nitride-based HEMTs are considered to be one of the most promising devices to realize the high power, high frequency, and high-temperature applications due to their inherent material properties such as wide bandgap, high saturation velocity, and high thermal stability [18–21]. Advances in materials and device technology in the past years have led to a remarkable rate of progress in the performance available from GaN-based HEMTs. High breakdown electric field (4 MV/cm) and excellent thermal conductivity make GaN an attractive candidate for high-temperature systems [22]. In comparison to others III–V semiconductor material, GaN material devices can operate at a much higher temperature (750 °C), without degradation in performance. The capability of high-temperature operations enables the device to be useful in applications in extreme environments [23]. An assortment of chemicals, gases, and biosensors has allowed HEMT technology to be operated at lifted temperatures [24, 25]. Long-term stability and high-temperature operation are significant prerequisites for gas sensors. For instance, in a hydrogen gas sensor, hydrocarbons are dissociated by catalytic metals only at elevated temperatures. The Si-based gas sensor has a point of confinement to be operated underneath 250 °C, prohibiting them from being utilized as hydrocarbon detectors or for other applications requiring a high-temperature operation. Therefore, semiconductor devices based on AlGaN/GaN heterostructure have great potential for sensor applications in harsh environments. Experiments are designed to validate the hypothesis that supervised machine learning can be used to extract useful information, i.e. model equation parameters from measured data. To expedite development of the methodology, I-V characteristics data generated by computer simulations is considered as reference and used as training data for a supervised machine learning algorithm. Simple mathematical functions are chosen as hypothesis functions. Using hypothesis functions and extracted values of function parameters, an attempt is made to replicate I-V characteristics so that root mean squared error (RMSE) between training data and replicated characteristics is acceptable. Additionally, RMSE between first-order derivatives of training data and model outputs is also evaluated. For ease of comparison, graphs of experiment outcomes are also provided. Outcomes of experiments expose flaws and bottlenecks in this methodology. Supervised machine learning algorithms chosen can replicate only a small subset of I-V characteristics. Due to the simplistic nature of hypothesis functions and the optimization algorithm, experiments were often hindered by long adaptation times. This chapter is a feasibility study for the proposed approach and can serve as a starting point for future research endeavors.

It is always necessary to accurately predict the device electrical characteristics by simulation before fabrication. It reduces the development cost and time required for long and tedious growth and processing steps. This helps in the interpretation of the fundamental device physics. Many series of simulations have been executed on various methods of GaN-based HEMTs such as transport and mobility properties, [26–28] self-heating effects, [29] carrier lifetime, trapping effects [30,31], etc. The temperature-dependent models for some physical characteristics have been reported by Islam and Huq [32]. A combined investigation on a simulation study of GaN-based devices for their performance at high temperatures is greatly needed. The circuit simulations always play a key role while testing of designed circuits and predicting device performance. The integrated circuits fabrication technique is very costly. That is why modeling and simulation of characteristics of semiconductor devices is vital prior to fabrication [33,34]. Several models were reported for GaN HEMT devices [35–38]. These models are empirical and followed the induction process for development. in which the user needs to give an initial guess of the empirical parameters and apply some assumptions that might be erroneous. Additionally, these models were ordinarily generated for low-temperature values, e.g., RF applications. Researchers are giving an extraordinary consideration to the GaN HEMT models that can be developed from gathered information [39,40]. The best method to model the extracted device data is machine learning. Through this method, the user can identify the parameters that can be with the desired data with the least interference. Machine learning techniques are more accurate compared to traditional modeling techniques and additionally can direct to a model that is not just useful for interpolation but also for extrapolation [41]. The work related to short-range temperature modeling using a machine learning technique was reported in Marinković [42]. Nonetheless, reported work focused on neural network modeling for a shorter temperature range (25 °C to 80 °C). This range is extremely constrained for GaN devices, which could be operated at considerably higher temperatures [43]. The GaN HEMT device modeling techniques using a machine learning algorithm are hardly observed in the reported research with respect to high temperatures. There are very few reports available in the literature on the modeling of GaN HEMT devices based on AI in relation to high temperatures.

Machine learning has become one of the most promising modeling techniques recently for device modeling [44]. In Braha and Shmilovici [44], an intelligent algorithm for device modeling based on a DT is used to predict the lithographic process parameters for semiconductor manufacturing. The exceptional model performance obtained shows the remarkable ability of this modeling technique. This case study reports on the nonlinear data modeling for the output characteristics of the GaN-based HEMT device using the machine learning regression technique, specifically the DT. The aim of this study is to develop a device model that can play with measured device characteristics over wide temperature ranges.

9.6.1 Experimental Setup

A single gate finger device of 100 μm unit gate width was used for simulation, as shown in Figure 9.3. The epitaxial structure comprised of a 25 nm Al0.25 Ga0.75N

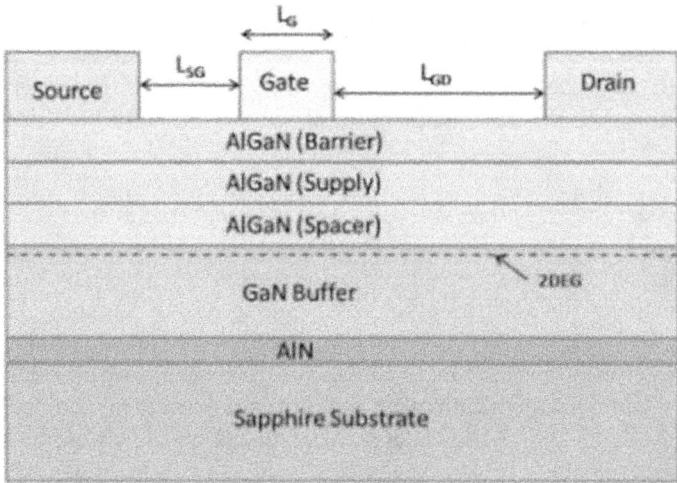

FIGURE 9.3   Schematic representation of simulated device.

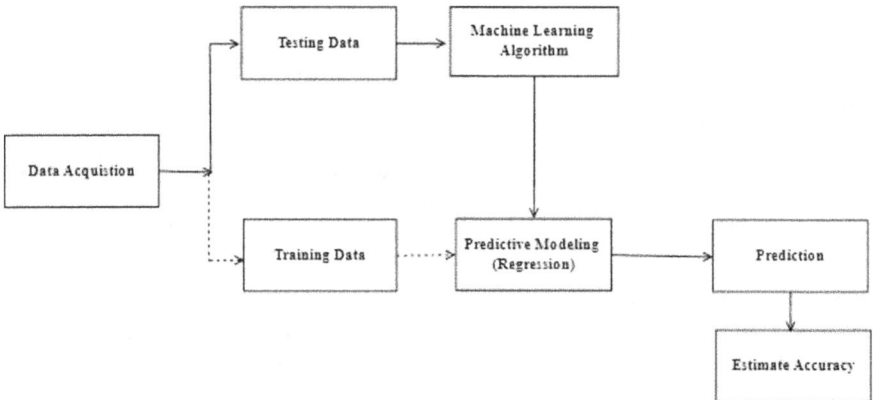

FIGURE 9.4   Standard flowchart of device modeling using machine learning.

layer, a 2.7 μm GaN buffer, and an AlN layer with sapphire as a substrate. The gate length (LG), source-to-gate length (LSG), and gate-to-drain length (LGD) of the device are 1 μm, 1 μm, and 2 μm, respectively.

The data extracted from the device simulations using the TCAD tool were utilized to train the machine learning techniques to develop the model for the GaN HEMT device (Figure 9.4).

The machine learning technique used to model the device characteristics is the DT algorithm. It is a widely used algorithm for inductive inference by which it approximates the target function [45]. A DT is a flow-chart-like arrangement. Here, every single node signifies a test on a variable. All branches denote a test outcome, and the cost of the class variables are located at the leaves of the tree [46]. This algorithm uses training examples (based on historical data) to construct a classification model, which describes the connection between classes and attributes. Once it has learned, the regression model can order new, obscure cases. The benefit of the DT models is that they are robust to noisy data and efficient in learning disjunctive expressions [47]. In this work, an exceptionally efficient and extensively utilized classification-based algorithm named C4.5 [48–50] is employed. C4.5 employs a top-down, greedy construction of a DT. This technique utilizes a top-down, greedy approach to the DT. This approach starts with inspecting which characteristic (input variable) ought to be tried at the root of the tree. For all instances, the feature is assessed via a statistical assay (information gain) to decide performance level for classification of the training examples. The best attribute is chosen and employed as a test at the root node of the tree. Afterward, a relative of the root hub is formerly made for every viable estimate of this characteristic (discrete or continuous), and the training models are arranged to the fitting relative nodes. The whole cycle is then continued utilizing the training models related to every relative node to choose the best ascribe to test by then in the tree. The developed DT could be employed to group the specific data element. It could be done with beginning the root and progressing across the leaf node, demonstrating how a class came across. Each nonterminal node represents a test or decision to be carried out on a single attribute value (i.e. input variable value) of the considered data item, with one branch and subtree for each possible outcome of the test. All the nodes indicate a test that is executed on a specific parameter. These tests could be executed with the chosen data element besides branches for every viable outcome of the test. A decision was taken while reaching a terminal node. At each non-leaf decision node, the attribute (i.e. input variable) specified by the node is tested, which leads to the root of the subtree corresponding to the test's outcome. By every nonleafy decision node, the characteristic is determined by the node being examined. This progress prompts the root of the subtree relating to the test's result. After putting the attributes by superior information gain nearest to the root, the algorithm prefers choosing shorter trees over longer ones. The algorithm can likewise retrace to reexamine prior decisions by utilizing a pruning strategy called rule post-pruning [1,49,50]. The "rule post-pruning" technique is used to conquer the overfitting issue, which regularly emerges in learning errands.

9.6.2 RESULTS

We have done the simulation of I-V characteristics for GaN-based HEMT devices for temperature range from 300 K to 600 K using silvaco-ATLAS. The machine learning models were trained on the dataset with the different gate to source voltage (V_{gs}) and temperature oscillating from 300 K to 600 K. The temperature values higher than 300 K were used to test the model's capability to extrapolate outside the input range. The hyperparameters optimization of the DT function gives three different optimized parameters, the MinParentSize = 1, the MaxNumSplits = 2000, and the NumVariablesToSample = 3. The output curves related to the GaN HEMT device in a temperature range varyings from 300 K and 600 K are presented in Figure 9.5. The applied voltage between gate to the source is 2 V. The data extracted from the DT model are compared with simulated outcomes and a mean

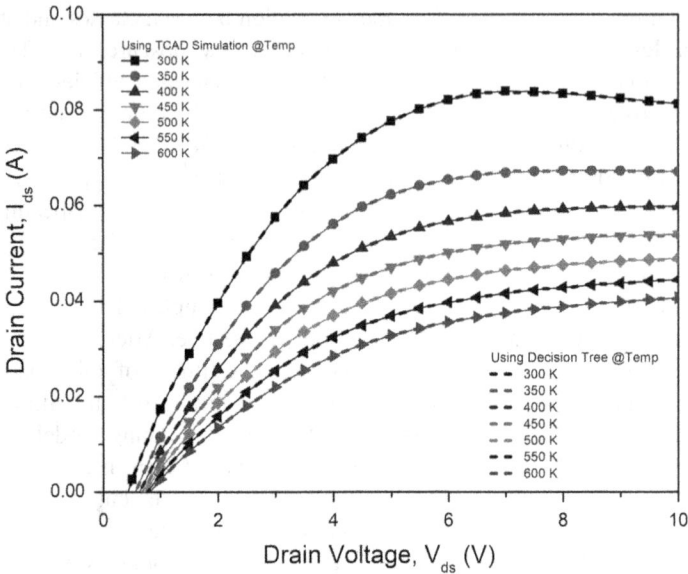

FIGURE 9.5   Comparison of drain current at different Temp (300 K to 600 K) at V_{gs} = +2V.

square error between these two was calculated as 5.32×10^{-8}. The temperature dependence of the maximum current and transconductance of different devices was further studied. The transconductance (gm) versus gate-source voltage (V_{ds}) for drain bias V_{ds} = +10 V of the simulated AlGaN/GaN HEMT is shown in Figure 9.6.

We observed a reduction in I_{ds} (≈52%) with an increase in temperature from 300 K to 600 K. The gate bias was ramped from –6V to 2 V for each of the drain biases. The peak current density was reached of V_{gs} = +2 V, V_{ds} = +10 V @ 300 K, with a current density of 0.84 A/mm.

The high current density could be attributed to the extremely high charge that was accumulated in the channel because of the polarization effects and high saturation peak velocity of electrons in GaN.

The output drain current becomes degrading with the temperature hike from 300 K to 600 K. To comprehend the explanation, we have tested for different parameters, for example, electron concentration and mobility. Figure 9.7 describes the behavior of electric concentration versus depth (μm) at different temperatures. Peak electron concentration along the channel was about $1.3 \times 10^{19}/cm^3$ @ 300 K corresponded to the highest I_{ds} = 0.84 A/mm when the device was turned on. Such an effect could be explained, looking at the differences between the electron mobility values at different temperatures, as displayed in Figure 9.8. The electron mobility in the channel was 1181 cm^2/Vs @ 300 K. When the temperature increases, electron mobility degrades along the channel due to phonon and impurity scattering. The predicted I_{ds} values at 300–600 K are close to the actual simulations. These approximations show the mean square error of 2.17×10^{-6}. The predicted characteristics utilizing the DT algorithm almost overlap alongside the simulation's ones.

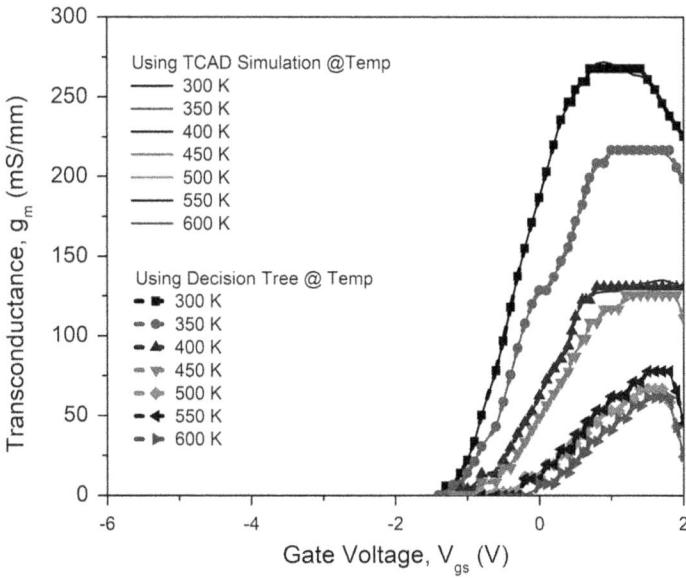

FIGURE 9.6 Comparison of transconductance at different Temp (300 K to 600 K) at $V_{ds} = +5V$.

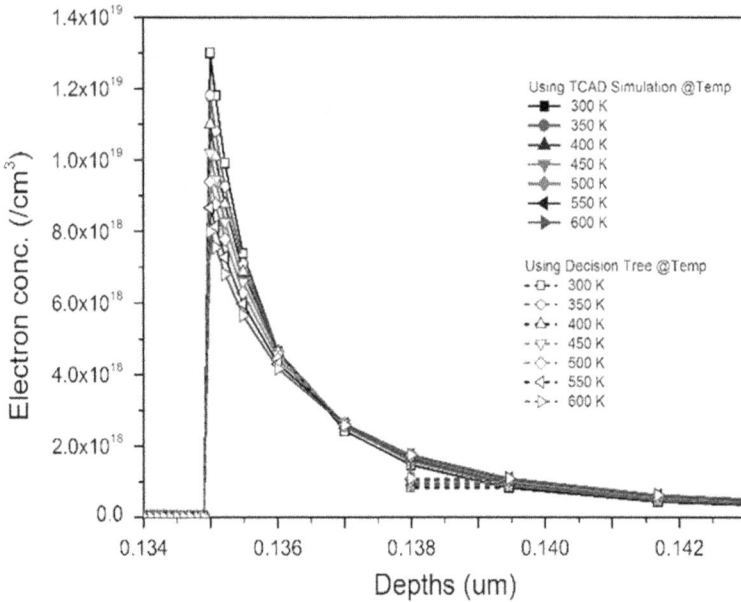

FIGURE 9.7 Comparison of electron concentration at different temp.

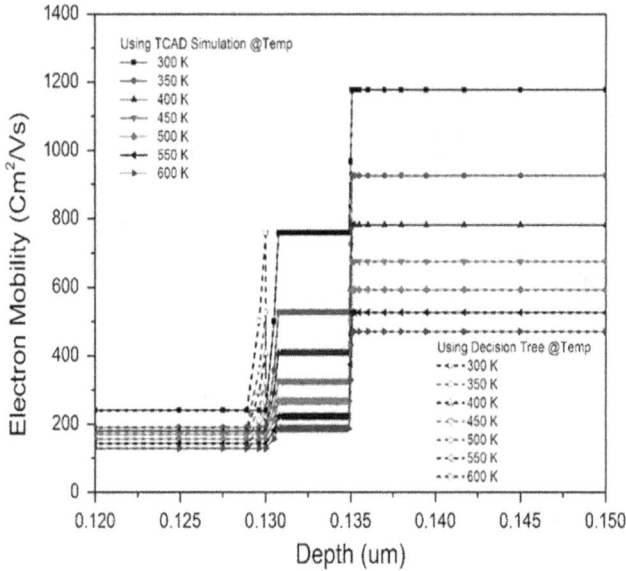

FIGURE 9.8 Comparison of electron mobility at different temp.

9.7 CONCLUSION

This chapter studies machine learning algorithms from the perspective of semiconductor device modeling. It aims to discuss and review major machine learning algorithms for semiconductor device modeling. We have done the measurements of I-V characteristics for GaN-based HEMT devices for temperature range from 300 K to 600 K. The proposed data modeling method for the I-V characteristics is the DT machine learning algorithm. The inputs to this model are V_{gs}, V_{ds}, and temperature. The predicted Ids values at 300–600 K are close to the actual simulations. These approximations show the mean square error of 2.17×10^{-6}. The predicted characteristics utilizing the DT algorithm almost overlap alongside the simulation's ones. Our next future work will commence increment in the number of inputs with including transistor size as an input parameter. We have also planned to increase the temperature range by training the model over larger temperature range, beyond 600 K to improve model performance.

ACKNOWLEDGMENTS

The authors acknowledge the fund provided by TEQIP III- RTU (ATU) CRS project scheme under the sanction no. "TEQUIP-III/RTU (ATU)/CRS/2019-20/19".

REFERENCES

[1] Gore, Chinmay Chaitanya. Study of machine learning algorithms for their use in semiconductor device model development. Master's thesis, 2015.

[2] Susto, Gian Antonio, Terzi, Matteo, and Beghi, Alessandro. Anomaly detection approaches for semiconductor manufacturing. *Procedia Manufacturing* 11, 2018–2024 (2017).

[3] Ding, Duo, Wu, Xiang, Ghosh, Joydeep, and Pan, David Z. Machine learning based lithographic hotspot detection with critical-feature extraction and classification. In 2009 IEEE International Conference on IC Design and Technology, pp. 219–222. IEEE, 2009.

[4] Luo, Rui. Optical proximity correction using a multilayer perceptron neural network. *Journal of Optics* 15, (7), 075708 (2013).

[5] Litovski, V. B., Radjenovic, J. I., Mrcarica, Z. M., and Milenkovic, S. L. MOS transistor modelling using neural network. *Electronics Letters* 28, (18), 1766–1768 (1992).

[6] Murphy, Kevin P. *Machine learning: A probabilistic perspective*. Cambridge, Massachusetts, United States: MIT Press, 2012.

[7] Tsividis, Yannis, and McAndrew, Colin. *Operation and modeling of the MOS transistor*. Oxford, England: Oxford University Press, 2011.

[8] Khusro, Ahmad, Hashmi, Mohammad S., and Ansari, Abdul Quaiyum. Exploring support vector regression for modeling of GaN HEMT. In 2018 IEEE MTT-S International Microwave and RF Conference (IMaRC), pp. 1–3. IEEE, 2018.

[9] Hari, Nikita, Chatterjee, Soham, and Iyer, Archana. Gallium nitride power device modeling using deep feed forward neural networks. In *2018 1st Workshop on Wide Bandgap Power Devices and Applications in Asia (WiPDA Asia)*. IEEE, 2018.

[10] Cai, Jialin, Yu, Chao, Sun, Lingling, Chen, Shichang, Su, Guodong, Liu, Jun, and Su, Jiangtao. Machine learning based pulsed IV behavioral model for GaN HEMTs. In 2019 IEEE MTT-S International Wireless Symposium (IWS), pp. 1–3. IEEE, 2019.

[11] Ambacher, O., Foutz, B., Smart, J., Shealy, J. R., Weimann, N. G., Chu, K., Murphy, M., Sierakowski, A. J., Schaff, W. J., and Eastman, L. F.: Two-dimensional electron gases induced by spontaneous and piezoelectric polarization in undoped and doped AlGaN/GaN heterostructures. *Journal of Applied Physics* 87, 334–344 (2000).

[12] Ren, F., and Pearton, S. J. Semiconductor device-based sensors for gas, chemical, and biomedical applications. Boca Raton, FL, USA: CRC Press, 2011.

[13] Sharma, N., Joshi, D., and Chaturvedi, N. An impact of bias and structure dependent LSD variation on the performance of GaN HEMTs based biosensor. *Journal of Computational Electronics* 13, (2), 503–508 (2014).

[14] Sharma, N., Mishra, S., Singh, K., Chaturvedi, N., Chauhan, A., Periasamy, C., Kharbanda, D. K., Prajapat, P., Khanna, P. K., and Chaturvedi, N. High resolution AlGaN/GaN HEMT based electrochemical sensor for biomedical applications. *IEEE Transactions on Electron Devices* 66, 1–7 (2019).

[15] Lalinský, T. et al. AlGaN/GaN based SAW-HEMT structures for chemical gas sensors. *Procedia Engineering* 5, 152–155 (2010).

[16] Sharma, N., Dhakad, S. K., Periasamy, C., and Chaturvedi, N. Refined isolation techniques for GaN-based high electron mobility transistors. *Materials Science in Semiconductor Processing* 87, 195–201 (2018).

[17] Dhakad, S. K., Sharma, N., Periasamy, C., and Chaturvedi, N. Optimization of ohmic contacts on thick and thin AlGaN/GaN HEMTs structures. *Superlattices Microstructures* 111, 922–926 (2017).

[18] Osvald, J. Polarization effects and energy band diagram in AlGaN/GaN heterostructure. Applied Physics A—Materials Science and Processing 87, 679 (2007).

[19] Gelmont, B., Kim, K. S., and Shur, M. Monte Carlo simulation of electron transport in gal-lium nitride. *Journal of Applied Physics* 74, 1818–1821 (1993).

[20] Pearton, S. J., Zolper, J. C., Shul, R. J., and Ren, F. GaN: Processing, defects, and devices. *Journal of Applied Physics* 86, 1–78 (1999).

[21] Levinshtein, M., Rumyantsev, S., and Shur, M. *Properties of advanced semiconductor materials*. New York: Wiley, 2001.

[22] Vitanov, S., Palankovski, V., Maroldt, S., and Quay, R. High-temperature modeling of Al-GaN/GaN HEMTs. Solid-State Electronics 54, 1105–1112 (2010).

[23] Luther, B. P., Wolter, S. D., and Mohney, S. E. High temperature Pt Schottky diode gas sen-sors on n-type GaN. *Sensors and Actuators B* 56, 164–168 (1999).

[24] Ryger, I., Vanko, G., Kunzo, Lalinsky, P. T., Vallo, M., Plecenik, A., Satrapinsky, L., and Plecenik, T. AlGaN/GaN HEMT based hydrogen sensors with gate absorption layers formed by high temperature oxidation. *Procedia Engineering* 47, 518–521 (2012).

[25] Lalinsky, T., Ryger, I., Vanko, G., Tomaska, M., Kostic, I., Hascik, S., and Valloa, M. Al-GaN/GaN based SAW-HEMT structures for chemical gas sensors. *Procedia Engineering* 5, 152–155 (2010).

[26] Albrecht, J. D., Wang, R. P., and Ruden, P. P. Electron transport characteristics of GaN for high temperature device modelling. *Journal of Applied Physics* 83, 4777–4781 (1998).

[27] Cordier, Y., Hugues, M., Lorenzini, P., Semond, F., Natali, F., and Massies, J. Electron mobility and transfer characteristics in AlGaN/GaN HEMTs. *Physica Status Solidi (c)* 2, 2720–2723 (2005).

[28] Turin, V. O., and Balandin, A. A. Electrothermal simulation of the self-heating effects in GaN-based field-effect transistors. *Journal of Applied Physics* 100, 054501–054508 (2006).

[29] Islam, S. K., and Huq, H. F. Improved temperature model of AlGaN/GaN HEMT and de-vice characteristics at variant temperature. *International Journal of Electronics* 94, 1099–1108 (2007).

[30] Sharma, Niketa, Periasamy, C., Chaturvedi, N., and Chaturvedi, N. Trapping effects on leakage and current collapse in AlGaN/GaN HEMTs. *Journal of Electronic Materials*, 60, 1–6 (2020).

[31] Sharma, N., and Chaturvedi, N. Design approach of traps affected source–gate regions in GaN HEMTs. *IETE Technical Review* 33,(1), 34–39 (2016).

[32] Galup-Montoro, C. *MOSFET modeling for circuit analysis and design*. Singapore: World Scientific, 2007.

[33] Deng, W., Huang, J., Ma, X., and Liou, J. J. An explicit surface potential calculation and compact current model for AlGaN/GaN HEMTs. *IEEE Electron Device Letters* 36, (2), 108–110 (2015).

[34] Sharma, N., Joshi, D., and Chaturvedi, N. An impact of bias and structure dependent Lsd variation on the performance of GaN HEMTs based biosensor. *Journal of Computational Electronics*, 13,(2), 503–508 (2014).

[35] Oishi, T., Otsuka, H., Yamanaka, K., Inoue, A., Hirano, Y., and Angelov, I. Semiphysical nonlinear model for HEMTs with simple equations. In Integrated Nonlinear Microwave and Millimeter- Wave Circuits (INMMIC), pp. 20–23. IEEE, 2010.

[36] Sang, L., and Schutt-Aine, J. An improved nonlinear current model for GaN HEMT high power amplifier with large gate periphery. *Journal of Electromagnetic Waves and Applications*. 26, (2–3), 284–293 (2012).

[37] Linsheng, L. An improved nonlinear model of HEMTs with independent transconductance tail-off fitting. *Journal of Semiconductors* 32, (2), 024004–024006 (2011).

[38] Gunn, S. R. Support vector machines for classification and regression. *ISIS Technical Report* 14, 85–86 (1998).

[39] Huque, M., Eliza, S., Rahman, T., Huq, H., and Islam, S.: Temperature dependent analytical model for current–voltage characteristics of AlGaN/GaN power HEMT. Solid-State Electronics 53, (3), 341–348 (2009).

[40] Chang, Y., Tong, K., and Surya, C. Numerical simulation of current–voltage characteristics of AlGaN/GaN HEMTs at high temperatures. *Semiconductor Science and Technology* 20, (2), 188–192 (2005).

[41] Breiman, L. Statistical modeling: The two cultures. *Statistical Science* 16, (3), 199–231 (2001).

[42] Marinković, Z. et al. Neural approach for temperature-dependent modeling of GaN HEMTs. *International Journal of Numerical Modelling: Electronic Networks, Devices and Fields* 28, (4), 359–370 (2015).

[43] Neudeck, P. G., Okojie, R. S., and Chen, L.-Y. High temperature electronics-a role for wide bandgap semiconductors. *Proceedings of the IEEE* 90, (6), 1065–1076 (2002).

[44] Braha, D., and Shmilovici, A. On the use of decision tree induction for discovery of interactions in a photolithographic process. *IEEE Transactions on Semiconductor Manufacturing* 16, (4), 644–652 (2003).

[45] Quinlan, J. R. Induction of decision trees. *Machine Learning* 1, (1), 81–106 (1986).

[46] Witten, I. H., Frank, E., Hall, M. A., and Pal, C. J. *Data mining: Practical machine learning tools and techniques.* Burlington, Massachusetts, United States: Morgan Kaufmann, 2016.

[47] Mitchell, T. M. *Machine learning.* Burr Ridge, IL: McGraw Hill, 1997.

[48] Braha, D. (Ed.). *Data mining for design and manufacturing: methods and applications.* Boston, MA: Kluwer Academic, 2001.

[49] Mitchell, T. M. *Machine learning.* New York: McGraw-Hill, 1997.

[50] Quinlan, J. R. Induction of decision trees. *Machine Learning*, 1, 81–106 (1986).

10 Securing IoT-Based Microservices Using Artificial Intelligence

Sushant Kumar[1] and Saurabh Mukherjee[2]
[1]Asst Prof, Research Scholar, Banasthali University,
 Vanasthali, Rajasthan
[2]Prof, Banasthali University, Vanasthali, Rajasthan

CONTENTS

10.1   INTRODUCTION: BACKGROUND AND DRIVING FORCES

The Internet of Things (IoT) is a technology that strongly enters people's reality. All environments are involved: urban, industrial, office, or home. The interest generated and the speed of adoption of the technology have produced a certain disorder and informality in the process. This has as a consequence that important elements were left aside; one of the most relevant is that of security [1].

In principle, IoT security does not have to be different from security in a typical computer network. In practice, however, there are environmental difficulties that further complicate the security problem. Many IoT devices are computationally limited, preventing the use of several known robust security mechanisms. The large number of devices that can be involved in an IoT network, as well as the exponential increase in the number of interactions, exacerbates the problem. The diversity of the equipment used, both in hardware and software, complicates the possibility of generalizing the proposed solutions [2]. There are a wide variety of methods and tools that can be used to undertake the work [3]. In this work, the intention was to use those techniques that, as well as the IoT, set trends and are used successfully in other related fields. Among these techniques, the most promising were microservices [4] and OAuth2 [5], which, when combined with techniques

DOI: 10.1201/9781003201038-10

and technologies, perhaps a little more traditional to the base, but equally successful as TLS [6] and MQTT [7], provided an appropriate structured environment.

The general objective was to provide the world of IoT with a safe architectural alternative adapted to new technological trends, which can be used generically in a multitude of situations. More specifically, first, special relevance was given to generating an alternative for smart homes. For this reason, the methodology encourages the clear division of functions, but takes care of the fluid integration and the availability of tools for the creation of new utilities. Second, among the many existing security problems, those related to client authentication when calling multiple services and confidentiality when transmitting information, especially when compromising the network, were attacked. Finally, we were aware of the hardware restrictions of many devices, especially sensors.

10.2 PREVIOUS WORK

In the ecosystem of IoT devices, these are largely unsafe, as they are small and energy-efficient devices; therefore, they also have limited computational resources. This last condition greatly affects trying to include complex security schemes [8]. At an industrial level, there are several enterprises applying IoT, for example, in intelligent transport [9, 10] and in agriculture [11]; however, one of the great current goals related to home and office automation is connected living. This objective requires important advances in the field of IoT, where it is necessary to provide answers to the problems related to the enormous increase in devices that must interact [12]. A particular case in the IoT is that of smart homes, since many times the solutions that are implemented are ad-hoc, by the users themselves, who tend to try to reduce costs and efforts as much as possible, which is generally reflected in a minimal and probably non-existent security scheme [13].

The devices involved require interconnection in a many-to-many scheme. To ensure the exchange of information, it is necessary to implement an identity management system that scales appropriately. In this sense, Lounis and Zulkernine [14] propose a system for the home that combines EAP, OAuth and DTLS. Also concerned with identity management and access control, Swamy and Kota [15] confirm the possibilities of OAuth and make an architectural proposal compatible with services.

Fernández-Caramés [16] analyze the problem of IoT security in the home and highlight how important it is to prevent sensors from indiscriminately capturing and distributing household data. As an example, they present the case of private conversations, which should not be published. Among the possible approaches, they mention as a viable alternative approaches oriented to services, to balance between centralization and distribution of control; furthermore, the trend in software and distributed applications seems to be generally directed towards the use of microservices [9,17–19].

Delving along this line [20], they highlight the benefits that a microservices architecture, additionally based on TLS/PKI, can have in IoT, by lightening development and maintenance tasks, beneficial for both providers and distributors as well as users, at the same time, reinforcing interconnection security. Case studies such as that of Wazid et al. [21] confirm in a practical way the possibilities of these techniques.

Although in most of the related works SSL/TLS is the preferred security and encryption mechanism, [22] what Sharma et al. propose is very interesting when analyzing the complications that may exist at a practical level in IoT with TLS. Their work analyzes the possibility of using SSH and highlights the advantages provided by the data compression issue included in said protocol, which is especially advantageous when working over HTTP. Zarca et al. [23] contribute to the IoT environment with a model-driven approach and propose an OAuth-oriented model with a strong UML inclination.

This proposal, through transformations, could be adapted to a specific architecture, offering the possibility of customizing it to the required environment. Another interesting architectural and security proposal is the one presented by Yi et al. [24], where instead of a traditional mechanism such as that presented by the SSL certification authorities, an approach of local certification authorities would be used, which more frequently, but also with a lighter process, would authenticate the IoT equipment. It can be considered that a middle point between the two proposals that we have just reviewed would be that of Sood et al. [25], which uses traditional certificates, but with close authentication, at the node level. They emphasize that this mechanism could be complemented with one of authorization, such as OAuth or similar.

Zhou et al. [26] emphasize OAuth, but above all, with the particularity of concentrating its security architecture on the gateway equipment, where the base station or sink node resides, which is in charge of processing heavy of authenticating, authorizing, and establishing the links between clients and resources. This approach is very relevant when we consider how susceptible those edge devices are linked to edge computing [27], through which an entire IoT system can be compromised. One of the high points of security in edge systems in IoT is usually related to the use of MQTT (or similar protocols), for which several works, such as that of Amato et al. [28], propose improvements over said protocol.

10.3   PROPOSED WORK

For this work, it was mainly considered that within the IoT ecosystem it is necessary to segment the location, scope, and access of the equipment involved in two layers: local or edge and centralized. In the local layer, represented schematically in Figure 10.1, we have those elements that will be invariably installed in the smart home or office, such as sensors, actuators, edge processors such as gateways and brokers, and mobile user devices.

Sensors are all equipment capable of capturing physical phenomena, virtual events, or periodic signals. Those sensors with the necessary capacity can communicate directly with the central broker; otherwise, they will interact with equipment in the edge processing subsystem. The sensors will be static, when they emit a constant signal that would generally be used by mobile equipment moving in the environment as Bluetooth beacons for positioning. The dynamic sensors will capture measurements of the environment, which will vary depending on the environmental conditions, such as luminescence, temperature, and humidity, among others.

The actuators will allow interaction with hardware- or software-generating events or actions. They will receive instructions either directly from the broker or

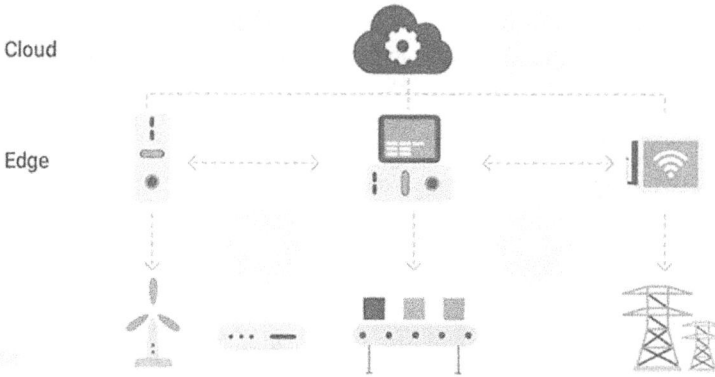

FIGURE 10.1 Components at the local or edge layer of the IoT system.

from a preprocessor when necessary. They are divided into premises, located in the intelligent environment, such as light or temperature controllers. They can also be remote, such as those capable of sending instructions, probably over the network, to a distant computer, but controlled from the home or smart office, such as when it is required to send an SMS, email, or tweet.

Finally, this layer contemplates the preprocessor equipment, which can also be called edge processors or brokers. These capture raw information from the sensors to forward it to the centralized broker or actuator when the sensor is incapable. The information can be sent as received from the sensor, or it can be preprocessed and this result sent. These computers can be Raspberry Pi or small computers such as tablets. In the centralized layer, presented in Figure 10.2, the teams in charge of the general coordination of all the components are considered, and it is considered that three

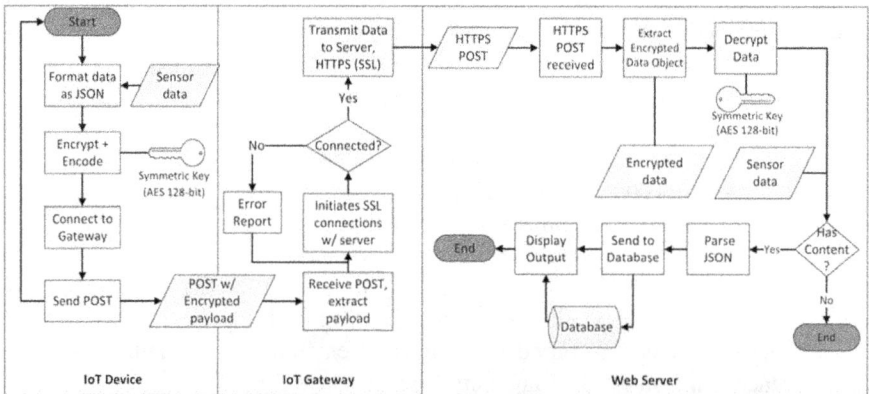

FIGURE 10.2 Components in the centralized layer of the IoT system.

subsystems are necessary: administration, processing, and persistence. These subsystems are linked to each other and also to the edge layer through a centralized broker.

The administration subsystem defines the parameters and configuration of the system and presents the web interfaces for the administrator users to interact with the entire system. In the case where central processing is done with multiple machines, it also manages the resulting cluster. The main part of this subsystem is then monitoring, which will allow all types of users to review the relevant information, preferably through dashboards and statistical tables. For all heavy information processing, the corresponding subsystem takes the data collected from the broker and processes it as defined by the specific applications or needs of the IoT system.

In general, the processing will be divided depending, above all, on the urgency of the processing, in real time, which processes the data in a continuous flow, as they arrive from the sensors; in memory, which collects the information in the cluster's memory, depending on the needs, and processes it in small batches; and batch, which interacts in general with the storage system, for processes where the amount of data is greater than what fits in the system's live memory. The information generated by the IoT system is directed to the persistence module, which safeguards the data for later use either in the development of models or in the generation of reports. Several alternatives must be considered, depending on the size of the information and the way it will be accessed.

Finally, the entire system, and more specifically the border and centralized layers, must connect and exchange information, which is achieved through a central broker, in charge of managing all the message queues, thus reducing the complexity of the interactions.

10.4 RESULTS

The resulting architectural design took into consideration, above all, the need to ensure the exchange of information of all the components of the system, trying to take care of the speed of calculation at all times. These elements require reconciling characteristics that are often incompatible. For example, more robust cryptography systems may require more computing power than many lightweight devices, such as sensors, provide. The final architecture designed, implemented, and tested is the one outlined in Figure 10.3, which will be described in detail below. In the first place,

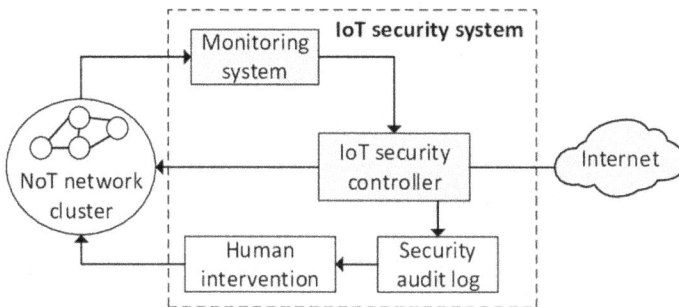

FIGURE 10.3 Secure architecture.

the components involved will be specified, and then the security functionality in general will be presented.

10.4.1 COMPONENTS

In this section we will try to define in a general way the types of components or equipment involved in the architecture proposed in Figure 10.3, in which there are basically those in charge of the registry service (Registry), the equipment providing authentication services for clients and users (UAA), and then, in a broad way, all the teams providing general services, as well as all the client teams (Services and Clients). Although for simplicity the components will be referred to in the singular, the architecture considers that, for each category or type of component, especially services, they can work in clusters.

a. Registry Services (REG)

The main task of the REG is to allow services to register through IP and aliases (service name) and thus make them available to customers, who will connect to the REG to request the information with which they will finally connect to the services of interest.

The REG also provides a load balancing service, by detecting that a service is registered in a cluster (multiple computers with the same service). The first point of contact for all other components of the system, whether these are services or clients, is the REG, which requires a static IP; all the other components of the system, however, can work with dynamic IPs, through a DNS.

b. Authentication Services (UAA)

The UAA takes its acronym from "user authentication and authorization". Within our architecture, it basically provides us with the authentication service, which works under OAuth2. The UAA stores the data of all clients in the system, including their roles. With this information, the UAA user services may or may not authorize the use of certain elements. Any component can connect with the UAA to, through its client credentials (user and password), request an access token. In the same way, any component of the system can request the UAA to validate a token received from a third party.

c. Generic Services (SRV) and Client Computers (CLI)

The last category of components accommodates all other services and all clients. In general, these components will interact with each other after having registered/authenticated in the system with the help of the REG and the UAA. The services can be very diverse, and it is up to the system administrator to decide which ones to require. However, in our architecture for IoT, there are some that are fundamental, for which they have been implemented in the test system, and they will be mentioned below.

To allow interconnectivity and, at the same time, reduce its complexity, the messaging service was implemented, which in the methodology is represented by the central broker (Figures 10.1 and 10.2).

The broker is able to receive and distribute all the messages circulating in the system and basically allows all services and clients to establish a single connection with the broker in general, to deposit messages and retrieve them from one or more queues. This broker can work with any communication protocol, or a combination of them. However, since in the world of IoT, at least today, the most widespread protocol is the Message Queuing Telemetry Transport (MQTT), it is the one used in the implementation presented here. Another service implemented for the proof of concept of the architecture is the one related to persistence, as a necessary support to subsequently implement batch processing.

For this, a transit service was implemented that takes the information from the broker and transfers it to a Hadoop cluster [29], where different types of tools of said ecosystem can be used to process the information. One of the cases that was worked on, given the nature of the IoT information, especially that coming from the sensors, was that of time series. For this, two data series services were built, thus providing graphing and trend analysis, among others. Regarding customers, all the sensors are considered here, which provide information to the system, the actuators, which react with the environment thanks to the information from the system, and all those devices, mobile or desktop, that allow the user to enter to configure the system, collect processed information, or even act also as sensors and actuators.

d. Security Schemes

The system's security architecture comprises three fundamental scenarios: the basic one, to which all elements must adhere to in their transactions, unless otherwise specified; the lightweight scheme, generally used only when starting a worker process on the system; and the strengthened one, for relationships of trust between services.

e. Basic Scheme

This is the default scheme that the system components will use in their transactions. This scheme is represented in Figure 10.3 by the dotted line that encompasses the system, and uses a combination of one-way TLS plus OAuth2. Every service must provide its public security certificate (PKI) to clients, who can then validate it with the certificate authority (CA). Likewise, every client must provide the services with an OAuth access token so that they can validate it with the authentication service. The use of TLS, in our scheme, is especially necessary to be able to encrypt the content of the information that is transmitted. It is used only on the services side to limit as much as possible the overhead that would imply, above all, at the administration level (but also of resources and processing), to use it in all the components. The security breach that appears is compensated with the use of OAuth2, through which the clients are in turn validated by the services.

f. Light Outline

This is a scheme that could be considered insecure, which is why it is provided only for those cases where an access token cannot yet be obtained, or when it is considered redundant to request it. Two cases exist at the moment in the work environment, which implement this scheme. When components enter the system in order to initiate their transactions, it is generally necessary to have the OAuth token, but since the UAA server IP may have changed, the first step is to contact REG to request the updated IP. For this connection, the client does not yet have the token, which is why it is not possible to work with the basic scheme. The second implementation of this scheme was applied to avoid unnecessary redundant connections and occurs when the service receives the token and must validate it with the UAA.

It is for this type of case that the lightweight scheme comes into play. The service provides its PKI, with which the communication is encrypted, but the client is not obliged to validate it (although it is recommended that they do so). The service provides an "insecure" access point for the client, which does not require the token. This is the scheme provided by REG exclusively to be able to deliver the UAA data.

g. Strengthened Schema

Similar to the problem worked in the light scheme, sometimes two services require interconnection, but at least one of them (who acts as a client) is unable to obtain its access token. When dealing with services, it is not convenient to open an insecure channel as is done in the lightweight scheme. In order to maintain the security standard, then, it was decided to implement a two-way TLS scheme, which is possible, without incurring greater overhead, since, being services, they already have their PKI anyway. Additionally, in general the services will be executed on equipment with greater processing capacity. This implementation also requires a dedicated channel to be able to execute this type of validation, and the example is given by the communication between the REG and the UAA. The UAA is the one that provides the access tokens and therefore should validate itself, which would generate a security hole. The REG, then, opens a dedicated channel so that a UAA service can register in this way at all times. By connecting the UAA with the REG, they exchange their respective PKIs, mutually validating each other over TLS, without reducing the system's security standard.

h. Functionality

Returning to Figure 10.3, the dotted line represents the scope of the basic security scheme, which encompasses the entire system. Internally, the numbers 1 (monitoring system), 2 (IoT security controller), and 3 (security audit log) can be seen, circled, which indicate the recommended starting order to guarantee the fluidity of the service. In practice, at least the services have in their base library the functionality to retry the connection when this starting order is not respected. However, this can be subject to unnecessary delay.

In the first place, the registration server, REG, is started, which will provide a central access point for the acquisition of contact information for the other services: every service will register in the REG its respective IP and its alias (service name), and every client will search here, by aliases, for the IP of the service required to be able to connect with it. REG offers three access points, each of which must handle a different security mode: the first, lightweight, allows any client to obtain the UAA's IP without any additional security; the second mode, strengthened, allows the connection of the UAA using two-way TLS; the last, basic one, requires OAuth2, which allows clients to request information from services and to record their contact information.

Second, an authentication server (UAA) is started, which will provide OAuth2 credentials to clients. The enhanced security mechanism, with two-way TLS validation, is used between the UAA and the REG. The UAA connects with the REG, as well as any other service, to give its IP and aliases and thus be available to the entire system. Once these two services, REG and UAA, are online, all other components, services, and clients can start their work.

Finally, then, as point 3, any other component, be it this service or client, will proceed as follows: first, using the lightweight security scheme, they will connect with the REG to request the IP of the UAA. They establish the connection with the UAA and request the access token, using their client credentials. With the access token in hand, the basic security scheme can already be used and, in the case of services, they will be registered with the REG, delivering IP and aliases, to wait for client requests, or act as a client from another service, as needed. In the case of a client, the next step is to use a service, where the basic security scheme will be applied; it connects to the REG and by means of an alias it requests the IP of the service of its interest, to then connect with said service.

10.4.2 Deployment and Testing

The implementation decision was mainly that each of the components can be executed on a variety of computers and with the least interdependence, for which we proceeded to work on a microservices architecture that allows their deployment either as independent processes, or within a containerization structure, such as Docker [30]. For desktop services and clients, Java was used with Spring Boot in general. For the central messaging service, it was decided to work using the MQTT protocol, and for the development of the broker, the Moquette library [31] was taken as a base, to which it was modified to add support for OAuth2 mainly. An HDFS cluster was used as a basis for persistence services [32]; time series services were built on the local network, which, according to the current interests of the project, were the most suitable for processing the data coming from the sensors.

It interacted with Open TSDB [33] and Prometheus [34]. As for customers, it was decided to build generic libraries for different types of systems, which facilitate the process of developing specific applications. A Java client library was developed, one for Android mobiles, one for Arduino MKR [35], and another for ESP32 [36], the latter three based on Eclipse Paho [37]. The Arduino libraries were intended exclusively for use in sensor and actuator controllers. System tests were conducted

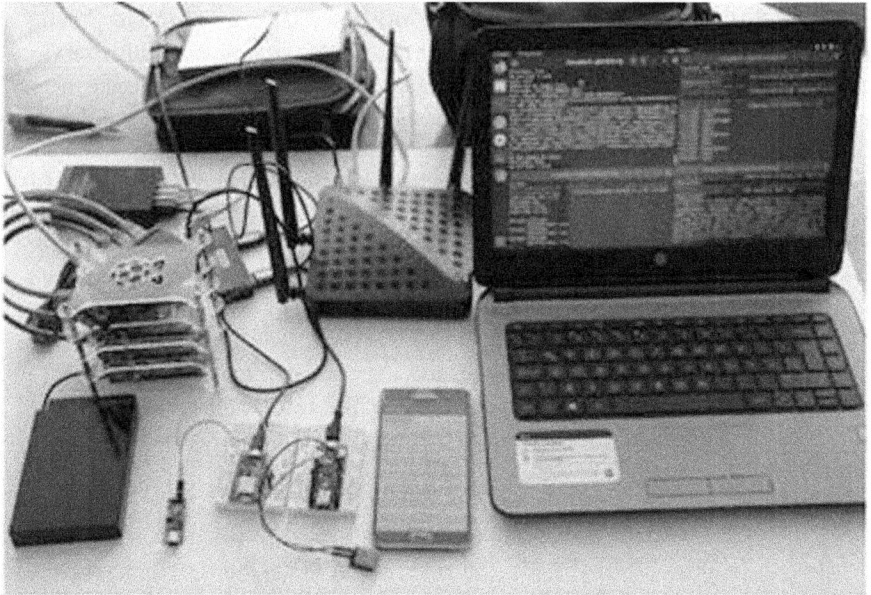

FIGURE 10.4 ESP8266 monitoring noise level with a KY038; a MKR1010 monitoring room temperature with a DHT11; an N9005 injecting random messages; and a laptop monitoring all services.

in a controlled office environment. The main equipment was an RPi 3B + that served as a Wi-Fi Gateway, providing DNS and NTP services, among others. MKR 1010 and ESP32 controllers were used simultaneously, which permanently received information from temperature, humidity, and noise sensors, such as the DHT11 and the KY038. The RPi3B + also hosted the REG, UAA services, and the MQTT messaging broker. Persistence services over HDFS as well as TSDB were installed on Linux on an i5-4210U with 8 GB of RAM. In this last equipment, generic services were also installed for sending and receiving messages, with which the fluidity of the interaction was verified.

As a mobile client, a N9005 was used, which had two functions: a dummy sensor, sending a large number of random numbers to the system, and a light actuator, warning each time that the measurements of the real sensors exceeded certain levels parameterized in the app. Figure 10.4 presents one of the setups of the testing system.

In the experimentation related to the security tests, it was decided to assume that a possible attacker was already connected to the local network and that he was, potentially, capable of making any request and capturing all the traffic. In the first test, Zap was used to perform both active and passive scanning, and it was verified that security was maintained (encrypted traffic), except in the case (documented in the architecture) of the lightweight scheme. Then, Wireshark was used for a rather manual cut verification, analyzing the packets by connection type or scheme, and again it was validated, as expected, that, except for the lightweight scheme, all traffic was kept encrypted, as presented in the Figure 10.4.

10.5   RESULT AND DISCUSSION

The process of including complex security schemes in lightweight IoT devices is not trivial, as stated by Khan and Salah [8], and in this work, this statement is agreed. Two notable cases were that the handling of TLS with auto-generated certificates for MKR 1010 required regenerating all the firmware to include the CA, and that it could not be carried out on ESP8266 controllers due to the unavailability, in practice, of open-source libraries sufficiently complete to guarantee the expected level of security. The approach adopted generally takes up the warnings made by Lin and Bergmann [13] and tries to provide a sufficiently complete and simple system so that with minimal technical support it could be implemented at home, and then managed directly by the user.

This, of course, has limitations given by the great variety of types of users that may wish to be included in the IoT. However, the tests carried out suggest that the heart of the prototype would allow us to provide this, if some facilities at the user interface, equipment, and installers level can be included. This work confirms what has already been stated by Lounis and Zulkernine [14] and by Fernández et al. [15] in relation to the problem of identity scaling and the benefit that can be obtained both with OAuth and with a service-based approach. Delegating authentication to a single point, then relying on temporary credentials, as OAuth2 allows, keeps the identity infrastructure light,

It is precisely this approach to services/microservices that allows working with TLS to not be too demanding at the maintenance level, as highlighted by some researchers. This proposal recognizes the importance that at the security level should be given to the edge equipment and following the line of Zhou et al. [26] and Park and Nam [27], it especially reinforces the gateway equipment in order to be able to implement improvements later in the MQTT protocol, in the style of what was done by Amato et al. [28], but mainly including the use of OAuth2 and TLS (Figure 10.5).

10.6   CONCLUSIONS

This work was motivated by a pressing current need: to secure IoT systems, especially those linked to a home context, and to do so without impairing the user's freedom of access to collect information and to modify the configurations of their system. This need, as it turned out, starts among other things from the relative informality of the IoT, especially in a smart home environment. It began with a methodological conception that stratifies the environment in layers: the one most closely related to the interaction with the space by collecting information and executing actions to modify the microenvironment, and the layer of centralized processing and analysis of the information. Both layers were interrelated with a centralized connection that unifies them while also maintaining their light coupling.

This methodology was aimed at allowing an implementation based on microservices, where, in addition to avoiding any type of monolithic structure as much as possible, a group of services and libraries were provided for clients that greatly facilitate the generation of new utilities and components. The tests carried out with the services, clients, and sensors generated under this infrastructure confirmed both

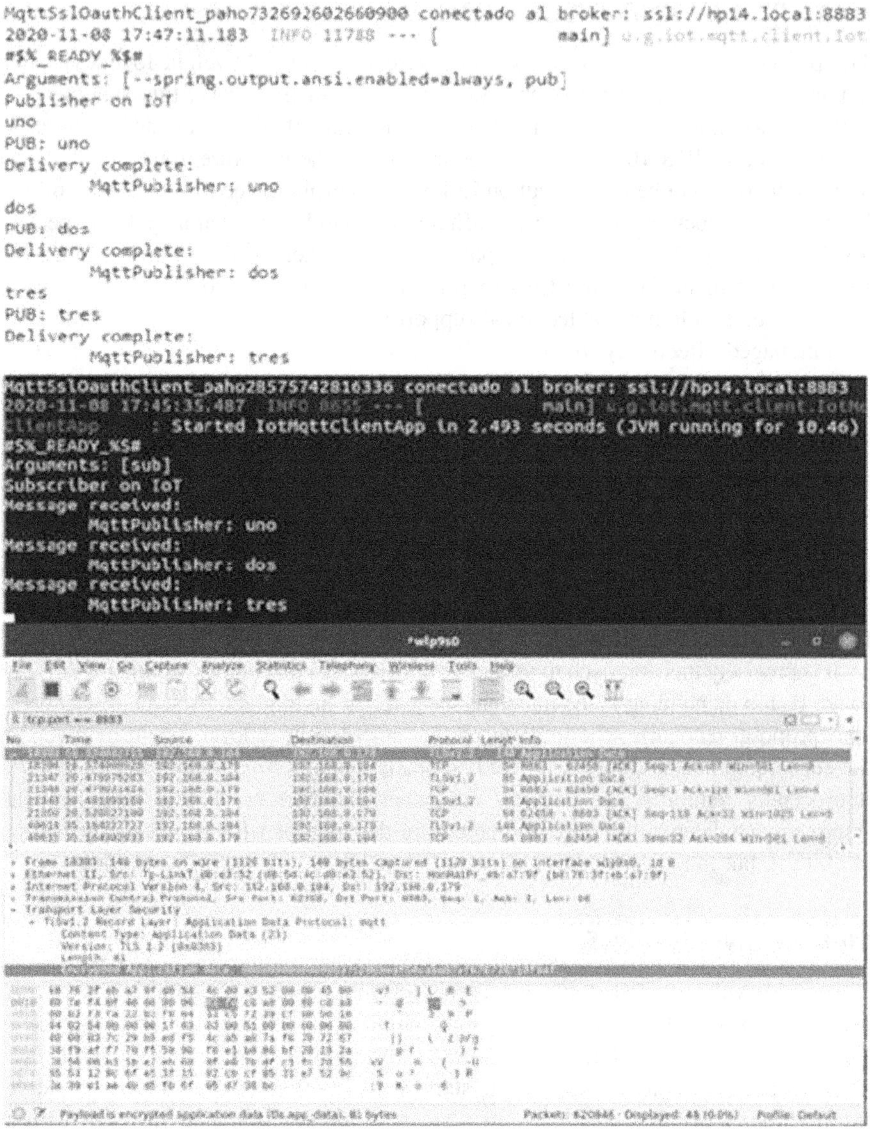

```
MqttSslOauthClient_paho732692602660900 conectado al broker: ssl://hp14.local:8883
2020-11-08 17:47:11.183  INFO 11788 --- [           main] o.g.iot.mqtt.client.Iot
#$%_READY_%$#
Arguments: [--spring.output.ansi.enabled=always, pub]
Publisher on IoT
uno
PUB: uno
Delivery complete:
        MqttPublisher: uno
dos
PUB: dos
Delivery complete:
        MqttPublisher: dos
tres
PUB: tres
Delivery complete:
        MqttPublisher: tres
```

FIGURE 10.5 Wireshark trace that captures the packets and verifies that, in transit, they are encrypted.

the robustness and the relative ease of use of the components. The main point is security. Although it was not the easiest to implement, once the three defined schemes for the different types of connections were debugged, it proved to be a robust choice, which withstood the tests of improper access. It should be noted, however, that a field-specific methodical testing scheme remains to be designed and implemented, which will be the subject of future work. We can, however, affirm that the combination of TLS, OAuth2, and MQTT, produced the expected results to a large extent.

As main contributions, we have the implementation of a solid secure architecture of IoT for the home, the stable and fluid combination of at least three high-level technologies for security management, and a functional reference implementation that can be made publicly available for free use.

REFERENCES

[1] W. Iqbal, H. Abbas, M. Daneshmand, B. Rauf and Y. A. Bangash, "An In-Depth Analysis of IoT Security Requirements, Challenges, and Their Countermeasures via Software-Defined Security," in *IEEE Internet of Things Journal*, vol. 7, no. 10, pp. 10250–10276, Oct. 2020. doi: 10.1109/JIOT.2020.2997651

[2] N. Neshenko, E. Bou-Harb, J. Crichigno, G. Kaddoum and N. Ghani, "Demystifying IoT Security: An Exhaustive Survey on IoT Vulnerabilities and a First Empirical Look on Internet-Scale IoT Exploitations," in *IEEE Communications Surveys & Tutorials*, vol. 21, no. 3, pp. 2702–2733, Thirdquarter 2019. doi: 10.1109/COMST.2019.2910750

[3] D. Shin, K. Yun, J. Kim, P. V. Astillo, J. Kim and I. You, "A Security Protocol for Route Optimization in DMM-Based Smart Home IoT Networks," in *IEEE Access*, vol. 7, pp. 142531–142550, 2019. doi: 10.1109/ACCESS.2019.2943929

[4] F. Meneghello, M. Calore, D. Zucchetto, M. Polese and A. Zanella, "IoT: Internet of Threats? A Survey of Practical Security Vulnerabilities in Real IoT Devices," in *IEEE Internet of Things Journal*, vol. 6, no. 5, pp. 8182–8201, Oct. 2019. doi: 10.11 09/JIOT.2019.2935189

[5] V. Hassija, V. Chamola, V. Saxena, D. Jain, P. Goyal and B. Sikdar, "A Survey on IoT Security: Application Areas, Security Threats, and Solution Architectures," in *IEEE Access*, vol. 7, pp. 82721–82743, 2019. doi: 10.1109/ACCESS.2019.2924045

[6] S. Siboni et al., "Security Testbed for Internet-of-Things Devices," in *IEEE Transactions on Reliability*, vol. 68, no. 1, pp. 23–44, March 2019. doi: 10.1109/TR.2 018.2864536

[7] M. A. Al-Garadi, A. Mohamed, A. K. Al-Ali, X. Du, I. Ali and M. Guizani, "A Survey of Machine and Deep Learning Methods for Internet of Things (IoT) Security," in *IEEE Communications Surveys & Tutorials*, vol. 22, no. 3, pp. 1646–1685, Thirdquarter 2020. doi: 10.1109/COMST.2020.2988293

[8] C. Choi and J. Choi, "Ontology-Based Security Context Reasoning for Power IoT-Cloud Security Service," in *IEEE Access*, vol. 7, pp. 110510–110517, 2019. doi: 10.1109/ACCESS.2019.2933859

[9] M. G. Samaila, J. B. F. Sequeiros, T. Simões, M. M. Freire and P. R. M. Inácio, "IoT-HarPSecA: A Framework and Roadmap for Secure Design and Development of Devices and Applications in the IoT Space," in *IEEE Access*, vol. 8, pp. 16462–16494, 2020. doi: 10.1109/ACCESS.2020.2965925

[10] F. Hussain, R. Hussain, S. A. Hassan and E. Hossain, "Machine Learning in IoT Security: Current Solutions and Future Challenges," in *IEEE Communications Surveys & Tutorials*, vol. 22, no. 3, pp. 1686–1721, Thirdquarter 2020. doi: 10.11 09/COMST.2020.2986444

[11] M. Frustaci, P. Pace, G. Aloi and G. Fortino, "Evaluating Critical Security Issues of the IoT World: Present and Future Challenges," in *IEEE Internet of Things Journal*, vol. 5, no. 4, pp. 2483–2495, Aug. 2018. doi: 10.1109/JIOT.2017.2767291

[12] B. Liao, Y. Ali, S. Nazir, L. He and H. U. Khan, "Security Analysis of IoT Devices by Using Mobile Computing: A Systematic Literature Review," in *IEEE Access*, vol. 8, pp. 120331–120350, 2020. doi: 10.1109/ACCESS.2020.3006358

[13] D. Wang, B. Bai, K. Lei, W. Zhao, Y. Yang and Z. Han, "Enhancing Information Security via Physical Layer Approaches in Heterogeneous IoT With Multiple Access Mobile Edge Computing in Smart City," in *IEEE Access*, vol. 7, pp. 54508–54521, 2019. doi: 10.1109/ACCESS.2019.2913438

[14] K. Lounis and M. Zulkernine, "Attacks and Defenses in Short-Range Wireless Technologies for IoT," in *IEEE Access*, vol. 8, pp. 88892–88932, 2020. doi: 10.11 09/ACCESS.2020.2993553

[15] S. N. Swamy and S. R. Kota, "An Empirical Study on System Level Aspects of Internet of Things (IoT)," in *IEEE Access*, vol. 8, pp. 188082–188134, 2020. doi: 10.1109/ACCESS.2020.3029847

[16] T. M. Fernández-Caramés, "From Pre-Quantum to Post-Quantum IoT Security: A Survey on Quantum-Resistant Cryptosystems for the Internet of Things," in *IEEE Internet of Things Journal*, vol. 7, no. 7, pp. 6457–6480, July 2020. doi: 10.1109/ JIOT.2019.2958788

[17] J. Wang et al., "IoT-Praetor: Undesired Behaviors Detection for IoT Devices," in *IEEE Internet of Things Journal*, vol. 8, no. 2, pp. 927–940, Jan. 2021. doi: 10.11 09/JIOT.2020.3010023

[18] X. Li, Q. Wang, X. Lan, X. Chen, N. Zhang and D. Chen, "Enhancing Cloud-Based IoT Security Through Trustworthy Cloud Service: An Integration of Security and Reputation Approach," in *IEEE Access*, vol. 7, pp. 9368–9383, 2019. doi: 10.1109/ ACCESS.2018.2890432

[19] S. Malani, J. Srinivas, A. K. Das, K. Srinathan and M. Jo, "Certificate-Based Anonymous Device Access Control Scheme for IoT Environment," in *IEEE Internet of Things Journal*, vol. 6, no. 6, pp. 9762–9773, Dec. 2019. doi: 10.1109/ JIOT.2019.2931372

[20] I. Farris, T. Taleb, Y. Khettab and J. Song, "A Survey on Emerging SDN and NFV Security Mechanisms for IoT Systems," in *IEEE Communications Surveys & Tutorials*, vol. 21, no. 1, pp. 812–837, Firstquarter 2019. doi: 10.1109/COMST.201 8.2862350

[21] M. Wazid, A. K. Das, V. Odelu, N. Kumar, M. Conti and M. Jo, "Design of Secure User Authenticated Key Management Protocol for Generic IoT Networks," in *IEEE Internet of Things Journal*, vol. 5, no. 1, pp. 269–282, Feb. 2018. doi: 10.1109/ JIOT.2017.2780232

[22] V. Sharma, I. You, K. Andersson, F. Palmieri, M. H. Rehmani and J. Lim, "Security, Privacy and Trust for Smart Mobile- Internet of Things (M-IoT): A Survey," in *IEEE Access*, vol. 8, pp. 167123–167163, 2020. doi: 10.1109/ ACCESS.2020.3022661

[23] A. M. Zarca, J. B. Bernabe, A. Skarmeta and J. M. Alcaraz Calero, "Virtual IoT HoneyNets to Mitigate Cyberattacks in SDN/NFV-Enabled IoT Networks," in *IEEE Journal on Selected Areas in Communications*, vol. 38, no. 6, pp. 1262–1277, June 2020. doi: 10.1109/JSAC.2020.2986621

[24] M. Yi, X. Xu and L. Xu, "An Intelligent Communication Warning Vulnerability Detection Algorithm Based on IoT Technology," in *IEEE Access*, vol. 7, pp. 164803–164814, 2019. doi: 10.1109/ACCESS.2019.2953075

[25] K. Sood, K. K. Karmakar, S. Yu, V. Varadharajan, S. R. Pokhrel and Y. Xiang, "Alleviating Heterogeneity in SDN-IoT Networks to Maintain QoS and Enhance Security," in *IEEE Internet of Things Journal*, vol. 7, no. 7, pp. 5964–5975, July 2020. doi: 10.1109/JIOT.2019.2959025

[26] W. Zhou, Y. Jia, A. Peng, Y. Zhang and P. Liu, "The Effect of IoT New Features on Security and Privacy: New Threats, Existing Solutions, and Challenges Yet to Be Solved," in *IEEE Internet of Things Journal*, vol. 6, no. 2, pp. 1606–1616, April 2019. doi: 10.1109/JIOT.2018.2847733

[27] C.-S. Park and H.-M. Nam, "Security Architecture and Protocols for Secure MQTT-SN," in *IEEE Access*, vol. 8, pp. 226422–226436, 2020. doi: 10.1109/ACCESS.2020.3045441

[28] F. Amato, V. Casola, G. Cozzolino, A. De Benedictis and F. Moscato, "Exploiting Workflow Languages and Semantics for Validation of Security Policies in IoT Composite Services," in *IEEE Internet of Things Journal*, vol. 7, no. 5, pp. 4655–4665, May 2020. doi: 10.1109/JIOT.2019.2960316

[29] S. Sathyadevan, K. Achuthan, R. Doss and L. Pan, "Protean Authentication Scheme – A Time-Bound Dynamic KeyGen Authentication Technique for IoT Edge Nodes in Outdoor Deployments," in *IEEE Access*, vol. 7, pp. 92419–92435, 2019. doi: 10.1109/ACCESS.2019.2927818

[30] M. Oh, S. Lee, Y. Kang and D. Choi, "Wireless Transceiver Aided Run-Time Secret Key Extraction for IoT Device Security," in *IEEE Transactions on Consumer Electronics*, vol. 66, no. 1, pp. 11–21, Feb. 2020. doi: 10.1109/TCE.2019.2959593

[31] S. Pérez, J. L. Hernández-Ramos, S. Raza and A. Skarmeta, "Application Layer Key Establishment for End-to-End Security in IoT," in *IEEE Internet of Things Journal*, vol. 7, no. 3, pp. 2117–2128, March 2020. doi: 10.1109/JIOT.2019.2959428

[32] S. Mandal, B. Bera, A. K. Sutrala, A. K. Das, K. R. Choo and Y. Park, "Certificateless-Signcryption-Based Three-Factor User Access Control Scheme for IoT Environment," in *IEEE Internet of Things Journal*, vol. 7, no. 4, pp. 3184–3197, April 2020. doi: 10.1109/JIOT.2020.2966242

[33] G. George and S. M. Thampi, "A Graph-Based Security Framework for Securing Industrial IoT Networks From Vulnerability Exploitations," in *IEEE Access*, vol. 6, pp. 43586–43601, 2018. doi: 10.1109/ACCESS.2018.2863244

[34] R. Sairam, S. S. Bhunia, V. Thangavelu and M. Gurusamy, "NETRA: Enhancing IoT Security Using NFV-Based Edge Traffic Analysis," in *IEEE Sensors Journal*, vol. 19, no. 12, pp. 4660–4671, June 2019. doi: 10.1109/JSEN.2019.2900097

[35] E. Dushku, M. M. Rabbani, M. Conti, L. V. Mancini and S. Ranise, "SARA: Secure Asynchronous Remote Attestation for IoT Systems," in *IEEE Transactions on Information Forensics and Security*, vol. 15, pp. 3123–3136, 2020. doi: 10.1109/TIFS.2020.2983282

[36] N. Ghosh, S. Chandra, V. Sachidananda and Y. Elovici, "SoftAuthZ: A Context-Aware, Behavior-Based Authorization Framework for Home IoT," in *IEEE Internet of Things Journal*, vol. 6, no. 6, pp. 10773–10785, Dec. 2019. doi: 10.1109/JIOT.2019.2941767

[37] Z. Deng, Q. Li, Q. Zhang, L. Yang and J. Qin, "Beamforming Design for Physical Layer Security in a Two-Way Cognitive Radio IoT Network With SWIPT," in *IEEE Internet of Things Journal*, vol. 6, no. 6, pp. 10786–10798, Dec. 2019. doi: 10.1109/JIOT.2019.2941873

11 Applications of the Approximate Computing on ML Architecture

Kattekola Naresh[1] *and Shubhankar Majumdar*[2]
[1]ECE Department, VNR VJIET, Hyderabad, India
[2]ECE Department, NIT Meghalaya, Shillong, India

CONTENTS

DOI: 10.1201/9781003201038-11

11.1 APPROXIMATE COMPUTING

11.1.1 INTRODUCTION

Now-a-days, there are a lot of data that need to be processed. Conversely, we only have a limited amount of resources to make this feasible. Inaccurate computing is an implementation that can always be utilized to achieve enormous energy conservations at an expense of minimal accuracy. It might not always be beneficial to answer up to eight or nine decimal places precisely. This is equivalent to a solution with three or four decimal places. It is very important to know that cost of the form of energy used for a more accurate solution possibility. At the same time, special attentiveness should be paid to the portability of the application. In previous studies, Isaac Newton's numerical method can assist us in a similar way. Also, Newton's calculations allow us to conserve energy by performing calculations for the desired accuracy rather than exact accuracy.

There exist various machine learning algorithms that might be used with inaccurate computing methods. This will reduce the accuracy a bit but can significantly reduce power consumption and required chip area, which may effect speed, which is advantageous. Neural networks (NN) primarily depend on approximate computing cognizance at the momentum of software program-degree and have various advantages versus comparison with premature methods. Primarily, NNs were proved that allow you to suit with continuous features, also consequently the approach would universally be adapted by using unique duties. Second, substantial parallelism inside NNs is utilized with the aid of the expeditious development of numerous neural community enhancements. The proper NN would be simply reserialized and established inside the cloud and with the threshold and thus will reap high speed. This chapter is based on the idea of how approximate computing changes the design constraints in implementation of artificial neural networks (ANNs).

ANNs are computationally expensive for data-intensive applications, which may contain many neurons and several orders of magnitude larger number of parameters. Consequently, a significant amount of research effort has been spent to implement hardware NNs to achieve high energy efficiency. Given the fact that emerging recurrent neural networks (RNN) applications are inherently error-tolerant with noisy input datasets and/or involving human interfaces with limited perceptual capability; *approximate computing* has been advocated for these applications. These works have shown that, by relaxing the computational exactness requirement for neurons in ANNs, we can still achieve minor accuracy loss at the application level while gaining significant energy savings.

Nevertheless, a simple/unique neural community cannot be secure to work as an accelerator because of the scarcity of the approximation high-quality controls. Diverse parameters can constitute the approximation best, e.g. the absolute error and imply-square errors among the approximated price and actual price, etc. Compelling the ones parameters can set rules that manipulate the approximation first class. Distinctive satisfactory manage mechanisms, together with linear fashions and statistical fashions, neural communities, and Bayes communities, are dynamically used as they are expecting whether the approximates can correctly approximate the yield that is fed to the input data. The ones un-secured enter information are despatched to CPU to actual calculations. On the contrary, predictions are also used to monitor the yield and finalize the approximations over the run time. Anticipated mistakes overwriting the mistake suffers a rollback of calculations. Therefore, the structure can proactively regulate the approximation over run time by putting forth a greater calculation attempt.

11.1.2 APPROXIMATION

The major concept of an approximate computation is simple. We intentionally decrease the accuracy to conserve energy, time, and/or memory. This will be a pattern that sees loss of accuracy as an opportunity, not a loss. In recent years, various approximation methods have emerged that explore possibilities, including unreliable hardware, NN accelerators, and numerical approximations. The extensive challenge is about conveying this important concept in a basic way to all failed programmers, thereby guaranteeing a precise level of accuracy.

Approximate computation differs from computations linked to concepts such as probability computing. The approximate computation doesn't include presuppositions about the probabilistic characteristic of the process driving the network system. Moreover, the computation uses statistical characteristics of algorithms and data to exchange efficiency for energy and power conservation.

Approximate computation was often used along with NN accelerator, and they are often used in error-prone applications that are tolerant. There are studies that suggest several estimation methods. NNs exhibit parallelism and could be accelerated by special hardware [1]. There exist various quality measures such as pixel differences in an image, categorization of data and clusters, ranking accuracy, etc., and these can be subject to incorrect computations. Other quality measures include overall image quality index and validation. For many applications, there exist several performance measures that can be utilized to calculate quality reduction, such as k-means clustering accuracy and average centroid distance, which are being used as performance metrics [2]. Other areas include image processing, face detection, and search engines. Approximate computing always contributes to a variety of devices and components such as analytic models, CPUs and GPUs, simulators and inaccurate computation techniques, and eventually SRAM cells and cache memory.

In a few more scenarios, the usage of AC is not avoidable. Either the possibility of AC comes from inheritance, or in few scenarios, the AC can be used very actively to maximize efficiency [2].

The incentives and opportunities for AC are as follows:

- Intrinsic scope or need for approximation
- Software tolerance and user error resilience
- Optimization of efficiency
- Quality configurability

To address the full potential of approximate computing, some of the challenges are as follows:

- AC's limited range and benefits
- Accuracy problems
- Finding a strategy for a specific application
- More scalability
- Provides high quality and adaptability

Using only an approximation of one processor component can lead to unintended consequences for other components, as well as erroneous or short-term conclusions. In the future, a comprehensive system-wide impact assessment of exchanges and the use of exchanges across multiple components will be required. As pre-existing systems use multiple control techniques such as pre-amplification, data compression, dynamic voltage/frequency scaling, etc., it is also important to ensure these schemes and ACT synergies for seamless AC integration in commercial systems [2].

The war between the growing call for enumerating and slow growth of hardware capacity activates the reheated amount of improvement in approximate computing that has brought large achievement to each study's community and industry. A large number of applications that do not require totally accurate calculations can obtain outstanding accelerations and a drastic discount in electricity intake by way of anchorage approximate computing, specifically in sectors that use real-time computation, rapid feedback, and calculative power intake, such as photograph processing and clinical computation and learning. Approximation computing may appear in one-of-a-kind hierarchies, including hardware, machine, and software ranges.

11.1.3 STRATEGIES OF APPROXIMATION COMPUTING

There are several strategies to approximate any model, including loop perforation, lossy compression, voltage scaling, precision scaling, memory access skipping, memory usage, load value approximation, NN distribution, data sampling, and more. A few of those techniques are outlined here.

Precision scaling: There exist many approximation computation schemes that change the precision of the input data width to decrease the amount of memory needed.

Loop perforation: There is a technique that uses loop perforation to basically skip some of the looping steps to reduce the cost of computations.

Load value approximation: Due to storage load loss, data are retrieved from main memories or storage to another level, resulting in higher latency.

The load value approximation uses the approximate characteristic of an application to estimate the load value, thereby obscuring the delayed storage.

Using memorization: The procedure of storing a result of a function for further usage and reusing it with the similar input is called memorization.

When designing hardware and software modules, there may be several strategies for making inaccurate calculations and using approximations. Several studies are currently underway to adjust the calculations using the approximation principle.

11.1.4 What to Approximate

There is a way to build an approximate model by finding precise critical sections or code segments, that is, the circulation of control in your program. It is important to identify widely accepted code, and application acceptance levels should be analyzed before adopting inaccurate calculations. The approximation of a single code segment will affect the others. Some of the general advantages of approximation are as follows:

- High performance/energy savings
- Flexibly adapt to your application's accuracy requirements
- Minimum hardware cost

Approximate computing is later performed in the following areas:

- Multimedia and Image processing
- Machine learning
- Scientific data processing
- Database searching
- MapReduce

There can be many other streams that can use approximation computing, and they can be studied through experiments and research work. For example, if any circuit needs to be optimized, the circuit Boolean expression needs to be optimized. Here, the respected K-Map of the circuit is tried to fold with maximum literals in such a way that there should not be more errors that affect the functionality of the circuit. The error analysis and the design requirement are the major functions of approximate computing.

For example,
K-Map for accurate and approximate 2-bit multipliers can be written as shown in Figure 11.1

B1B0 A1A0	00	01	11	10
00	000	000	000	000
01	000	001	011	010
11	000	011	1001	110
10	000	010	110	100

B1B0 A1A0	00	01	11	10
00	000	000	000	000
01	000	001	011	010
11	000	011	**111**	110
10	000	010	110	100

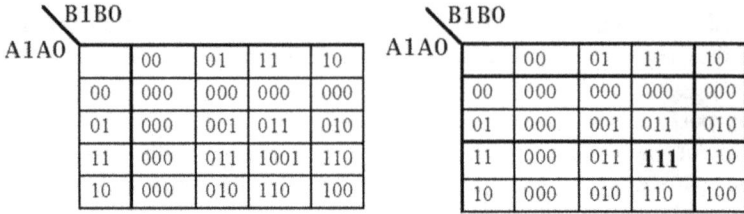

FIGURE 11.1   K-Map for 2-bit multiplier for accurate and approximate design [Source: V. Mrazek 2018].

From Figure 11.1, the implementation of a 2-bit multiplier for accurate design needs 4 partial products, whereas the approximate design can be optimized with 3 partial products, which has an error in 1 position of 16 combinations with the nearest value. Hence, the gate-level design can be optimized to a major extent with one error position.

11.1.5   ERROR ANALYSIS IN APPROXIMATE COMPUTING

Approximate computing evaluation and execution are dependant on gate-level design of the proposed arithmetic circuits, as well as the reported circuit scan coded with MATLAB for the calculation of the error matrices. For all the possible input combinations between 0 and $2^N - 1$, N stands for the count of inputs, which have been considered for the variations, and the outputs are observed. The output of the inexact arithmetic circuit is compared with the output of the exact/accurate arithmetic circuit, and the error metrics are examined. These error matrices can give us an idea about the accuracy of the proposed circuits.

The following performance metrics for error analysis (as defined in) have been evaluated for the comparison:

1. Error Distance (ED):

For such an inexact design, a parameter has been utilized to evaluate the inexactness in accordance with the exact yield; the pre-defined error distance is being proposed as shown with respect to the figure of merit for an inexact computing. For specified input, the error distance (ED), is explained as an arithmetic difference between the exact result (E) and the inexact result (I) [3–8].

$$ED(E,\ I) = |E - I_1| = \left| \sum_i [i] * 2^i - \sum_j [j] * 2^j \right| \tag{11.1}$$

Here, I and j are the indices for the bits in E and I, respectively.

2. Error Rate (ER):

ER is characterized as the level of incorrect yields among all yields.

$$ER = \frac{Total\ Number\ of\ Erroneous\ Output}{Total\ Number\ of\ Outputs} \times 100 \qquad (11.2)$$

3. Total Error Distance (TED):

It is the absolute sum of error distance.

$$TED = |\textstyle\sum ED| \qquad (11.3)$$

4. Mean Error Distance (MED):

MED is the average for a set of outputs.

$$MED = \frac{Total\ Error\ Distance}{Total\ Number\ of\ Outputs} \qquad (11.4)$$

5. Normalized Mean Error Distance (NMED):

NMED is the normalized value of MED.

$$NMED = \frac{MED}{Smax} \qquad (11.5)$$

Smax: It is the maximum magnitude of the output value of the precise adder.

11.2 MACHINE LEARNING

11.2.1 INTRODUCTION

Artificial intelligence comes under the broader framework of machine learning. Machine learning aims to create advanced systems or machines that can learn and train themselves automatically through experience, without any human interaction being expressly programmed or requested. Machine learning is a continually developing activity in this regard. It seeks to describe the data modeling of the dataset at hand and to incorporate the data into machine learning structures that businesses and organizations might utilize.

11.2.2 NEURAL NETWORKS

An NN is influenced by the framework of the human brain. It is fundamentally a model of machine learning that is used in unsupervised learning (more specifically, deep learning). An NN is a web of interrelated entities known as nodes in which a simplistic computation is responsible for each node. NN functionality is like the neurons that are present in the brain of the human body.

11.2.2.1 Architecture

NN's design is often referred to as "architectural design" or "configuration". It consists of basic groups of the count of layers. It also comprises the process of interconnect to change the weights varied. The selection of the design obtained the results that can be extracted. It is the most important aspect to a NN's implementation.

The easiest design is a design in which two layers of input and output are divided into units. Each block in the input layer has an input value and an output value equal to the input value. Due to the combination of function and transfer function, all input-level blocks are connected to the inputs in the output block. There exists greater than one output block. In such situation, there is a logistic regression or linear depending on whether the transfer function is linear or logistic. The regression coefficient is the system weight.

By summating one or more hidden layers among all the inputs and output layers and at such node, the output layers and blocks improve the predictive power of the NN. However, there should be as many hidden layers as possible. This allows the NN to generalize it without storing all the data in the training set, preventing abuse.

11.2.2.2 Abilities and Disabilities

- While NNs work well with data on linearity and nonlinearity, there is a general condemnation of NNs, especially robotics, which might require extensive practical implementation training. In such a case, any learning engine needs enough representative examples to understand the fundamental context, which will help expand into new cases.
- NNs also work when one or more websites are not responding to the network but take up a lot of processing and storage resources required to run large and efficient neural software networks. The brain has the right hardware for the task of processing signals through neural graphs, but even modeling the von Neumann technique in its simplest form can eliminate NNs designed to populate millions of database connections, a lot of RAM, and computer and hard disk memory.
- NNs learn through data analysis and do not need to be reprogrammed, but they are reprogramming devices. These are referred to as modeled black boxes, though this offers a very small amount of the further view into what these models will really do. The user has to specify the inputs, verify its practice, and wait for the outputs.

11.2.3 Machine Learning vs. Neural Network

- To discover meaningful patterns of interest, machine learning uses advanced algorithms that analyze, learn from, and use data. An NN is a gathering of algorithms, which utilizes data processing involved in neuron graphs in machine learning.
- Wherein schemes of machine learning make judgements with respect to what is already studied from the info, an NN organizes schemes in procedures that might make prescribed choices on their own. Even though schemes of machine

learning might develop from fundamentals, they may contain a few human interventions throughout the beginning levels.

    o As the nested layers within transversal data along hierarchies of various definitions, NNs do not really need human involvement, which finally enables us to learn through our failures.

- It is possible to classify machine learning models into two types: supervised and unsupervised learning modules. NNs, though, can be classified into NNs that are recurrent, feedforward, modular, and co-evolutionary.
- In a simple way, an ML model works – it is fed and learned from data. Each time that it constantly learns from the results, the ML model becomes more sophisticated and educated. The configuration of an NN, on the other hand, is very complex. In it, the data flows through multiple layers of interconnected nodes, through which each node classifies the previous layer's characteristics and data before transmitting the results to other nodes in subsequent layers.
- Although machine learning models are adaptive, learning from new sample data and interactions continuously evolves. The models can thus classify the trends in the details. Here and the only input layer is data. There are several layers, though, also in a basic NN model.

    o The first layer, preceded by a hidden layer, is the input layer, and then eventually the output layer. One or more neurons are found in any layer. You can improve its analytical and problem-solving ability by raising the number of hidden layers within such an NN model.

- Probability, analytics, and programming. Hadoop and Big Data, knowledge of ML architectures, algorithms, and data structures are skills needed for machine learning. NNs include skills such as modeling data, geometry, and graph theory; linear algebra; programming and statistics; and probability.
- In fields such as hospitals, banking, e-commerce (recommendation engines), financial services and insurance, self-driving vehicles, online video streaming, IoT, and transportation and logistics, to name a few, machine learning is implemented. In the other hand, NNs have been used to address various market problems, including, among other things, revenue forecasting, data analysis, consumer study, risk assessment, voice recognition, and character recognition.
- There are a few key difference between the machine learning and NNs. NNs are basically part of deep learning, which is a branch of machine learning. NNs are, however, few not but regularly sophisticated implementation of machine learning that is actually seeing applications in a variety of areas of interest.

11.2.4 CLASSIFICATIONS OF NEURAL NETWORKS IN MACHINE LEARNING

There are two important types of NNs that use machine learning.

  1. Artificial Neural Network – ANN
  2. Convolutional Neural Network – CNN

Here, the classifications of NN are CNN and ANN. The way we communicate with the globe is changing. These various forms of NNs are at the center of profound learning revolutions and powered technologies such as self-driving cars vehicles, voice recognition, unmanned aerials, etc.

11.2.4.1 Artificial Neural Network (ANN)

The human mind, which has been modeled beyond several millions of years and also has evolved, will simply outmatch the present day Von Neumann computing regarding several objects of applications that are involved with analytics, inference, and recognition. ANNs, which can make out a huge inventiveness from biological NNs, have also become very successful in a varied spectrum of applications that is regression analysis, function approximation, filtering, robotics, clustering, pattern and sequence recognition, etc. Some remarkable recent accomplishments within the domain of ANNs is the technique of deep learning networks (DLN). Practicing a large contents of computing power and data, DLNs can classify images and interprets, does a variety of deduction tasks recognize speech and read documents, into the bargain or as no else used or predefined algorithm. Microsoft's "Project Adam" is among such an enterprise that belongs to DLN with greater than 2 billion connections.

Recently, synthetic NNs (ANNs) have been evaluated as top-notch overall production in specific studies, consisting of scientific semantic segmentation, face detection, and image processing. In the latest, traits in imperative transforming gadgets (CPUs) and photograph-transforming gadgets (GPUs) yield good memory supplies and large performing speeds for instruction and performance of ANNs. However, due to having restricted memory for portable devices, the range of performing units, and therefore the battery potential of these ANNs, is not practical. The best situation is to scale back the ANN hardware architecture to increase the hardware precision.

Because of having huge computational necessities, the hardware administrations of such neuromorphic algorithmic architecture show inefficacious in the realms of region and strength intake. The provocation of hardware administrations of ANNs has become a great study in specific outlooks. A unique way for stalking effective hardware administrations of NNs was to make modifications to the schemes of the network systems. To utilize the graphical processing devices (GPUs), also to take advantage of the similarities of ANNs, has additionally been met. The alternative technique is utilization of rising tool technology to administer synapses and neurons with greater effectively. Usage of a hybrid skilled CMOS machine/crossbar array, section-alternate memory device, resistive RAM, and spin-primarily based devices in these circumstances is always explored. Luckily, NNs and related programs disclose intrinsic software flexibility to mistakes that make them desired applicants for approximate computations. Discovering the pre-existing error flexibility of a system, electricity effectiveness may be achieved with the aid of using a diffusion of hardware and software program strategies.

An ANN consists of a network of neurons that are related to one another. The training period where an error is applied between the required and actual answer is minimized by utilizing an iterative development algorithm. The bias and weighted values of ANN are calculated. The inputs are usually normalized between −1 and 1 during preparation. In contrast to the ANN instructed with unnormalized inputs, such normalization can reduce the instruct run-time and output of an ANN. Likewise, a lower count of layer sand neurons achieves similar precision. In addition, test results are used to provide an impartial assessment after the training phase of the final model, and the certainty is measured as a performance metric, or misclassification rate.

11.2.4.1.1 Feedforward ANN

A single neuron (or perceptron) may be thought of as a regression analysis. ANN is a multiple layer group of neurons and perceptron. As inputs are processed only in the forward direction, ANN is also known as a feedforward neural network. The below Figures 11.2 and 11.3 show the building blocks of neuron and feedforward neural network.

As it will be seen, ANN is comprised of three layers – hidden layer, output layer, and input layer. The input layer gives permission to the inputs. The hidden layer performs the operation on the inputs, and therefore the output layer gives the result. However, each layer tries to find out the certain weight of the data. ANNs try to solve problems related to the following:

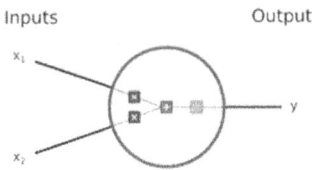

FIGURE 11.2   Building blocks of neuron [Source: M. E. Nojehdeh 2020].

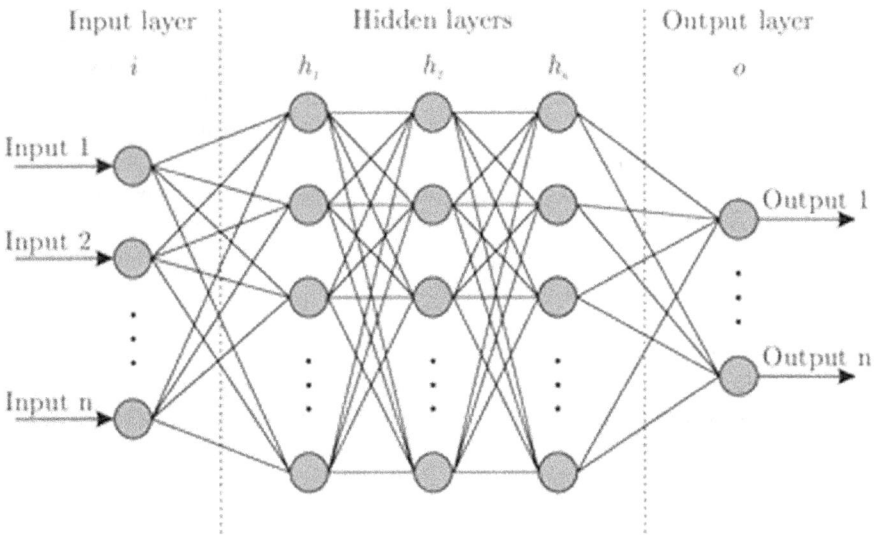

FIGURE 11.3   Feedforward neural network [Source: Miroslav Kubat 2017].

- Table format
- Visual format
- Words and text format

ANNs are often arranged in layers. The layers are made of several interdependent "nodes" containing "activation functions". NNs can be organized into three layers:

a. **Input Layer**
   Every input layer will give with purpose to have a descriptive attribute value entered in each and every order that will be observed. There are nodes present in the input layers that are usually sufficient for the number of independent variables. "Input layer" refers to a template for a network that interacts with high numbers of hidden layers, whereas the input layer nodes will be passive. In other words, it does not change the information. Get a unique value from the inputs and then duplicate it on multiple outputs. Duplicate each value in the input layer, and send it to some or all hidden nodes.

b. **Hidden Layer**
   Hidden layers specify the input value of the system. In this case, the primary arc comes from another hidden node or an input node associated with each node. Connect the outgoing arc with a hidden node or another hidden node. Regarding the hidden layer, specific transformation is completed with a weight network of "connection". It will have more than one hidden layer in the network if required. Values entering the hidden node will be multiplied with their respective weights, which are predefined groups that will be saved in the program. Then, add respective weights of the input layers to get the value.

c. **Output Layer**
   All hidden layers relate to the output layer to obtain data. The output layer acquires the connection from the input layer or the hidden layer. It replaces the initial value corresponding to which a variable value response is expected. Deployment issues usually only have single output layer nodes. Active nodes in the warehouse mix it up and vary information to provide output values. There are some advantages of the NN to perform useful data gestures depending on choosing the right pressure. This is often different from regular IP. ANN is considered a simple mathematical model to improve present analyzed data techniques. Although this is incompatible with the functioning of the neurons in the brain of the mammal species, it is still an important component for artificial intelligence.

11.2.4.1.2 Abilities of Artificial Neural Network (ANN)

ANNs can study any nonlinear function. Hence, these networks are often referred to as overall performance estimates. ANNs will give the weights that can be linked up with inputs and outputs.

One of the common reasons is the universal approximation of the activation functions. Nonlinear properties are introduced into the network through the activation function. This helps the network learn tough relationships in both the outputs and inputs. The output of each neuron occurs due to the activation of a weighted amount of input. But wait a minute! What if there is no activation function? Networks only learn linear functions and may not learn complex relationships. The activation function can be an ANN main function to activate the ANN.

11.2.4.2 Convolution Neural Network (CNN)

CNNs are a direct community for deep learning. These CNN models are used in a variety of applications and domains and are regularly seen in virtual processing data projects. Some blocks built for CNNs are filters, also known as cores. The core typically extracts its function from its input through acts of belief.

11.2.5 Novel Algorithm in ANN

11.2.5.1 Introduction

NNs are the most important model for many innovative applications today. Over the past decade, they have processed different visual formats and voice analysis, and have differentiated verbal communication. They are currently the leading technologies in computer science and are often used in various fields. The weight gain algorithm is the center of some or all NNs. When the architectures of NNs are stable, only the weight can be obtained as the function.

Therefore, it will be a tough process to get these weights. Flexible nature is present in the NN structure and hardness of the nonlinearity dynamics. The weight of an NN can sometimes not be determined. According to reality, to calculate the gradient, NN is iteratively upskilled using a gradient descent technique combined with backpropagation (BP). While techniques of gradient descent are generally efficient and work effectively with the latest NN, this has some disadvantages. The routine processes they use are time consuming. If you use modern accelerated hardware, such as a GPU or a tensor processor unit (TPU), the training process will take weeks or days to achieve desirable results. Second, it is not guaranteed that honest decisions will be made through the curriculum. This may cause gradient descent if structured for convex functions. Still, NNs tend to be very non-convex. Third, gradient descent gradients require gradients to be used, and whether the NNs being upskilled are different, these gradients can explode or disappear when calculated in BP. Gradient descent is also overly sensitive, so small configuration changes like learning rate or momentum can make a major contrast in the outcomes. It is task-specific to set certain hyperparameters and relies on personal experience.

On the other hand, using gradient descent, an end-to-end NN trained is a common recording equipment model that is difficult to interpret. You may be exposed to NNs that have been condemned for a long time [9].

11.2.5.2 Weights of Neurons

In order to understand how the weight works, it's really beneficial to visualize a conceptual NN. There is an input layer inside the NN, which takes the input signals and passes them to the next layer.

Next, there is a sequence of hidden layers in the NN that applies modifications to the data input. It is within the hidden layers' nodes that the weights are applied. For instance, the input data can be taken by a single node and multiplied by any predefined

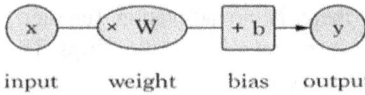

FIGURE 11.4 Weight and bias of neural network [Source: M. E. Nojehdeh 2020].

weight value, thereby adding a bias ahead of processing the data to the further layer. The NN's final layer is also known as the output layer. To generate the correct numbers in a specified range, the output layer often tunes the inputs from the hidden layers. Figure 11.4 shows the ANN process for a single neuron.

11.2.5.3 Weight vs. Bias

Weights and bias are all factors within the network that are necessary to learn. Until learning eventually starts, an instructive NN can randomly select the weight and bias values. Both parameters are changed to the appropriate values and to the right performance as training proceeds. In the scope of their effect on the input results, the two parameters vary. Simply put, bias shows how far from their expected value the forecasts are. The disparity between the output of the function and its desired outcome is made up of biases. A low bias means that more assumptions are made by the network about the output source, while a high bias value produces less predictions about the output form. Weights, with respect to another hand, could be brought up as the connection's power. Weight affects the major contents of impact. A difference in the input would make an impact upon the output. Less weight value would not change the input, and a greater weight value can change the input more substantially.

$$Y = \sum (\text{weight} * \text{input}) + \text{bias} \qquad (11.6)$$

11.2.5.4 Neuron (Node)

It is the fundamental block of an NN. A specified count of inputs and bias values will be obtained. As a signal (value) comes, the value of the weight is compounded. It has four weight values if a neuron has four inputs, which can be changed throughout the training phase. Figure 11.5 shows the operations at one neuron of an NN.

$$Z = x_1 * w_1 + x_2 * w_2 + \ldots + x_n * w_n + b * 1 \qquad (11.7)$$

$$\hat{Y} = a_{\text{out}} = \text{sigmoid}(z) \qquad (11.8)$$

$$\text{Sigmoid}(z) = 1/(1 + e^{-z}) \qquad (11.9)$$

11.2.5.4.1 Bias (Offset)

It's an external neuron input, and it's always 1. It has its own relation weight. This helps ensure that there will be an activation in the neuron, even though all inputs are none (all 0's).

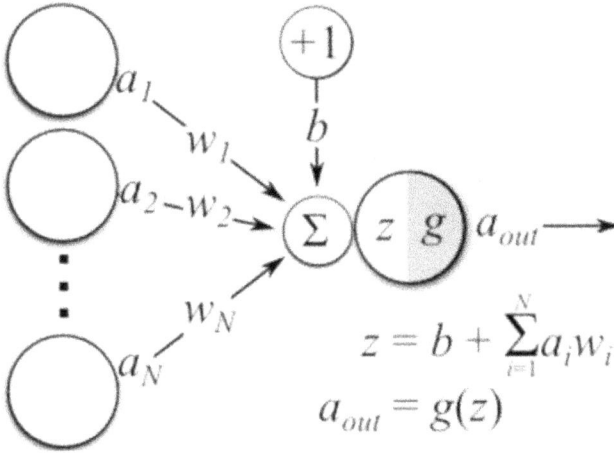

FIGURE 11.5   Operations at one neuron of a neural network [Source: Miroslav Kubat 2017].

11.2.5.4.2 Activation Function (Transfer Function)

To apply non-linearity to NNs, activation functions are used. In a narrower set, it squashes the values. The activation feature of Sigmoid shatters values in the 0 to 1 size. In the deep learning market, there are several activation functions used and ReLU, SeLU, and TanH are favored over the role of sigmoid activation.

11.3   APPROXIMATE MACHINE LEARNING ALGORITHMS

11.3.1   INTRODUCTION

In multiple domains, error-resilient applications will make way for trading off among power and quality of service (QoS), reductions in area, and time latency. Machine learning is one commonly used group of these domains. Despite providing high precision, machine learning algorithms always are familiar with their computational complexity [10]. The usage of approximate computing techniques (ACTs) was also suggested to resolve the complexity. As an efficient way to achieve real-time and effective low power enactment of machine learning algorithms, actions are increasingly being adopted.

The problems and reasons due to the usage of the approximate computing were raised by H. Younes in Younes et al. [10]. The major problem was the targeting of error resilience of systems and users. In addition, in visual and other computer applications, the discernment limitations of humans encourage the scope of approximation. An outline of approximate computing techniques for embedded machine learning was discussed in Venkataramani et al. [11]. ACTs will be implemented over the level of software, circuit, and design. When applying approximate computing, algorithmic- or software-level techniques mimic the top-most option taken. This regards the fact of reducing the complexity of the data/algorithm right from the beginning would inevitably lead to reductions at both the levels of the circuit and design.

For embedded systems, decrements include energy/power, hardware resources, and time latency.

A perspective of relating algorithmic-level actions on machine learning algorithms is proposed in the work discussed in Sivanantham [12]. Achieving an efficient classification with reduced complexity is the key inspiration behind this work. In addition, each ACT will obey as a valued composition knob for trading off the quality of latency in the proposed approach [10]. For this reason, when monitoring the performance time latency versus quality, each ACT was applied alone. Subsequently, two separate applications, i.e. image classification and touch modality classification, have been applied to the proposed approach [10]. Two machine learning algorithms under supervision have been used: support vector machine (SVM) and K-nearest neighbor (KNN).

11.3.2 Approximate Computing Techniques

As shown in Figure 11.6, the algorithmic-level method is subsequently divided into two classifications. The data-oriented group includes altering the properties of the dataset. Data-format (DFM) adjustment and dataset reduction (DsR) are included in this group. Also, to apply DsR, start from an accessible dataset, though the size of the dataset can be decreased by randomly eliminating a few samples at a rate of k. DsR could be applicable by either downscaling or downsampling by the prescribed way of producing a dataset. Downsampling sets the signal sampling frequency to processing interface, whereas by using downscaling the value proportions in matrix-based/vector program are reduced. DFM, precisely known as the precision scaling,

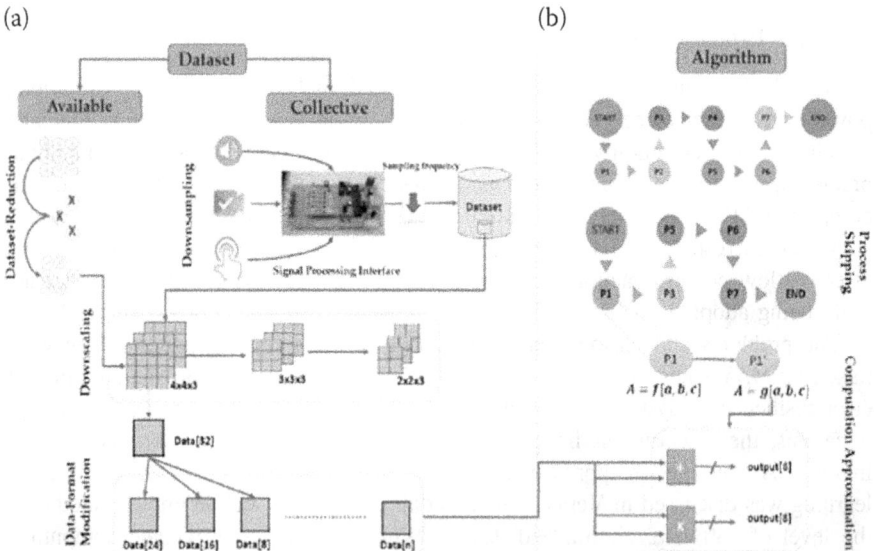

FIGURE 11.6 Implementing techniques of algorithmic level approximate computing: (a) data-aligned, (b) processing-aligned [Source: H. Younes 2019].

involves the usage of representation of fixed-point data. Bit-width can be decreased by DFM, which makes an ease of the arithmetic operations' complexity in return. The group of processing attacks the mathematical algorithm itself. DFM involves methods like computation of approximation (CA) and computation skipping (CS) [10]. CS can be used in a multiple-step or multiple-stage algorithm by fully escaping the task, or by escaping a definite numerous task within the iterative algorithm process and is known as loop-perforation. CA substitutes an equivalent version for a computationally costly function. A comparable performance with an appropriate error margin must be given by an equivalent feature. For DFM, CA is an interrelated step, because it establishes fixed-point arithmetic functions [10].

11.3.3 APPROXIMATE ALGORITHMS FOR MACHINE LEARNING

Two macine learning algorithms were used to implement the suggested solution outlined in Figure 11.6, KNN and SVM. The two binary categorization concerns, i.e. image considerations and connect modality recognition, respectively SVM and KNN, are used. The raw data handled are defined in tentorial form for both issues. Especially, a tensor T (4 * 4 * 30,000) replicates a simple contact mode captured with 2 KHz sampling frequency through a 4 * 4 tactile sensory array over 10 seconds and tensor M (15 * 15 * 3) is produced with resizing an image with RGB channels with a 15 * 15 rectangle [10].

Downsampling is applied to the touch modality classification issue by decreasing the skilled set size using a random removal process by 10%, 20%, and 30%, respectively. During the downscaling, every recording of the touch modality is deleted or cut off from 3.5 to 7 seconds. Thus, it conducts an average in-time domain. This resulted in a 4 * 4 * 1 size tensor for every contact; 24 and 16 bits technique-quantization, respectively with <6,18> and <6,10> accuracies, was then implemented. Two processes are involved in the KNN algorithm: distance calculation and sorting processes. In the distance computation method, processing-oriented actions have been implemented. Next, the arithmetical square-root operation was skipped. Consequently, two testing points were compared using the distance-squared metric. Secondly, their 16- and 24-bit fixed-point complements replaced floating-point adders and multipliers.

Downsampling will be considered in the consecutive manner as applied to KNN for the image classification issue. CS and CA have been introduced as well. First, to decrease the count of repetitions carried in Jacobi algorithm, which is needed for singular value decomposition (SVD) computations, loop-perforation by a skipping factor of sf{2; 3; 4} was utilized in Younes et al. [10]. An architecture of variable accuracy was then adopted. The architecture involves 24-bit quantization for SVD symmetrization, and unfolding blocks with <12,12> accuracy while retaining a floating-point depiction for a computation module of the kernel.

11.3.4 RESULTS AND ANALYSIS

With accordance of the average precision loss and speed relative with respect to exact classifiers, also the effect of every ACT over the output of the classifiers is

TABLE 11.1

Implementation Results

Approximate Computing Algorithm	Classification Problem			
	Touch Mobility		Image	
	Accuracy Loss (%)	Speedup	Accuracy Loss (%)	Speedup
Downsampling	≤3%	1.7X	<6%	1.15X
Downscaling	6.4%	2X	Not Applicable	Not Applicable
DFM (24-bit)	3%	1.14X	4.7%	3.2X
DFM (16-bit)	8%	1.3X	>>10%	Not Applicable
Loop Perforation	NA	NA	4%	2X

[Source: S. Sivanantham 2013]

shown in Table 11.1. In the two categorizations, the problems with tolerable precision loss and the proposed approximate SVM classifiers and KNN (K = 3) have demonstrated their advantages.

For example, recognition of connect modality is increased by a multiple of 1.7 over an exact loss, which is less than 3%, where a speed of 3.2 could be achieved for image classification with an accuracy loss of less than 5% [12], which is shown in Table 11.1.

11.4 CASE STUDY 1: ENERGY-EFFICIENT ANN USING ALPHABET SET MULTIPLIER

11.4.1 INTRODUCTION

The primary power famished additives of ANN is the multiplier blocks within the neurons that will multiply loaded data and corresponding weights (synapses). Labeling the difficulty by the characteristics of alphabet set multiplier (ASM) is advised, which is approximate. Use the idea of computation sharing multiplication (CSHM) [12–14] for conspiring the energy efficiency of ASM. In ASM, novel multiplication is reconfigured with generalized shift and upload operations. The ASM includes an in-laptop bank that produces a few of the "alphabets", which might decrease the ordered multiples to the entry. Based totally at the weighted cost, an accurate mixture of alphabetical picks, addition, and shift operations are derived to make the result, i.e. product. Also attaining electricity advantages, the wide variety of "alphabets" utilized inside the proposed ASM are less than vital for perfect (correct) operation. Also, the result may not aid all the combinations of multiplication. To assure right working of the neural community, and make sure that the unsupported multiplication mixtures no longer will lead to huge mistakes throughout the testing [15], limitations are kept on the weights that are acquired from conventionally skilled networks. Those limitations are the same for quantizing, which would reduce a few records. As a result, accuracy loss is suffered. But, to acquire the quality of output, we observe retraining of NN with limitations in the location.

Existing ASM can be replaced by an conventional multiplier in the artificial neurons to reduce the intake of energy and also increase the various benefits, that is, as increase in processing speed and reduction of area [15]. Last, we impose a more complex neuron design, which will not constitute any of the pre-processing bank, and a multiplier-less neuron, emerging to huge improvements with less accuracy decrement in energy usage [16].

The primary operation of these ANNs includes levels, such as training and testing. First, the training method is normally done offline so it is not an electricity intake issue. The skilled ANN is being utilized to test the random records of data, that is performed on the chip. Huge networks, which have millions of neurons, and the trying out system, even though less computed in depth than schooling, may additionally require great computation. The checking out method is generally ahead in propagation, that includes activation operations, summation, and multiplication. The vast electricity consuming calculations amongst those will be multiplication, through which some distance over-weighs the activation and summation. Hence, the fundamental recognition is to alleviate this trouble by way of presenting a solution that is energy efficient. On these paintings, by using the approximate ASM, we primarily mirror the traditional multiplier inside the neurons. Sooner or later we take some time to produce a synthetic neuron without a multiplier block. Notice that, introducing an NN along the approximate multiplier may lead to minimum reduction of accuracy but achieve good-sized strength discount.

In the multiplication operation, from smaller bit sequences, products will be generated and will be the lower multiples of input "I_n" (multiplier). The putrefaction depends upon the multiplicand "W", in such an instance is representing the synaptic weight. Table 11.2 shows the decomposition of two multiplication operations; they are W1 × I and W2 × I.

Note that if I, 3I, 5I, 7I, 9I, 11I, 13I, and 15I are pre-existing, then total multiplication will be decreased to a smaller number of shifts and additions. Some of the small bit sequences are called alphabets. In ASM [12–14], certain pre-specified alphabets are shifted and added instead of directly multiplying the multiplier and multiplicand. These alphabets are collectively referred to as alphabet sets, which consist of lower order multiples and generate these alphabets in a pre-computer library is needed.

TABLE 11.2

Decomposition of Multiplication Operation

Weights	Decomposition of Product
$W_1 = 01101001_2$ (105_{10})	$W_1 \times I = 2^5.(0011).I + 2^0.(1001).I$
$W_2 = 01000010_2$ (66_{10})	$W_2 \times I = 2^6.(0001).I + 2^1.(0001).I$

[Source: S. S. Sarwar 2016]

ASMs consist of four stages:

  i. Produce characters.
 ii. Choose characters.
iii. Shift the respected characters.
 iv. Summate the shifted characters.

11.4.2   8-bit 4 Alphabet ASM

In this, synaptic weights have been treated as 8 (12) word length and further can be spitted into three (two) quartets of ASM. Its final adding requirement after selection and shifting operations is done. Depending on the multiplicand, multiple various combinations of selection, addition, and shifting can take place, and these combinations are administered by a control block. To include all desired sets of combinations and to compute a conventional multiplication functionality for the bit words with a width of 4 bits, 8 alphabets {1,3,5,7,9,11,13,15} are required. The count of characters is straightforward converted to consume the power from the front computer block, and the number of alphabets is proportional to the number of communication buses [16].

Therefore, using the errors resilient capability of neural computing, a reduction of the number of present alphabets gains a the low level in complex of routing and power consumption. Figure 11.7 shows the 8-bit 4-alphabet ASM [16].

The multiplier "I" (input) is induced with the front end of the computer, where it gives four characters. By this example, a character set is {1,3,5,7}. The multiplicand "W" is partitioned into two fixings, and these will be fed as the inputs to the "control block" circuit. In accordance with "W" (multiplicand), for "shift" and "select" blocks an appropriate control logic is generated. The appropriate alphabet is selected by the selection unit and passes this alphabet to the shift block. Then, the shift units will enter the set of number of shifts. Finally, two disparate values will be summated by an "adder" unit to produce the eventual multiplied result. Consider an example of a multiplier unit "M" and the multiplicand 01001010_2. From this we must generate 1010_2 M (10 M) and 0100_2 M (4 M) $\times 2^4$ (MSB is shifted by 4 bit), and summate them

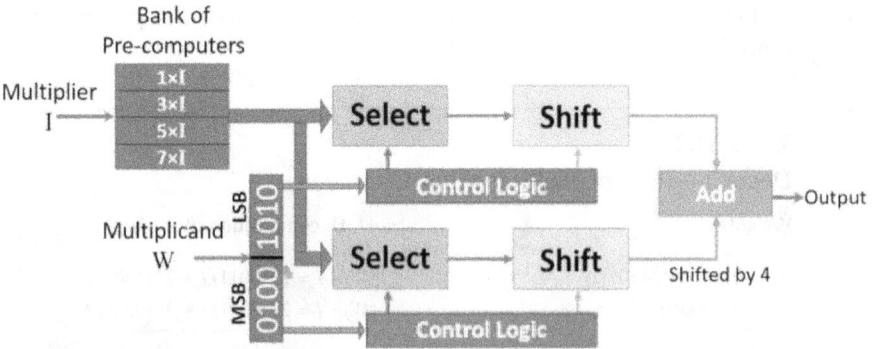

FIGURE 11.7   8-bit 4 alphabet ASM [Source: S. S. Sarwar 2016].

both. You could generate 10 M by transferring the character 5 M by 1. You can generate 4 M by shifting the character 1 M by 2.

The eventual summation value is proved by the below expressions:

$$01001010_2 \times M = (4M) \times 2^4 + (10M) \times 2^0$$

11.4.3 Four Alphabet ASMs Using CSHM Architecture

These ASMs are only helpful when they can be used with a smaller number of alphabets in a distributed manner, that is, when they share alphabets with different multiplication units. CSHM architecture achieves this goal. Figure 11.8, which consists of a common computer front storage area, is shared among multiple multiplication units.

In the feedforward ANN, you can share the system front storage library, and every input will be multiplied with several distinct weights to get various neurons (Figure 11.3). Processing blocks that can computes four neurons in succession are implemented so that four ASM blocks can share the alphabet from the public system library, as shown in Figure 11.8.

The effectiveness of ASM depends largely on the count of characters that can be utilized to include the range of selection, addition, and shift combinations. When the bit sequence decomposed with a multiplier computation contains 4 bits, a character

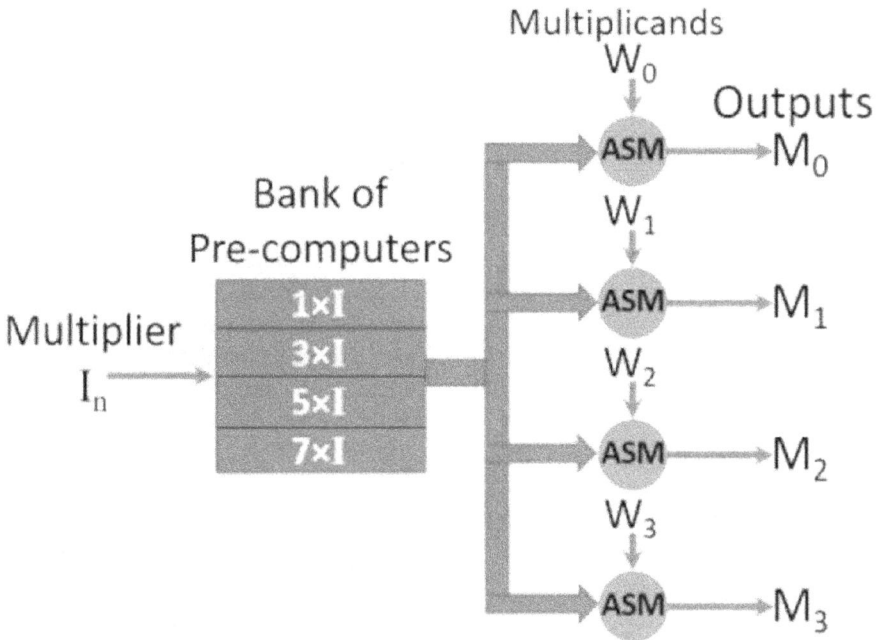

FIGURE 11.8 Four alphabet ASMs using CSHM architecture [Source: S. S. Sarwar 2016].

with a set of eight characters {1,3,5,7,9,11,13,15} will be enough to use selection, shift to generate any product, and then add operation. To obtain a significant performance improvement, we recommend using a minimum number of alphabets because all combinations cannot be covered, resulting in a multiplicative approximation.

For example:

Consider 4 characters {1,3,5,7}. From that we could produce 12 (inclusive of 0 (0000_2)) from 16 possible 4-bit combinations through a shift operation (for example, from 1 (0001_2)) to get 2 (0010_2)), 4 (0100_2), and 8 (1000_2)). In this case, the unsupported quaternary value of bits is {9,11,13,15}. Therefore, since the alphabet set used does not support LSB 1001_2 (9_{10}), the product $01101001_2 \times I$ cannot be generated with any combination of selection, shift, and addition. To overcome this problem, constrained training of ANNs is introduced such that the non-supported set of combinations will not happen. Also, the ANN application is error-proof. We can use this method and obtain an appropriate set of weights, simultaneously at the same time imposing constraints on the network to cause minimal or no loss of network accuracy. Compared with the original training, the cost of retraining is small.

An algorithm to limit the weight of 12-bit ASM will be explained below with an example as follow:

The 12-bit synaptic weight is regarded as 3-bit quartet P, a series version of Q and R, where R is the LSB and P is the MSB. Figure 11.9 shows the 12-bit weight value is divided into three quartets. Since, the 2's complement binary number system is used, the primary bit of P is the sign bit; so we don't have to take the sign bit into consideration because only the absolute value is multiplied. Therefore, P has 8 combinations, from 0 (000_2) to 7 (111_2), and Q and R have 16 combinations, from 0 (0000_2) to 15 (1111_2). If only 2 alphabets {1,3} are used, out of 16 the maximum number of combinations supported is only 8. In this case, 5 and 7 for P, while 5, 7, 9, 10, 11, 13 Q and R are 14, 15 are not supported. Therefore, to minimize the loss of precision, we change these non-supported values to the closest supported values. Algorithm 1 is a weight constraint mechanism of a 12-digit, 2-alphabet {1,3} multiplier.

11.4.3.1   Rounding Logic

In order to perform multiplications correctly, we need to round down/up non-supported values to the closest supported values to ensure minimal information loss. For every two consecutive support values, consider their average value as a rounding threshold point. Consider two consecutive support values 8 and 12 (utilizing only the characters {1,3}); therefore, the threshold value will be $(8 + 12) / 2 = 10$. When the non-supported value 9 appears, it is converted to 8; otherwise if 10 or 11 appear, it is converted to 12. For different unsupported values, the rounding threshold is different.

Figure 11.10 introduces and explains the training and testing methods, an overview of ANN design methods. The input is an NN, and its commensurate

| P (MSB) | Q | R (LSB) |

FIGURE 11.9   12-bit weight value decomposed into three quartets [Source: S. S. Sarwar 2016].

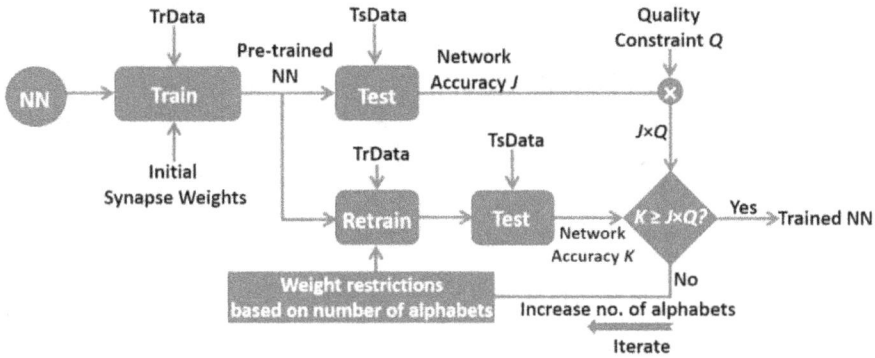

FIGURE 11.10   Overview of the ANN design methodology [Source: S. S. Sarwar 2016].

training data set (TrData), test data set (data), and quality constraint (Q), which indicates the tolerable quality degradation in the implementation. Quality specifications are for specific applications.

In order to check the validity of our model, we use it in a face detection program, which detects whether there is a face based on the input image data. Here, just 2 is the number of final output neurons. Input neurons of 1024 are used and 100 hidden layer neurons. We first created 8-bit and 12-bit synaptic weights for unsupported (for periodic multipliers) and restricted conditions using the training data collection (for ASM). Then, using the test data collection, we checked the network and obtained better results with a maximum precision drop of 0.47%. Table 11.3 displays the results.

After the success, we used the MNIST dataset to solve a more complex "handwritten digit recognition" problem. As earlier, to stimulate synaptic weights (in this, the final number of output neurons is 10) a similar method was used. Then, we used these synaptic weights to test the accuracy of the system in the designed processing engine.

The accuracy results are shown in Table 11.4.

TABLE 11.3
For Face Detection Accuracy Results of NN

Width of Synapse	No. of Characters	Accuracy (%)	Accuracy Loss (%)
8-bit	traditional NN	90.7	–
	4 {1,3,5,7}	90.5	0.22
	2 {1,3}	90.3	0.39
	1 {1}	90.2	0.47
12-bit	traditional NN	90.7	–
	4 {1,3,5,7}	90.6	0.12
	2 {1,3}	90.5	0.19
	1 {1}	90.5	0.24

[Source: S. S. Sarwar 2016]

TABLE 11.4
Digit Recognition Accuracy Results of NN

Width of Synapse	No. of Characters	Accuracy (%)	Accuracy Loss (%)
8-bit	traditional NN	97.5	–
	4 {1,3,5,7}	97.4	0.04
	2 {1,3}	97.4	0.06
	1 {1}	97.1	0.35
12-bit	traditional NN	97.6	–
	4 {1,3,5,7}	97.6	0.03
	2 {1,3}	97.4	0.19
	1 {1}	97.4	0.25

[Source: S. S. Sarwar 2016]

11.4.4   MULTIPLIER-LESS NEURON

By using ASM in artificial neurons, the results are obtained, and from this we noticed that even if there is only a single alphabet {1} with all layers, within ~0.5% of the traditional implementation we can achieve network accuracy. The additional benefit for utilizing only a single alphabet, especially {1}, because we do need not to produce and use any other character set, is that the input is at most enough to meet the need of 1 {1} alphabet. This means that we do not need to do multiplication, just shift and add. This can eliminate the need for pre-computer storage areas and an alphabet "selection" unit, shown in Figure 11.11. Therefore, the circuit will be faster, and it is more compact and consumes minimum power, resulting in "multiplier-free" neurons.

11.4.5   RESULTS AND ANALYSIS

Simulation is done by using multilayer perceptron model and CNN in the experiment, with the support of C++ and MATLAB. A multilayer BP network is realized using

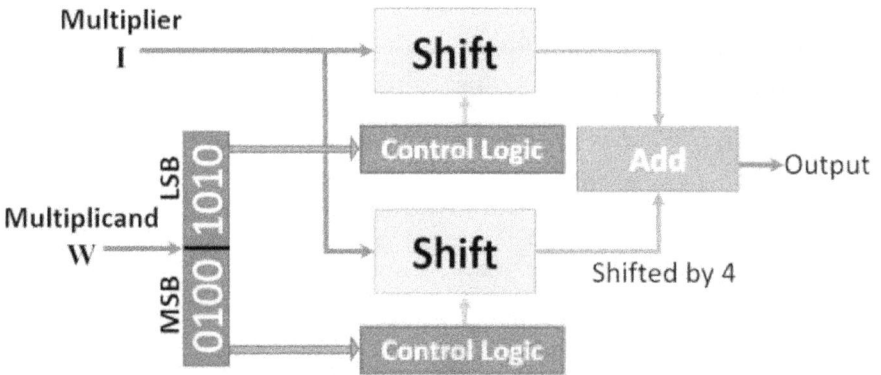

FIGURE 11.11   8 bit 1 alphabet {1} ASM [Source: S. S. Sarwar 2016].

TABLE 11.5

Lists the Benchmark Metrics Used

Application	Dataset	NN Model	Number of Layers	Number of Neurons	Number of Trainable Synapses
Digit Recognition (8-bit)	MNIST	MLP	2	110	103,510
Digit Recognition (12-bit)	MNIST	CNN (LeNet) [17]	6	8010	51,946
Face Detection (12-bit)	YUV Faces	MLP	2	102	102,702
House Number Recognition	SVHN	MLP	6	1560	1,054,260
Tilburg Character Set Recognition	TICH	MLP	5	786	421,186

[Source: S. S. Sarwar 2016]

these tools. The NN using the corresponding training data set is trained here. Then, during the retraining of the NN, a restriction on weight update imposed for the minimum count of characters in the ASM-based neuron can be utilized. The synaptic loads and test patterns are used as input from the trained NN to our processing engine. In Verilog at the register transfer level (RTL), a processing engine is implemented and is linked to Synopsys Design Compiler Ultra of IBM at 45 nm technology. This is always utilized to compute the exhaustion of energy and area under constant velocity circumstances. Table 11.5 shows the number of typical layers for the benchmarks used.

11.5 CASE STUDY 2: EFFICIENT ANN USING APPROXIMATE MULTIPLY-ACCUMULATE BLOCKS

11.5.1 INTRODUCTION

When the total count of neurons in ANN raises, the consumption for the area of similar design rises using a different approach. Hence, restriction on applications on design plan of action, such as FPGA, will limit the number of computing resources [18]. To minimize the complexity of the ANN hardware, it was shown the weighted ANNs can be calculated to involve the small count of non-zero digits in instructs, and thus a small count of subtractions and adders can be used to realize their multiplication by input variables [19]. Also, floating-point weights are dynamically quantified in every layer, and also the fixed-point loads in binary characterization are presented in a delay efficiency framework using accumulators, and carry-save adders are implemented in Nedjah et al. [20] to decrease the relative latency of a MAC module. MAC modules are recently used in the implementation of neuro-morphic cores by utilizing a double model, namely dendritic-based and axonal. The structured evaluation of ANN designs by using MAC modules over FPGAs is described in Nedjah et al. [20]. In Venkataramani et al. [11], a post-preparation models and a design without the usage of multiplier technique are provided, which can decrease the design complication of a time-multiplexed ANN.

Considering that the area used by floating-point addition and multiplication operations is larger than that of integer counterparts, and requires a lot of energy, they are converted to integers, then the offset value is added, and the floating-point weight processing is found in the startup. This transformation is achieved purely by multiplying the offset and weight of each floating point by 2^q. Here, q represents the quantized digit, therefore finding a lowest number higher than the product or near value to it.

11.5.2 SMAC Neuron's Architecture

The neuron in Figure 11.12 uses the MAC block to perform neural calculations on the kth layer of the ANN. In the kth layer of ANN and a typical control module, m MAC blocks are used for calculation, where m and n represent output (or neuron) and input counts in each layer. The control module multiplies the corresponding

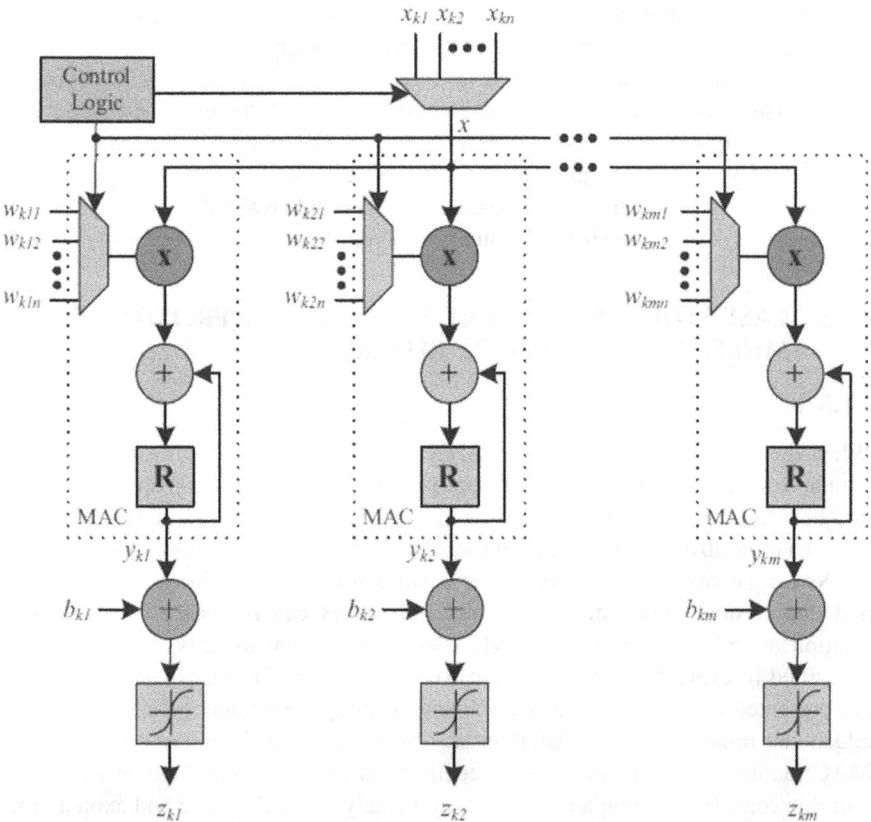

FIGURE 11.12 Neuron calculations at the kth layer using MAC blocks with ANN [Source: M. Esmali Nojehdeh 2020].

weights of the input variables. If an ANN contains λ in each layer of neurons, where the number of layers is represented by $1 \leq I \leq \lambda$ and λ, the appropriate value of MAC modules is $\sum_i^\lambda \eta_i$, which is the total count for neurons. There is a difficulty for the MAC block calculation, and register is calculated by the output and input counts of each layer of each neuron and each layer of the weight value. The complexity of the control module is evaluated by the input count on each layer. As the neuron calculation is obtained, after the calculation of the previous layer is completed, the neuron calculation of the next layer is started, layer by layer. Therefore, it can be achieved only by providing an output signal. After $\sum_i^\lambda (ii + 1)$ clock cycles, the calculation of the entire ANN for each layer input with λ layer and I is obtained, where $1 \leq I \leq \lambda$. The SMAC neuron architecture is shown in Figure 11.12.

11.5.3 THE ARCHITECTURE OF SMAC ANN

Figure 11.13 presents the ANN design with one MAC design and shows the process of ANN building a single MAC module. For consistency reasons, the reset signal and clock are not included. The control block contains three counters in the figure. In addition to the offset value for each input output, it also uses input variables to synchronize the weight multiplication and apply the activation function. The counter is related to the count of layers, the count of inputs, and outputs in every layer. Record X1, X2,... During the calculation of the primary hidden layer, the width of multiplies the preliminary inputs in the ANN, and all variables have associated weights. The multiplexer size of the input layer is calculated by the maximum input count of the entire layer, and the multiplexer size of the offset value and weight is determined by the number of offset value and weight, respectively. By using the MAX bit width of the MAC module, the width of the multiplier is calculated. Therefore, weights in the entire ANN are multiplied by the input variables. The MAX count of outputs of every layer is calculated by the number of registers, which are used for storing the results of each layer. We noticed that all ANN calculations together with the λ layer, the input of every layer and the "I" neurons of every layer, where $1 \leq I \leq \lambda$, are resulted ahead the $\sum_i^\lambda (\iota i + 2)\eta_i$ cycle [19].

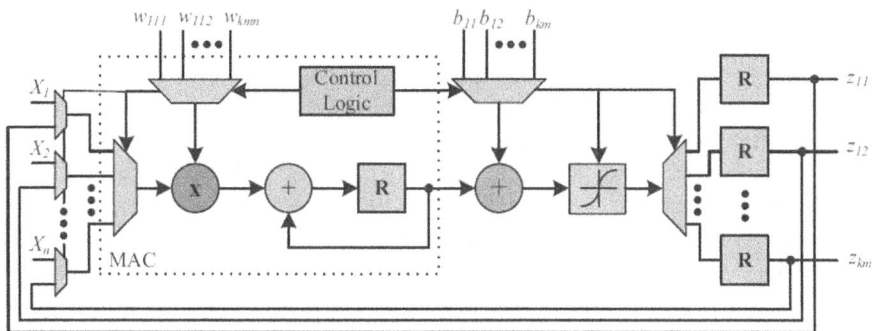

FIGURE 11.13 Designing a ANN by using simple MAC block [Source: M. Esmali Nojehdeh 2020].

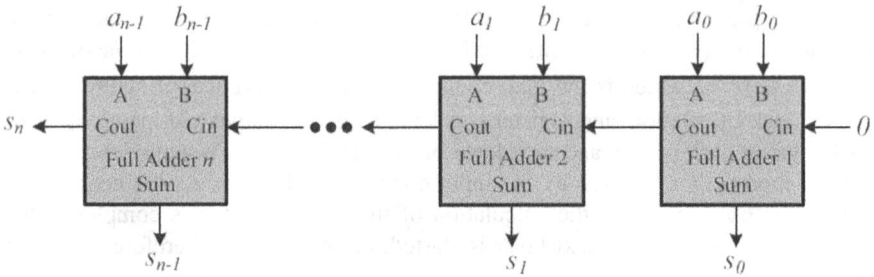

FIGURE 11.14   n-bit Ripple Carry Adder [Source: M. Esmali Nojehdeh 2020].

11.5.4   APPROXIMATE ADDER

The n-bit ripple carry adder consisting of n 1-bit full adders as shown in Figure 11.14. In this diagram, the input bits of FA are represented by carry (Cin). A and B, its output bits, are represented by sum and carry (Cout). Table 11.6 presents the 1-bit FA reality table. In the relevant analysis on the estimated ripple adder [21,22], it is presumed that the simultaneous errors of the FA's sum and cout outputs are likely to result in significant errors. The output of the adder is the result, not one output error. Nevertheless, the statement ignores the reality that although an error in among the outputs of the FA block, it raises the error that is in the output of the adder; the error which is in the output of the adder can be decreased by another error in the other output. As seen in Table 11.6, for example, on the ABCin = 010 entry of the approximately 1-bit APAD1 adder, the two errors on the FA output only produce a magnitude 1 error [19]. Thus, changing errors on the cout and sum, output will allow simplifying the convolution of approximately 1-bit adder hardware. In accordance with this reality, four approximate 1-bit adders (APAD) of distinct error digits and different hardware complications were involved in this. The truth table of all the pads is given in Table 11.6. In order to acquire the n-bit approximate ripple carry adder, an integrated procedure was also proposed in Nojehdeh et al. [19], which replaces the precise FA with APAD with different approximate levels [19,23].

11.5.5   APPROXIMATE MULTIPLIER

The realization of the precision multiplier includes two stages, namely, the use of "AND" gates to generate partial products and the use of half adders (HAs) 1 and FA to accumulate these partial products, shown in Figure 11.15, where the rectangular modules of 3 and 2 words represent FA and HA accordingly, the same 4-bit unsigned multiplier module is shown. Based on the likelihood of logic 0 and logic 1 occurring on the output of each HA and each FA, when constructing an approximate multiplier, the synthesis tool [19] takes note of the error on the output of the multiplier and substitutes the actual HA and FA blocks with their approximate forms in the exact multiplier. It create an estimated multiplier called an approximate multiplier based on chance (PBAM). Similarly, an estimated multiplier is created by the CGP method [24], which is obtained from the accurate multiplier.

TABLE 11.6
Truth Tables of Exact and Approximate 1-Bit Adders

Inputs A BCin	FA			APAD1				APAD2				APAD3				APAD4			
	Cout	Sum	Decimal	Cout	Sum	Error	Decimal	Cout	Sum	Error	Decimal	Cout	Sum	Error	Decimal	Cout	Sum	Error	Decimal
0 0 0	0	0	0	0✓	0✓	0	0	0✓	0✓	0	0	0✓	0✓	0	0	0✓	0✓	0	0
0 0 1	0	1	1	0✓	1✓	0	1	0✓	1✓	0	1	0✓	1✓	0	1	0✓	0⊗	−1	0
0 1 0	0	1	1	1⊗	0⊗	+1	2	0✓	1✓	0	1	0✓	1✓	0	1	0✓	1✓	0	1
0 1 1	1	0	2	1✓	0✓	0	2	0⊗	1⊗	−1	1	0⊗	1⊗	−1	1	0⊗	1v	−1	1
1 0 0	0	1	1	0✓	1✓	0	1	1⊗	0⊗	+1	2	1⊗	0⊗	+1	2	1⊗	0⊗	+1	2
1 0 1	1	0	2	1✓	0✓	0	2	1✓	0✓	0	2	1✓	0✓	0	2	1✓	0✓	0	2
1 1 0	1	0	2	1✓	0✓	0	2	1✓	0✓	0	2	1✓	1⊗	+1	3	1✓	1⊗	+1	3
1 1 1	1	1	3	1✓	1✓	0	3	1✓	1✓	0	3	1✓	1✓	0	3	1✓	1✓	0	3

[Source: M. Esmali Nojehdeh 2020]

$$
\begin{array}{cccccccc}
 & & & & a_3b_0 & a_2b_0 & a_1b_0 & a_0b_0 \\
 & & & a_3b_1 & a_2b_1 & a_1b_1 & a_0b_1 & \\
 & & a_3b_2 & a_2b_2 & a_1b_2 & a_0b_2 & & \\
 & a_3b_3 & a_2b_3 & a_1b_3 & a_0b_3 & & &
\end{array}
$$

$$
\begin{array}{cccccccc}
a_3b_3 & a_3b_2 & s_4 & a_3b_0 & s_2 & s_1 & a_0b_0 \\
 & a_2b_3 & c_3 & s_3 & c_1 & & \\
 & c_4 & & c_2 & & &
\end{array}
$$

$$
\begin{array}{cccccccc}
a_3b_3 & s_9 & s_8 & c_6 & s_6 & s_1 & a_0b_0 \\
c_9 & c_8 & c_7 & s_7 & & & \\
c_{12} & c_{11} & c_{10} & & & &
\end{array}
$$

$$
c_{13} \quad s_{13} \quad s_{12} \quad s_{11} \quad s_{10} \quad s_6 \quad s_1 \quad a_0b_0
$$

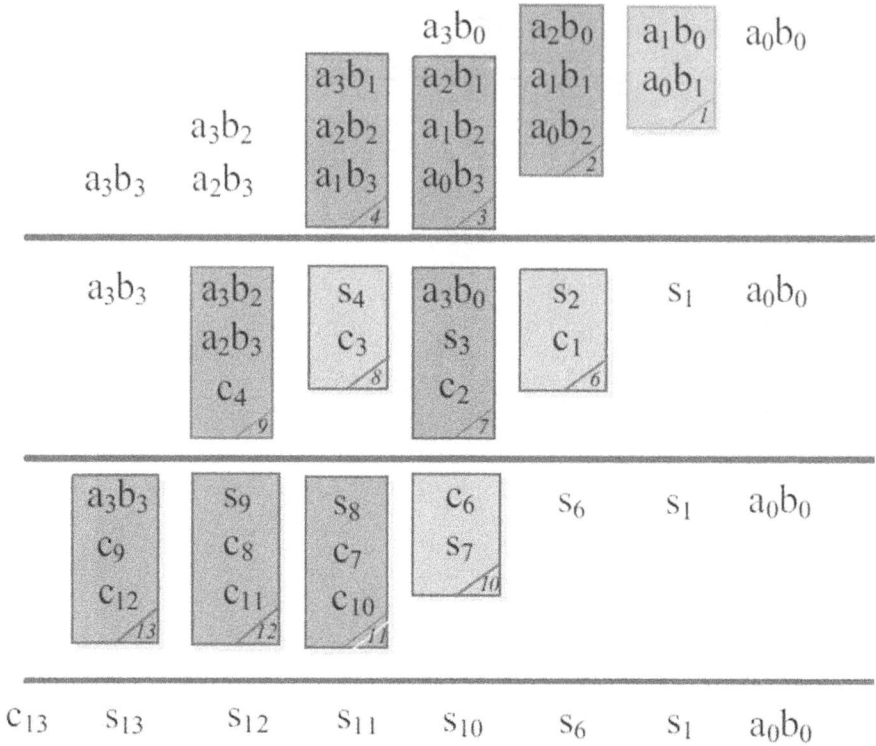

FIGURE 11.15 Exact 4-bit unsigned multiplier [Source: M. Esmali Nojehdeh 2020].

Along with the approximate multipliers, another method is also proposed called LEBZAM [19], which is achieved by setting the r least effective output of the accurate multiplier to zero. Here, r represents their approximate level. There are different algorithms proposed for the implementation of approximate multipliers and compressors used in multipliers [25–27].

The synthesis procedure is as follows:

i. Adjust the r least significant outputs of the accurate multiplier to 0.
ii. Remove every FA and HA module needed to achieve the lowest effective output of the precision multiplier.

Figure 11.16 shows the implementation of a 4-bit approximate multiplier where the value of r represents 3. Therefore, under the architecture introduced in Section 11.3, by utilizing approximate multipliers along various widths and approximate amounts in the MAC modules designed by the ANN, hardware accuracy can be considered, thereby leading to greatly reduced hardware complexity of the ANN. Likewise, by utilizing the approximate adder, the hardware complexity of ANN will be reduced ahead [19].

$$a_3b_0$$

$$a_3b_1 \quad a_2b_1$$

$$a_3b_2 \quad a_3b_2 \quad a_2b_2 \quad a_1b_2$$

$$a_3b_3 \quad a_2b_3 \quad a_1b_3 \quad a_0b_3$$

$$a_3b_3 \quad a_3b_2 \quad s_2 \quad a_3b_0$$
$$a_2b_3 \quad c_1 \quad s_1$$
$$c_2$$

$$a_3b_3 \quad s_5 \quad s_4$$
$$c_5 \quad c_4 \quad c_3$$
$$c_7 \quad c_6$$

$$c_8 \quad s_8 \quad s_7 \quad s_6 \quad s_3 \quad 0 \quad 0 \quad 0$$

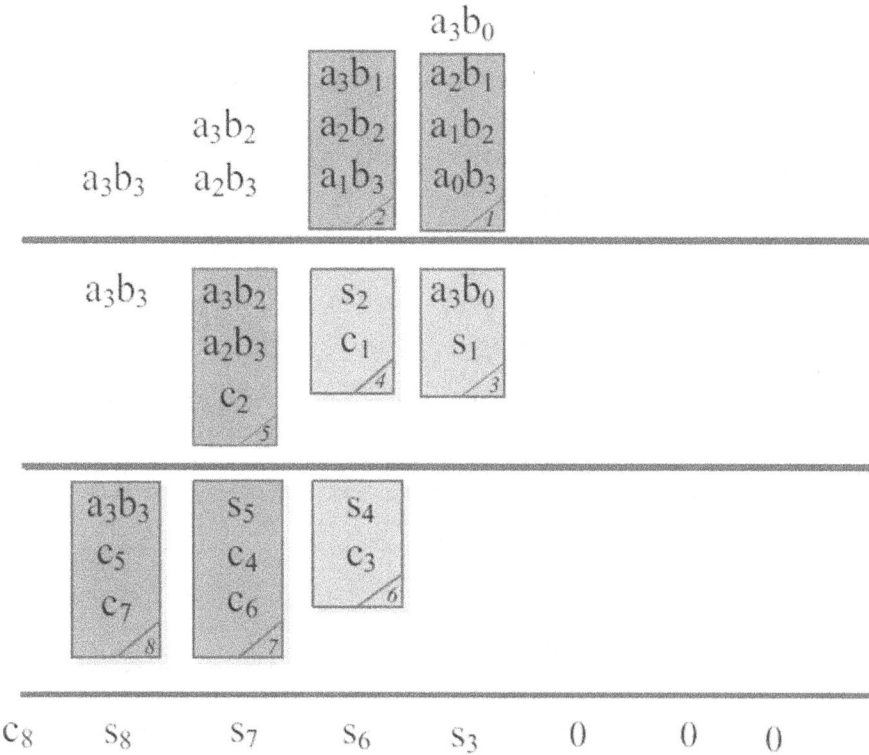

FIGURE 11.16 Approximate 4-bit unsigned multiplier with the least significant 3 bits are set to logic value 0 [Source: M. Esmali Nojehdeh 2020].

11.5.6 RESULTS AND ANALYSIS

We addressed the issue of virtually penned text detection as an application. We introduced a feedforward ANC with 16 inputs, a hidden layer consisting of 16 neurons, and an output layer consisting of 10 neurons in the CNN architecture of the MAC. The activation functions in the hidden layer and the output layer are respectively symmetric saturated linear and SoftMax. ANN is instructed by utilizing MATLAB's deep learning toolbox. Here, the preaching and the test inputs are averaged between 1 and −1, and loads are actuated and modified at random to reduce the real response when it is wrong and to reduce the predicted response error [19] using a BP-based learning method. The ANN is designed with 7494 values and verified with 3498 data. Ahead training, the misclassification rate is 4.85% [19].

Ahead converting the floating-point weights and summing rate to numbers, when the quantization rate q, which will be fixed at 8, the ANN architecture using precise multipliers and adders is the hardware misclassification rate (HMR) is found to be 5%, and defined in a behavioral way. The design is to use estimated adders and multipliers in this research without reaching the HMR limit (5.5%). Using the approximate adder of the approximate multiplier of and our LEBZAM approximate

TABLE 11.7

Results of SMAC Neuron Architecture Using Approximate Multipliers and Adders

Multiplier Type	Approximation Level				Area	Delay	Power	Energy	Area Gain	Energy Gain
	Hidden Mul	Output Mul	Mul Add	Add						
Behavioral (Exact)	0	0	0	0	15,327	3.58	1.44	174.7	0%	0%
mul12s2NM [24]	NA	10	NA	14	11,854	3.92	0.59	78.7	22%	55%
mul12s2KM [24]	NA	9	NA	15	13,133	3.95	0.69	92.4	14%	47%
PBAM [21]	7	7	12	11	10,226	3.66	0.61	76.2	33%	57%
PBAM [21]	7	7	12	12	9798	3.64	0.61	75.7	33%	57%
PBAM [21]	7	7	12	13	9534	3.66	0.62	77.2	39%	56%
LEBZAM [19]	6	10	9	13	10,392	3.58	0.58	70.1	32%	60%
LEBZAM [19]	7	12	10	13	8801	3.61	0.55	67.3	43%	61%
LEBZAM [19]	7	11	10	14	8989	3.61	0.52	63.6	41%	64%

[Source: M. Esmali Nojehdeh 2020]

multiplier [18], ANN is carried out under SMAC ANN and SMAC neuron architectures. The12s 2nm and mul12s 2KM approximate multipliers provide 12-bit input, and among other multipliers, the minimum area consumption and minimum error are chosen. Please note that the estimated amounts of the multipliers and adders on the hidden layer and the output layer are calculated manually, taking into consideration the HMR limit rate, and give the ANN design yield with the desired computational overhead and the important values. Then, explain the specification in Verilog, and synthesize it with the Cadence Genus platform and TSMC 40 nm model library [19].

Tables 11.7 to 11.8 show the gate-level yield of the ANN design. Here, the power, delay, and area represent the total area (μm^2), the delay over critical path (unit is ns), and the total consumption of power (mW). The latency represents the time (in ns), and it is essential to obtain the ANN output after the input is applied. It is set on as the multiple of the clock width multiplied to obtain the ANN output. To acquire an ANN output by using SMAC ANN and SMAC neuron, the number of clock cycles that are required are calculated as 34 and 468 of ANN, respectively. Furthermore, energy represents consumption of energy expressed in pJ, which is the product of waiting time multiplied by power consumption. We noticed that using the technique of retiming in the synthesis tool can iteratively improve the clock cycle. The test data in the simulation was used to generate the switching activity data needed to calculate the power consumption [19].

To check the ANN specification, this test data set is also used. Under the SMAC neuron architecture, Table 11.7 lists the gate-level effects of the ANN design, where the exact multiplier in the MAC block is only replaced by estimated values. It is noted that the estimated multipliers have been designed for a fixed scale, so an ANN architecture containing all the multipliers can have more energy consumption, delay, and area parameters than an ANN that uses specific differential multipliers

TABLE 11.8

Results of SMAC ANN Architecture Using Approximate Multipliers and Adders

Multiplier Type	Approximation Level		Area	Delay	Power	Energy	Area gain	Energy gain
	Mul	Add						
Behavioral (Exact)	0	0	3180	3.52	0.35	569.3	0%	0%
mul12s 2NM[24]	NA	13	2908	3.40	0.25	391.6	9%	31%
mul12s 2KM[24]	NA	13	3140	3.68	0.26	451.5	1%	21%
PBAM [21]	7	10	2972	3.55	0.26	426.6	7%	25%
PBAM [21]	8	9	2978	3.59	0.25	421.9	6%	26%
PBAM [21]	7	11	3029	3.84	0.25	448.5	5%	21%
LEBZAM [19]	6	14	3046	3.53	0.28	469.8	4%	17%
LEBZAM [19]	7	12	3041	3.62	0.26	440.2	4%	23%
LEBZAM [19]	7	13	3021	3.53	0.26	426.7	5%	25%

[Source: M. Esmali Nojehdeh 2020]

relative to these multipliers. This is also since optimized precision multipliers and adders are used in the logical synthesis tool. Using the estimated multiplier of, on the other side, will reduce the hardware complexity of the ANN by finding the required degree of approximation of the multiplier in the output layer and the hidden layer. Furthermore, the largest reduction in area, latency, and power usage comes from our estimated multiplier. Note that the trade-off between computational overhead and SMAC exact could also be transverse by changing the estimated level of multipliers [19].

Under the SMAC neural structure, Table 11.7 displays the gate-level effects of the SMAC neuron design, in which the exact multipliers and adders in the MAC block are displaced by approximate multipliers and adders. An approximate adder and an approximate multiplier are used to significantly minimize the complexity ANN's hardware. The maximum increases in area and consumption of energy are 43% and 64%, respectively, using the proposed approximate multipliers. The gate-level effects of the ANN specification by using the SMAC ANN architectonic are seen in Table 11.8, in which only the exact multiplier is substituted with an estimated multiplier in the MAC block. We require a single multiplier be reacquired, and the estimated multiplier preferred results in the highest region of gain and power usage. In addition, the use of the approximate adder will minimize the complexity of hardware, as seen in Table 11.8. It should be noted that using the approximate multipliers and adders will also expand the hardware, which could be noticed from the outputs [19].

Figure 11.17 shows the graph to represent the SMAC neuron to SMAC ANN with gain in percentages for replacing accurate designs with approximate design. From Figure 11.17, by replacing accurate adders and multipliers with approximate adders and multiplier designs, respectively, the gains in area, delay, power, and

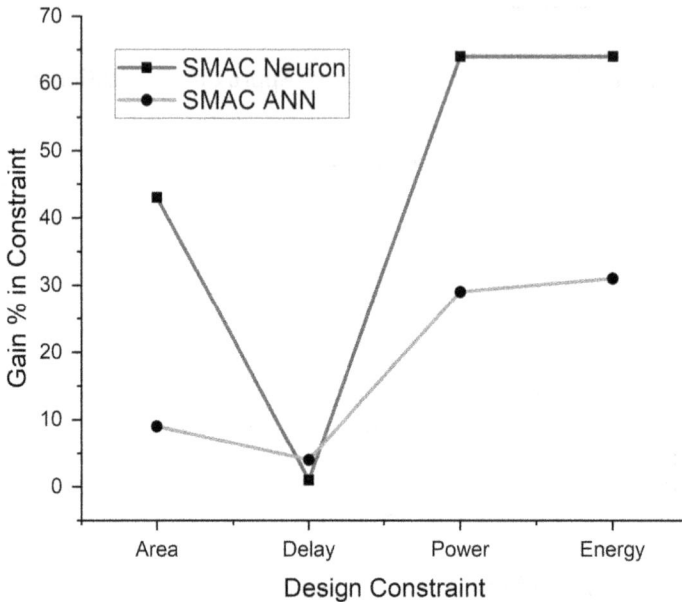

FIGURE 11.17   Graph to represent gain in constraint for area, delay, power, and energy.

energy for SMAC neuron were from 1% to 64%, and gains in SMAC ANN were from 4% to 31%, respectively. Hence, efficient design of ANN was chosen by approximate arithmetic circuits.

11.6   CONCLUSION

ANNs are one of the most well-established machine learning techniques that have a great scope of applications in approximate or estimated fields. Here, approximate computing is embedded to implement an NN model to increase the efficiency of the design. Hence, some of the machine learning algorithms are classified for better latency using approximation techniques. The arithmetic circuits are replaced with approximate modules to design an efficient SMAC ANN and SMAC neuron where the results show that the area, delay, power, and energy can be effective from 4% to 64% with the proposed models. Two types of arithmetic blocks were explained, which are multiplier less design of ANN and approximate multiply accumulate block. The results of alphabet set multiplier were compared with the size of synapses with the size of CNNs with 4,2,1 alphabet NN, respectively. One more result shows the comparison of SMAC neurons with different approximate multipliers and adders, which are effective in energy saving as per the earlier values. SMAC ANN and SMAC neuron implementations are energy efficient in different approximate levels of adders and multipliers. As large-scale NN gains great interest with complexity, there should always be a reduced or optimized technique to represent the efficient design of ANNs.

REFERENCES

[1] V. Kumar, R. Kant. 2019. "Approximate Computing for Machine Learning," in C. Krishna, M. Dutta, and R. Kumar (eds), *Proceedings of 2nd International Conference on Communication, Computing and Networking. Lecture Notes in Networks and Systems*, vol. 46. Springer, Singapore. doi: 10.1007/978-981-13-121 7-5_59.

[2] S. Mittal. 2016. "A Survey of Techniques for Approximate Computing," *ACM Computing Surveys*, vol. 48, no. 4, pp. 1–33, Mar. doi: 10.1145/2893356.

[3] Z. Yang, A. Jain, J. Liang, J. Han, F. Lombardi. 2013. "Approximate XOR/XNOR-Based Adders for Inexact Computing," *13th IEEE International Conference on Nanotechnology (IEEE-NANO 2013)*, Beijing, pp. 690–693. doi: 10.1109/NANO.2 013.6720793.

[4] H. Junqi, T. N. Kumar, H. Abbas, F. Lombardi. 2017. "Simulation-Based Evaluation of Frequency Upscaled Operation of Exact/Approximate Ripple Carry Adders," *2017 IEEE International Symposium on Defect and Fault Tolerance in VLSI and Nanotechnology Systems (DFT)*, Cambridge, pp. 1–6. doi: 10.1109/DFT.2 017.8244437.

[5] P. J. Braspenning, F. Thuijsman, A. J. M. M. Weijters. 1991. "Artificial Neural Networks," Heidelberg, Germany: Springer. doi: 10.1007/BFb0027019.

[6] M. E. Nojehdeh, M. Altun. 2020. "Systematic Synthesis of Approximate Adders and Multipliers with Accurate Error Calculations," *Integration*, vol. 70, pp. 99–107. doi: 10.1016/j.vlsi.2019.10.001.

[7] Gopinath Rebala, Ajay Ravi, Sanjay Churiwala. 2019. "An Introduction to Machine Learning," Heidelberg, Germany: Springer International Publishing, Aug. doi: 10.1 007/978-3-030-15729-6.

[8] Miroslav Kubat. 2017. "An Introduction to Machine Learning, Second Edition," Heidelberg, Germany: Springer International Publishing, Sep. doi: 10.1007/ 978-3-319-63913-0.

[9] Zhentao Gao, Yuanyuan Chen, Zhang Yi. 2020. "A Novel Method to compute the weights of Neural Networks," *Neurocomputing*, vol. 407. ISSN 0925-2312, doi: 10.1016/j.neucom.2020.03.114.

[10] H. Younes, A. Ibrahim, M. Rizk et al. 2019. "Algorithmic Level Approximate Computing for Machine Learning Classifiers," *26th IEEE International Conference on Electronics, Circuits and Systems (ICECS)*, Genoa, Italy, pp. 113–114. doi: 10.1109/ICECS46596.2019.8964974.

[11] S. Venkataramani, S. T. Chakradhar, K. Roy et al. 2015. "Approximate Computing and the Quest for Computing Efficiency," *52nd Annual Design Automation Conference on - DAC '15*, San Francisco, California, pp. 1–6. doi: 10.1145/2744 769.2751163.

[12] S. Sivanantham. 2013. "Low Power Floating Point Computation Sharing Multiplier for Signal Processing Applications," *International Journal of Engineering and Technology (IJET)*, vol. 5.2, pp. 979–985. doi: 10.1.1.411.8328.

[13] G. Karakonstantis, K. Roy. 2007. "An Optimal Algorithm for Low Power Multiplierless FIR Filter Design using Chebychev Criterion," *IEEE International Conference on Acoustics, Speech and Signal Processing - ICASSP '07*, Honolulu, HI, pp. II-49–II-52. doi: 10.1109/ICASSP.2007.366169.

[14] Jongsun Park, Hunsoo Choo, K. Muhammad et al. 2000. "Non-adaptive and Adaptive Filter Implementation Based on Sharing Multiplication," *IEEE International Conference on Acoustics, Speech, and Signal Processing. Proceedings (Cat. No.00CH37100)*, Istanbul, Turkey, vol. 1, pp. 460–463. doi: 10.1109/ ICASSP.2000.862012.

[15] V. Mrazek, Z. Vasicek, L. Sekanina. 2018. "Design of Quality-Configurable Approximate Multipliers Suitable for Dynamic Environment," *NASA/ESA Conference on Adaptive Hardware and Systems (AHS)*, Edinburgh, pp. 264–271. doi: 10.1109/AHS.2018.8541479.

[16] S. S. Sarwar, S. Venkataramani, A. Raghunathan et al. 2016. "Multiplier-less Artificial Neurons Exploiting Error Resiliency for Energy-Efficient Neural Computing," *Design, Automation & Test in Europe Conference & Exhibition (DATE)*, Dresden, pp. 145–150. doi:10.5555/2971808.2971841.

[17] Y. Lecun, L. Bottou, Y. Bengio et al. 1998. "Gradient-Based Learning Applied to Document Recognition," *Proceedings of the IEEE*, vol. 86, no. 11, pp. 2278–2324, Nov. doi: 10.1109/5.726791.

[18] L. Aksoy, S. Parvin, M. E. Nojehdeh et al. 2020. "Efficient Time-Multiplexed Realization of Feedforward Artificial Neural Networks," *IEEE International Symposium on Circuits and Systems (ISCAS)*, Sevilla, pp. 1–5. doi: 10.1109/ISCAS45731.2020.9181002.

[19] M. Esmali Nojehdeh, L. Aksoy, M. Altun. 2020. "Efficient Hardware Implementation of Artificial Neural Networks Using Approximate Multiply-Accumulate Blocks," *IEEE Computer Society Annual Symposium on VLSI (ISVLSI)*, Limassol, Cyprus, pp. 96–101. doi: 10.1109/ISVLSI49217.2020.00027.

[20] N. Nedjah, R. M. da Silva, L. M. Mourelle et al. 2009. "Dynamic MAC-Based Architecture of Artificial Neural Networks Suitable for Hardware Implementation on FPGAs," *Neurocomputing*, vol. 72, no. 10, pp. 2171–2179. doi: 10.1016/j.neucom.2008.06.027.

[21] A. Bernasconi, V. Ciriani. 2014. "2-spp Approximate Synthesis for Error Tolerant Applications," *Euromicro Conference on Digital System Design*, pp. 411–418. doi: 10.1109/DSD.2014.21.

[22] Z. Yang, A. Jain, J. Liang et al. 2013. "Approximate XOR/XNOR-Based Adders for Inexact Computing," *2013 13th IEEE International Conference on Nanotechnology (IEEE-NANO 2013)*, Beijing, pp. 690–693. doi: 10.1109/NANO.2013.6720793.

[23] D. De Caro, N. Petra, A. G. M. Strollo et al. 2013. "Fixed-Width Multipliers and Multipliers-Accumulators With Min-Max Approximation Error," *IEEE Transactions on Circuits and Systems I: Regular Papers*, vol. 60, no. 9, pp. 2375–2388, Sept. doi: 10.1109/TCSI.2013.2245252.

[24] V. Mrazek, R. Hrbacek, Z. Vasicek et al. 2017. "EvoApprox8b: Library of Approximate Adders and Multipliers for Circuit Design and Benchmarking of Approximation Methods," *Design, Automation & Test in Europe Conference & Exhibition (DATE)*, Lausanne, pp. 258–261. doi: 10.23919/DATE.2017.7926993.

[25] C. Lin, I. Lin. 2013. "High Accuracy Approximate Multiplier with Error Correction," *2013 IEEE 31st International Conference on Computer Design (ICCD)*, Asheville, NC, pp. 33–38. doi: 10.1109/ICCD.2013.6657022.

[26] N. N. Petra, D. De Caro, V. Garofalo et al. 2011. "Design of Fixed-Width Multipliers With Linear Compensation Function," *IEEE Transactions on Circuits and Systems I: Regular Papers*, vol. 58, no. 5, pp. 947–960, May. doi: 10.1109/TCSI.2010.2090572.

[27] Chip-Hong Chang, Jiangmin Gu, Mingyan Zhang. 2004. "Ultra Low-Voltage Low-Power CMOS 4-2 and 5-2 Compressors for Fast Arithmetic Circuits," *IEEE Transactions on Circuits and Systems I: Regular Papers*, vol. 51, no. 10, pp. 1985–1997, Oct. doi: 10.1109/TCSI.2004.835683.

12 Hardware Realization of Reinforcement Learning Algorithms for Edge Devices

Shaik Mohammed Waseem and Subir Kumar Roy
VLSI Systems Research Group, IIIT Bangalore, Bengaluru, Karnataka, India

CONTENTS

12.1 INTRODUCTION

12.1.1 REINFORCEMENT LEARNING AND MARKOV DECISION PROCESS

A computational approach to modeling the interaction of an agent (such as a robot) with its environment (such as its surroundings in which it funtions or operates) and the use of these interactions to modify the agent's functional behavior (called its actions) through the maximization of a notional metric termed a "return", is often the basis of reward-based learning. An interesting real-life example is that of a child trying to wave hands, stand up, walk, and learn according to the surroundings while trying to maximize his or her reward/learning with time. In a way, reinforcement learning (RL) is goal-focused learning of an agent while interacting with an

DOI: 10.1201/9781003201038-12

environment. A more formal definition of "reinforcement learning" has been stated by R. Sutton and A. Barto in [1] – *"Reinforcement learning is learning what to do – how to map situations to actions – so as to maximize a numerical reward signal. The learner is not told which actions to take, but instead must discover which actions yield the most reward by trying them out"*.

In some interesting applications of RL, such as in game theory – for example, in GO (a board game) – actions of an agent might also impact subsequent future rewards rather than only the immediate ones. In order to predict such behavior, a delayed reward strategy is additionally considered while designing RL algorithms through trial-and-error search, and this unique feature distinguishes it from other methods and modes of learning in the context of artificial intelligence (AI) and machine learning (ML) [2]. In the context of RL, the *agent* could be generally referred to as a learner or a decision maker, whereas the *environment* is a platform where an agent learns and decides what actions it needs to perform. For each action performed by an agent in the environment, it changes its *current state* to a *next state* and also concomitantly collects an appropriate *reward* (usually a numerical value) corresponding to its new state. The concept of *Environment-Agent-State-Transition-Reward* can be elegantly and more formally modeled by a finite Markov decision process (MDP), as illustrated in Figure 12.1. As can be seen, the agent-environment interaction happens in discrete time steps. At each time step (t = 0, 1, 2........ n), the agent receives the environment's state S_t. This state information can be in the form of a numerical value or a signal derived from all the available states in "S" – a collection or a superset of the all states modeling the environment under consideration. After selecting an available action A_t corresponding to state S_t from the superset "A", the agent moves to next state in the environment S_{t+1}, which is again a part of superset "S" and collects reward R_{t+1}, which is part of superset "R". This

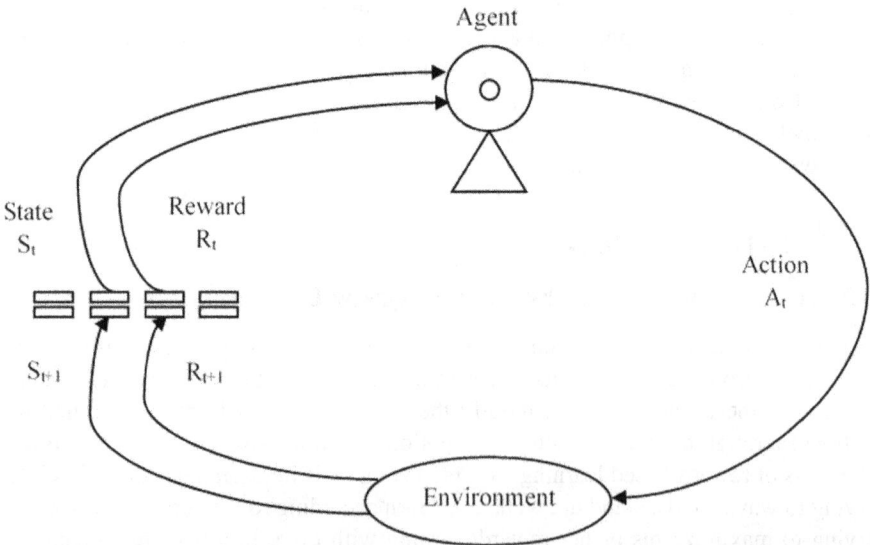

FIGURE 12.1 Markov decision process (agent-environment interface).

process continues in an iterative manner to generate a path given in equation (12.1). For an in-depth mathematical explanation on the use of finite MDP and its generalization to model RL, one can refer to Sutton and Barto [1] or Szepesvari [2].

$$S_t, A_t, R_{t+1}, S_{t+1}, A_{t+1}, R_{t+2}, S_{t+2}, A_{t+2}, R_{t+3}, S_{t+3}, A_{t+3}, R_{t+4}, S_{t+4}, A_{t+4} \ldots$$
$$(12.1)$$

12.1.2 HARDWARE FOR REINFORCEMENT LEARNING AT THE EDGE

While targeting implementation of AI and ML algorithms as specialized hardware accelerators has added advantages by way of higher throughput, lower power consumption, and ability to meet real-time constraints, there are concomitant complex design and implementation issues that need to be addressed, especially in the context of deploying algorithms that rely on deep reinforcement learning (DRL). DRL combines the power of both RL and deep neural networks (DNNs). Due to the presence of DNNs, any DRL algorithm will involve an extremely large number of parameters (generally referred to as the weights and biases in DNNs) that need to be trained on a large set of data as a part of a continuous form of learning. Updation of such a huge set of parameters periodically is indeed a formidable challenge and involves calibrated design methodologies for reading and writing memories that host these parameters. With real-time applications involving processing of extremely large datasets from numerous sensory sources (for example, in an Internet of Things, or IoT), this problem is further exacerbated and grows in complexity. To realize such applications on hardware, many techniques have been proposed in the literature. These will be discussed later in Section 12.2. To facilitate implementation of a DRL algorithm as a hardware accelerator, it is necessary to cast the various conceptual components of RL and its model in an MDP to hardware blocks in the implementation – for example, what could be an agent, an environment, and a reward in hardware scenarios/implementations. The agent can be best represented as a controller; the environment can be viewed as a plant (controlled system) and the reward as a control signal [1]. In subsequent sections, we provide details of how the environment has been represented as a collection of states with an underlying state transition graph, an agent as a hardware circuit, and a reward as read-only data stored in memory. We also give detailed architectural-level information and data flow explanations to enable the design, implementation, and validation of hardware accelarators implementing RL algorithms.

With recent advancements in portable electronic devices and IoT-based devices, the challenge has become much wider and even greater to enable the implementation of machine intelligence at the edge of these networked IoT devices. This arises due to the exceedingly large requirements of computational resources needed to accelerate DRL algorithms that cannot be provided by current VLSI/SoC technologies, design architectures, and design methodologies [3]. In such scenarios, specialized hardware architectures can play a key role in demonstrating feasibility of implementing such computational and memory-hungry algorithms.

Currently, the most commonly used hardware to accelerate DRL algorithms is graphics processing units (GPUs). While they are very good at compatibility and performance, GPUs unfortunately consume an inordinately high amount of power and energy. One good example of such GPUs could be NVIDIA's Turing Architecture based GPU [4]. Application-specific integrated circuit (ASIC) alternatives in the form of Google Edge Tensor Processing Unit (Edge TPU) coprocessor, available as part of Coral Development Board [5], Intel's Neural Compute stick 2 hosting Intel Movidius Myriad X Vision Processing Unit (VPU) [6], and Nvidia's Jetson Nano developer kit [7] are a few popularly available choices on the market with the capability to accelerate deep learning algorithms at the edge. Some other hardware we generally come across as part of our mobile electronic devices for edge computing include the famous System on Chips (SoCs), such as the Snapdragon 8 series from Qualcomm [8], Exynos 9820 from Samsung [9], Helio P60 from Mediatek [10], and Ascend series from HiSilicon [11].

Though there are several edge-based computing frameworks that have been developed by companies like Baidu (OpenEdge) [12], Microsoft (Azure IoT Edge) [13], and Amazon (AWS IoT Greengrass) [14], the work is continuously expanding to accommodate more flexible and compatible frameworks and make possible computing at the edge for several day-to-day applications and create a considerable impact on solving several societal issues and problems. There exist several deep learning libraries with support for edge computations. A few of the popular ones include PyTorch from Facebook [15], Apple's CoreML [16], Qualcomm's SNPE [17], Paddle-Mobile from Baidu [18], and MACE from XiaoMi [19]. Google has two versions as part of deep learning libraries, Tensorflow [20] is heavy weight and therefore unsuitable and generally not recommended for the edge. A lighter version called Tensorflow Lite [21] has been developed by Google for mobile/edge devices. Though these libraries have established themselves as potential choices, there exist several other open-source deep learning libraries that can serve equally well and therefore are available for consideration.

Field-programmable gate arrays (FPGAs), on the other hand, continue to remain an important and feasible choice amongst academic and industry researchers, as they have the ability to develop specialized hardware architectures targeted towards the edge without incurring the Silicon technology processing and fabrication costs needed for a fully ASIC implementation. Though FPGAs offer higher energy savings and distributed computing resources, they however lag and lack in compatibility when compared to GPUs. This is because the RL algorithms rendered in high-level programming languages can be directly compiled into central processing unit (CPU) and GPU instruction sets, the same needs to be behaviorally synthesized into gate-level netlists for implementations in an FPGA fabric. This is a time-consuming process in current generation FPGA design flows. However, modern FPGA platforms and fabrics with multi-core processors fabricated along with unconfigured core logic (for implementation of specialized hardware accelerators) can render multi-functional embedded systems with real-time processing units, higher data transfer capabilities, and the ability to implement advanced communication protocols. For our proposed approaches to the design and implementation of hardware accelerators for RL algorithms, we will present extensive data and results for Xilinx FPGA fabrics and platforms.

The chapter consists of several sections organized as follows. Section 12.2 gives a brief background about the algorithms, techniques, and hardware architectures available in the literature for RL. Section 12.3 illustrates our proposed hardware architecture for the simple reinforcement learning (SRL) algorithm, with architecture-level description and data flow descriptions across the modules. Implementation and simulation results, along with performance data considering a few key metrics obtained from our proposed hardware architecture, are presented in Section 12.4. In this section, we also provide a comparative analysis of the available hardware implementations for Q-learning as reported in the literature. Section 12.5 discusses several applications of the Q-learning and SRL algorithms. As part of future work, illustration of an autonomous robot for agriculture/farming industry is provided at both the hardware architectural and application level in Section 12.6. Section 12.7 concludes the chapter.

12.2 BACKGROUND

Designing hardware to accelerate RL algorithms has been an active area of research, with many engineers and scientists continuously proposing several architectures and finding different ways to make possible their implementation at the edge. After a proposal in 1989 by C. Watkins on Q-learning, a technical note by C. Watkins and P. Dayan further provided an in-depth view on Q-learning in the early 1990s [22], by detailing a convergence theorem and showing how the Q-learning algorithm provides an optimal path, as long as all discrete actions are repeatedly sampled in all states while discussing the Markov environments. Liu and Elhanany [23] proposed a pipelined hardware architecture that significantly reduces delay caused by action selection and value function updates. They provided a set of formal proofs related to reduced delays due to their approach. The proposed approach enabled the authors to mainly focus on application of Q-learning to large-scale or continuous action spaces.

Hwang et al. [24] proposed a hardware realization for a multilevel behavioral robot that can execute complex tasks in an autonomous fashion and called it Modular Behavioural Agent (MBA) with learning ability. The realization was made by considering a template that embeds an RL mechanism with a critic–actor model, and the proposed architecture was implemented on an FPGA that hosted a CPU core. Their hardware setup demonstrated with examples the ability of goal-seeking robots to reach their destinations in unstructured environments. In the literature, much of current research work continues to focus on RL algorithms and ways to accelerate them using specialized hardware architectures for edge applications. Though this comes with certain challenges, a necessary vision is being framed by both academia and the industry to address them.

Shi et al. [25] provided an introduction to the definition of edge computing along with several case studies to materialize the concept. Several challenges and opportunities were discussed as part of processing the data at the edge of the network. It is indeed essential to look at the advantages that come up with edge deployment, ranging from privacy, data security, data bandwidth savings, response time, etc.

Several applications that are part of the DRL paradigm (a combination of both RL and DNNs) are very difficult to deploy at the edge as they require huge computational resources, demand higher power, and are not portable in the majority of cases. As such, a convergence for both deep learning and edge computing has emerged as a topic of consideration.

Wang et al. [26] provided a comprehensive survey of how smart devices that generate huge amounts of data need to be processed at the edge without much delay in response. As part of the survey, the authors illustrated application scenarios of realizing an intelligent edge with a customized computing framework. With the aim of facilitating AI in the day-to-day life of humans, efforts are being made by both the hardware and software communities to minimize the cost of edge devices and make them more efficient.

Designing hardware architectures for DRL algorithms targeted at FPGAs has been a major area of interest with researchers proposing novel solutions for model-free environments and policy optimization-based algorithms. One such solution is proposed by Hyungmin Cho et al. [27]. They presented an FPGA-based asynchronous advantage actor–critic (A3C) DRL platform called FA3C in which they demonstrated its advantages with respect to performance and energy efficiency when compared to an implementation based on a NVIDIA Tesla P100 GPU platform. In both the implementations, the A3C agents were programmed to learn the control policies of selected Atari games.

With diverse problems that arise in robotics, distributed control, and a variety of other domains, multi-agent systems (MAS) could be used to propose solutions, wherein the tasks, depending upon the necessity, are handled by multiple agents. This, when done in conjunction with RL algorithms, allows researchers to propose solutions based on strategies like fully cooperative, fully competitive, and general (neither competitive nor cooperative) tasks. A complete overview of MAS and Multi Agent Reinforcement Learning (MARL) is provided as part of the technical report by Busoniu et al. [28] along with an example of coordinated transportation of an object by two robots while discussing cooperative robotics. Though realization of hardware for MARL algorithms comes with a challenge of parallelization of computing resources and fast and efficient data sharing between several on-chip modules, a considerable effort has been made by researchers in this direction. There is also a proposed algorithm by Matta et al. [29] on how the standard Q-learning algorithm has been extended as part of MAS to target applications like swarm robotics, which allows a knowledge sharing mechanism between multiple agents.

Several works providing information on how MARL works in different scenarios involving real-life/social dilemma and allows agents to learn and behave accordingly to accomplish tasks are available in the literature. Some of the research work reported in the literature has proven the multi-agent behavior with simulations of environment for complex tasks, while other works have experimentally shown the behavior on real hardware. Policy optimization algorithms involving multi- agent behavior in the form of actor–critic strategy are also extensively studied as part of knowledge transfer in order to mimic traditional student–teacher behavior. This indeed demands parallel computations to be handled for multi-thread/multi-core processing while allowing faster learning abilities and accomplishments.

Hardware architectures developed for such algorithm implementations must be very efficient in handling concurrent flow of data from multiple agents. Realization of hardware for MARL has been an active area of research, with challenges involving the efficient data transfer capabilities between different agents at an application level. When MARL algorithms are to be executed at the edge of the network, the challenges grow multi-fold, especially in terms of hardware resources.

Theoretical and algorithm approaches for MARL are presented in Zhang et al. [30] along with information on representative frameworks like Markov/Stochastic and extensive forms of games in accordance with scenarios involving full cooperation, full competition etc. An in-depth view of literature on Multi-agent Deep Reinforcement Learning (MDRL) is provided in Hernandez-Leal et al. [31] with an analysis on key components of MAS and RL. The practical challenges for implementing MDRL are also discussed to highlight computational resources, real-hardware implementations, etc.

A more practical approach of automating the control of a real robotic arm is discussed in Shao et al. [32], where the entire approach has been explained in two steps, with the first being the design space exploration techniques developed to enhance performance of an FPGA-based accelerator for RL policy training (Algorithm-Trust Region Policy Optimization (TRPO)) and the second being the transfer of trained RL policy to a real robotic arm. The feasibility of the approach was demonstrated by validating the entire experiment with a real-time task of picking pre-defined objects. Comparison of the approach was also made by authors, with state-of-the-art GPUs and CPUs to establish the advantages in terms of speed. Further, a good note is provided in Schwartz [33] on swarm intelligence and the evolution of personality traits, with discussion on simple simulation scenarios like "Robot leaving a room", and challenging ones like "several robots moving around a target in which the position of the target keeps changing from one simulation to the other". In such a scenario, the robots perceive the environment to be like a Gaussian potential field, where each single robot is able to recognize the target potential field along with the remaining robot fields and the base field. Approaches for multi-agent robotics (behavior based, collective, and evolutionary) along with different case studies on MARL, adaptation in multi-agent scenarios, and multi-agent self-organization are presented in Liu et al. [34].

12.3 HARDWARE REALIZATION OF SIMPLE REINFORCEMENT LEARNING ALGORITHM

12.3.1 ARCHITECTURE-LEVEL DESCRIPTION

As discussed in the earlier sections, for Markov decision processes (MDPs) such as shown in Figure 12.1, the agent in a particular state takes action in the environment and moves to the next state and thereby collects the rewards associated with the new state. In order to carry out general software-based simulations, there are several predefined open-source environments to run, simulate, and verify RL agent behaviors. One such example of an open-source environment is the OpenAI-Gym [35]. In this chapter, to simulate the simple reinforcement learning agent (SRLA), an environment

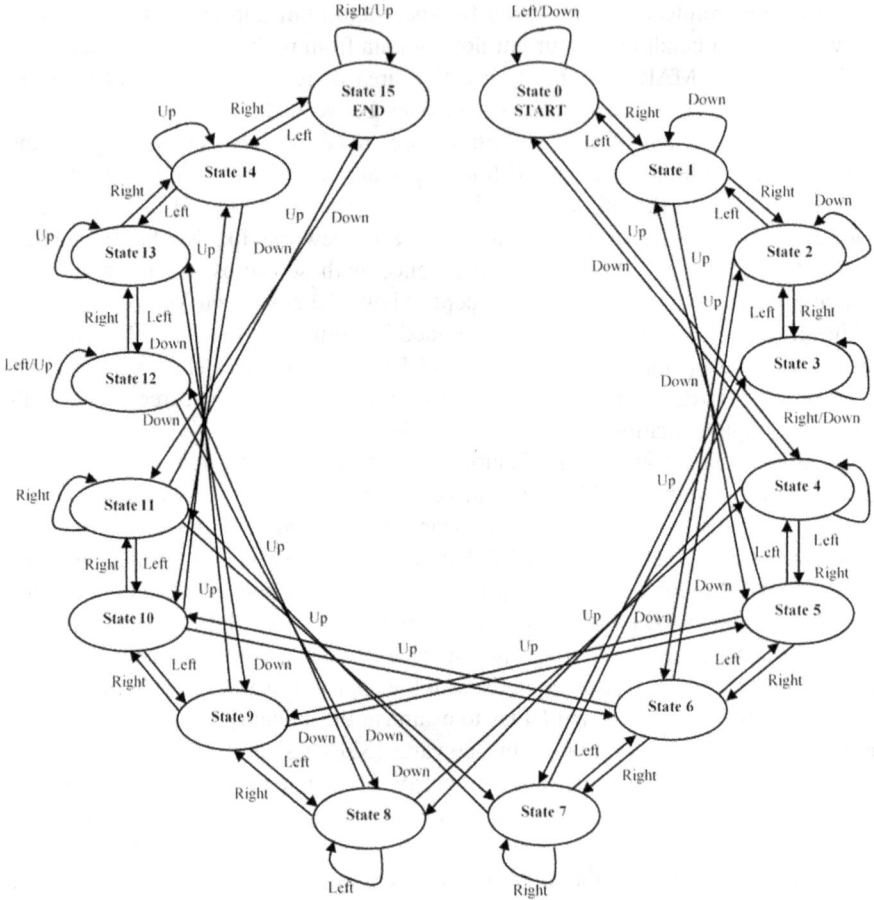

FIGURE 12.2 State diagram representing an environment with 16 states and 4 actions.

has been designed using a state transition diagram in which each state is a register entity holding a specific value according to its rewarding identity.

As shown in Figure 12.2, the environment has 16 states, with the start/initial state being *state0* and the end-state/goal being *state15*. In each state, the agent can take any of the available four actions – up, down, right, and left. The state transiton diagram models a grid-based environment in which border states such as *state0*, *state1, state2, state3, state12, state13*, etc., have only a subset of the above four actions that allows the agent to enter or transit into another state, possibly a new state. However, it is possible that the agent, with some of the actions in the subset, continues to remain in the same state it was occupying earlier. For example, consider *state0* in Figure 12.2. In this state if the agent takes either action *up* or action *right* it will end up occupying states *state4* and *state1*, respectively, while it will continue to remain in *state0* whenever it takes actions *left* or *down*. Similarly, while in *state12*, taking actions *right* and *down* will move the agent into *state13* and *state8*, respectively, whereas taking actions *up* and *left* will leave the agent in the same *state12*. Some states like *state5* will allow the agent to go into newer states for every action it takes – for example action *up* taken in *state5* will move the agent into *state9*, action *down* into *state1*, action *right* into *state6*, and action *left* into *state4*, respectively. Some other states, where every action taken will allow the agent to enter newer states, are the ones located at the center of the grid (modeling the environment), for example, *state6* and *state10*. Though a start state and an end state are provided in the state transition diagram for reference, it has been designed in such a way that it holds good even if the agent starts or ends, respectively, for any start and end state pairs.

As part of providing rewards to the SRLA for the states it traverses, a reward matrix has been designed by allotting each state with their respective reward points or values, as shown in Figure 12.3. The main aim of any RL algorithm-based agent is to collect positive reward points and maximize its cumulative reward while reaching the goal. Depending upon the traversal path taken (a collection of different states starting with the start state) for the actions executed by the SRLA in different states, those actions that provide positive rewards to the agent when a particular new next state is reached would be mapped with the new state. As the objective is to maximize the cumulative reward, this leaves room for only a single or unique action to be taken by an agent to reach a new state from any specific current state, thereby reducing the memory required to store the values for other actions in a state-action table (which is generally part of the Q-learning algorithm while populating the Q-table) as discussed in Notsu et al. [36].

The hardware architecture depicted in Figure 12.4 has two major components – SRLA and environment. The main objective of the hardware architecture targeted for rendering the SRL algorithm is to make its deployment possible in edge devices. As discussed in earlier sections and in Notsu et al. [36], the SRL algorithm is very compact as it results in the smallest memory footprint as well as a very low logic hardware footprint when implemented with the architecture given in Figure 12.4. However, one

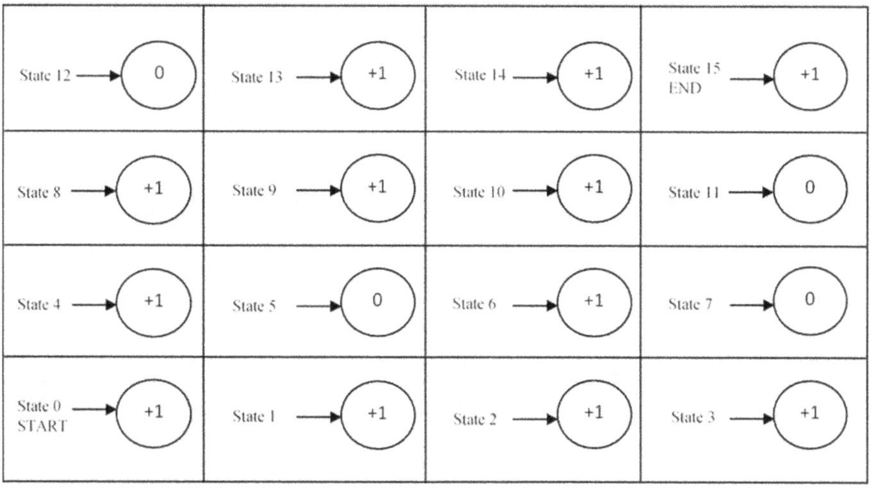

FIGURE 12.3 State-reward mapping as part of reward distribution in environment.

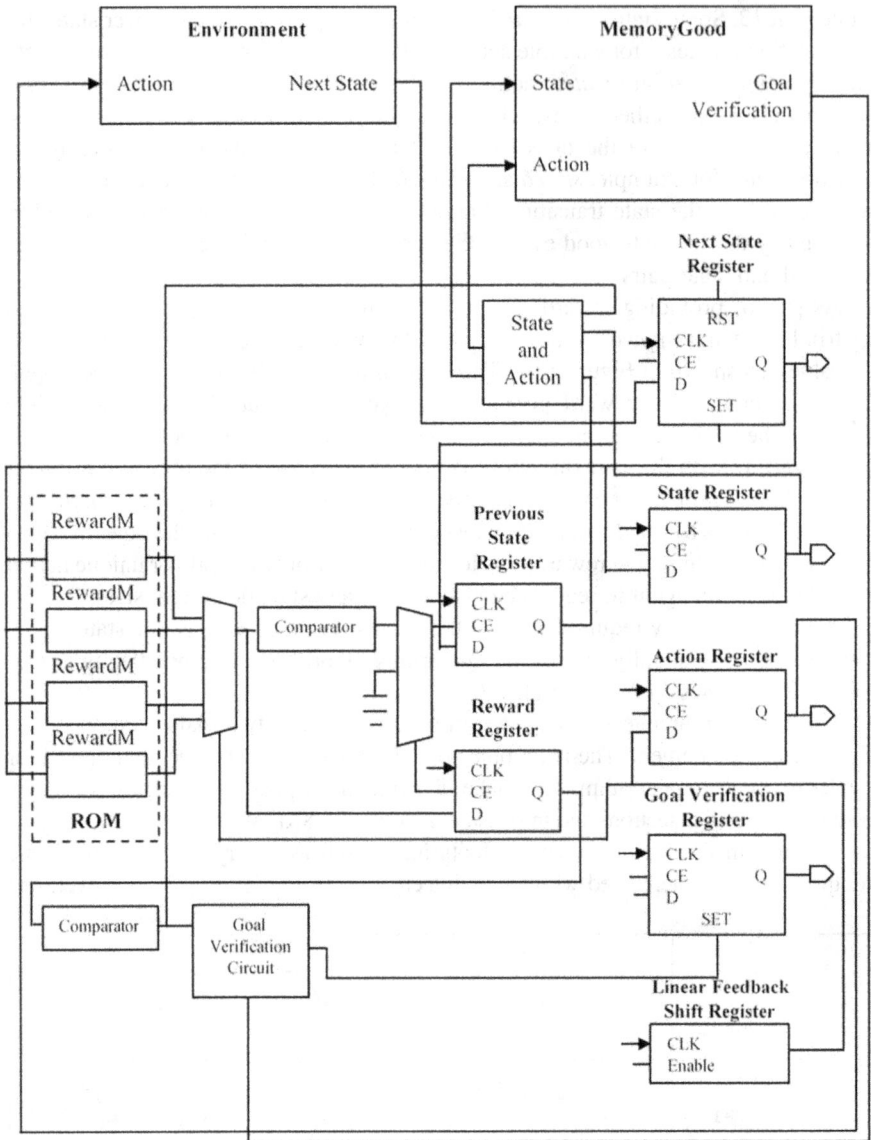

FIGURE 12.4 Hardware architecture for simple reinforcement learning algorithm.

drawback that has been reported in the literature is that it does not necessarily result in an optimal solution or path always, unlike the Q-learning algorithm. While this is certainly a disadvantage, SRL is good at providing sub-optimal solutions or paths for the RL agent that enables the goal to be reached with a lesser amount of resources. This is proven by our results and analysis given in the sections to follow.

The hardware architecture is designed with a pool of registers, Read Only Memory (ROM), memory array, and some comparators as and when necessary to mimic the functionality of the algorithm discussed in Notsu et al. [36]. The *State Register* holds value of the current state of the agent. The *Next State Register* holds the value of the new state reached by the agent for a specific action taken by it when it is in the current state. The *Action Register* is responsible for storing the value of any action that the agent chooses randomly from the set of actions applicable in the current state while moving around in the environment. In this particular implementation, four actions were considered to navigate in the environment (up, down, left, and right). A linear feedback shift register (LFSR) is employed to generate the actions randomly, as shown in Figure 12.4. The *Previous Register* shown in the architecture stores the value of the current state, once the agent moves to a new next state. This is needed to facilitate comparison and reward-based action assignment to the respective current state. *Reward Register* holds the respective reward value the agent has collected while moving to a new next state. *RewardM* is a ROM that is responsible for storing pre-determined reward values associated with each state of the environment. *Goal Verification Circuit* is responsible for verifying the correct state-actions mappings. *State and Action Circuit* is responsible for writing the respective state and action mappings to the *MemoryGood* memory array in a controlled and conditional way depending upon the types of rewards achieved by the agent while navigating in the environment. *Goal Verification Register* is a synchronous register that gets set if the conditional mapping of state and action is met as per the rewarding criteria of state and vigilance.

12.3.2 Flow of Data in the Hardware Architecture

The flow of data in the entire architecture is in the way that the action that is responsible for the agent to move from a current state to a new next state is first generated as part of the LFSR circuit from where the action register is populated and the respective value drives the state transition diagram (environment) to generate the next-state value that an agent has reached. The new next state is recorded in the *Next State Register*. The *Next State Register* value is provided as reference to the *RewardM (a ROM)* to collect the associated reward into the *Reward Register*. The reward is compared using a comparator, which populates the *Previous Register* with the current state value in case the reward collected is positive. The *State and Action Circuit* is a control logic that drives the respective state and action pair with associated positive rewards to be written in *MemoryGood*. The vigilance states that are responsible for verification are read from *MemoryGood* into the *Goal Verification Circuit* to set the *Goal Verification Register* if the conditions are met.

12.4 RESULTS AND ANALYSIS OF SRL HARDWARE ARCHITECTURE

For the architecture shown in Figure 12.4, behavioral simulations were carried out to verify the correctness of the implemented functionality. Behavioral simulations also allow estimation of hardware resource utilization along with power details. These were generated by targeting the implementation for Avnet Ultra96 v2 platform from Xilinx using the Xilinx Vivado 2019.2 software tool. Results captured as

TABLE 12.1

Comparison of Proposed Hardware Architecture for SRL with Silva et al. [37] and Spano et al. [38]

Parameter Under Consideration	Proposed Hardware Architecture for SRL (States = 16, Actions = 4)	Silva et al., Hardware Architecture for Q-Learning [37] (States = 12, Actions = 4)	Spano et al., Hardware Architecture for Q-Learning [38] (States = 16, Actions = 4)
Clock Frequency (MHz)	100	22	74
LUTs	24	3387	142
Flip-Flops/REG	28	1029	188
DSP/Multipliers	0	58	3
BUFG	1	0	0
Power (mW)	25	6	15

data in Table 12.1 prove that the proposed hardware architecture enables an implementation to be deployed as part of edge devices.

Silva et al. [37] and Spano et al. [38] proposed hardware architectures for implementation of Q-learning algorithms targeted for Xilinx Virtex-6 platforms and Xilinx Zynq Ultrascale+ MPSoC ZCU106 Evaluation kits, respectively. The authors present data related to the FPGA resource utilization, power consumption, and other important design attributes for different data formats and different number of states and actions. A tabulated comparison is provided in Table 12.1 to get an understanding and to know the advantages of the proposed SRL hardware architecture in comparison to those proposed in Da Silva et al. [37] and Spanò et al. [38]. It is worth mentioning that the implementation results provided as part of the proposed hardware architecture also include the environment as a state transition diagram, where to the best of our knowledge both the architectures being compared are concerned only about the agent's computational, memory, and other associated resource requirements. As seen, our proposed hardware architecture clearly has an advantage with respect to resource utilization. It is weaker on the power consumption parameter, which is seen to be slightly higher. This discrepancy is primarily due to the I/O pins, which were needed and realized for verification of hardware. These I/O pins consume power, which amounts to almost 96% (23 mW) of the power consumed by the entire implementation, as clearly given in the table. This could be easily reduced by design management of I/O pins for the target FPGA [39]. Hence, the proposed architecture nearly outperforms both the hardware architectures for the considered parameters, as shown in Table 12.1. The main aim for the comparison is to prove the usefulness of SRL hardware architecture to be beneficial for implementation on edge devices. The proposed FPGA-based hardware implementation also results in a good execution speed in comparison to the software-based execution of the SRL algorithm with Desktop CPU (x86-64 architecture). The execution speed is also at par with that of the architecture reported in Da Silca et al. [37]. As stated earlier, the SRL algorithm in itself doesn't guarantee an optimal path, as guaranteed by the Q-learning algorithm. This should be considered its shortcoming. From the perspective of hardware implementation on edge devices, the proposed architecture appears to do better compared to other architectures. It is the cost of non-optimality of the path generated by the agent for different applications, which needs to be taken into consideration while deciding on the choice of the hardware architectures for edge devices implementing these applications.

12.5   Q-LEARNING AND SRL ALGORITHM APPLICATIONS

Being one of the most famous RL algorithms, Q-learning has seen wide applications over a period of almost two decades, during which it has established its usefulness in domains such as robotics, financial services, medical maintenance facilities, game theory, and statistics, to name a few. This was made possible mainly due to the re-advent of neural networks and allied techniques that helped the Q-learning algorithm to provide a wide range of solutions in the form of deep Q-learning (a combination of both Q-learning and neural networks). Further extension of Q-learning algorithms to multiple agents has resulted in newer research areas being pursued to develop applications related to swarm robotics, cooperative robotics, and solutions related to multi-modal problems. A clear explanation of how the Q-learning algorithm could be used to solve real-life problems is provided in Jang et al. [40]. Many of the applications presented in Jang et al. [40] could also be realized to a large extent with the SRL algorithm targeted towards edge devices. Some of the popular single-agent Q-learning algorithms and their applications include *HiQ: A hierarchical Q-learning algorithm* with applications in solving the reader collision problem in RFID systems [41]. *Incremental multi-step Q-learning* has applications in adaptive traffic signal control with the added advantage of achieving fewer delays for variable traffic conditions, as in Shoufeng et al. [42]. Minih et al. [43] proposed and tested a novel artificial agent called the *Deep-Q network,* and it was demonstrated that the agent had the ability to play classical Atari-2600 games with the same level of control as a human being, thereby proving that the Q-learning algorithm along with neural networks could work with higher dimensional sensory inputs and actions. Multi-agent-based Q-learning algorithms like modular Q-Learning [44,45], swarm-based Q-learning [46], and semi-competitive domain mult-agent Q-learning [47] have their applications in the areas of multi-agent cooperation for robot soccer, exchange of information amongst the agents, and solving the famous prisoner's dilemma problem. Further, the ability of robots to learn from raw-pixel data without any prior knowledge was discussed in Zhang et al. [48]. The authors demonstrated the ability of a three-joint robot manipulator with externally provided visual observation to perform target reaching with a trained deep Q network (DQN). Zhu et al. [49] proposed a framework for agents based on DRL that enables agents to take actions and interact with objects while addressing multiple key issues related to the DRL algorithm, such as lack of ability to generalize and data inefficiency to name a few.

12.6 FUTURE WORK: APPLICATION AND HARDWARE DESIGN OVERVIEW

As a group, we are working towards developing hardware accelerators that could enable the RL algorithms to be deployed at the edge. As part of it, work in the direction of study of diferent algorithms and analysis of these algorithms from the perspective of hardware implementations are being carried out with respect to model-free, off-policy, and on-policy based algorithms. Applications targeted towards agricultural and the farming industry are being developed to bring in technology as a catalyst to improve farm yield and reduce losses to farmers. The preliminary work includes plant disease detection, treatment, and autonomous path planning. This section illustrates the details of how this goal can be achieved at both the application level and the hardware design level.

With rapid changes in climate across the globe, environmental conditions are deteriorating and becoming unsuitable for normal agricultural practices. This has led to a rapid shift towards greenhouse environments, where farmers look to boost their crops and farm produce yields through precisely controlled measures. Consider a robot that is equipped with the necessary sensors that would allow it to navigate any greenhouse environment in an autonomous way. Besides, we assume the robot has been designed to store, hold, and apply fertilizers, chemicals, insecticides, and medicines needed to treat diseased plants. As the robot is fully autonomous, the amount of sensory data that gets generated for path planning and disease detection among plants will be huge and should be processed rapidly at the edge to reduce delay in identification of plant diseases and application of insecticides and medicines precisely to the diseased portions of the plant (leaves or branches). RL algorithms with either Q-learning or policy optimization [50] could be used for this application to derive the necessary path for traversal. Data provided to the RL algorithm in most cases needs to be pre-processed as and when received from the host sensors, with dedicated algorithms responsible for pre-processing being executed on a CPU such as an open-source RISC-V processor. The hardware accelerator that we are implementing is being designed as a co-processor communicating with the RISC-V CPU to receive sensory data and send actuating data to the physical robot agent. The hardware accelerator is being targeted for deployment on the Xilinx Zynq Ultrascale+ MPSoC FPGA platform [51] to achieve proper processing system–programmable logic (PS-PL) management to move the data to and from the co-processor and accelerate the computations in coordination with the sensory data arriving at the RISC core. The entire problem definition, as stated in earlier sections, again points to the design and development of specialized hardware architectures for accelerating RL algorithms at the edge.

The higher level functionality of the robot is explained with a state diagram, as shown in Figure 12.5. The robot operates in five different states, namely, Power ON, Working Mode, Chemical Filling Mode, Docking Mode, and Power OFF mode. Each state has a set of instructions that needs to be handled as part of operation, as described in the Table 12.2.

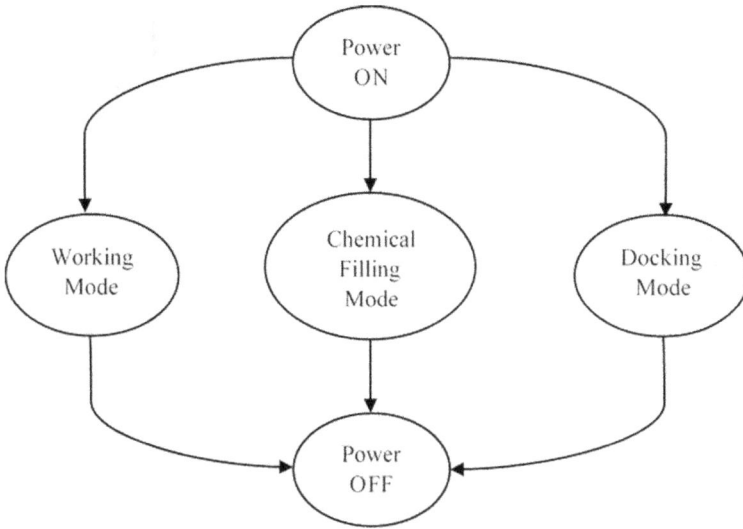

FIGURE 12.5   Autonomous robot-state diagram for different modes of operation.

12.6.1   HARDWARE DESIGN OVERVIEW

As has been the central focus of the chapter, the main aim would be to design a hardware architecture to accelerate the RL algorithm. The hardware architecture would be implemented on a FPGA fabric to which the post-processed sensory data that gives information about the environment details that surround the autonomous robot will be provided by the host processor on board. Figure 12.6 provides higher level details of the entire setup for the autonomous robot. We call the hardware architecture residing in the FPGA fabric, which is mainly responsible for the data computations and storage as part of RL algorithms, reinforcement learning agent (RLA). The data movement between the processing system (RISC-V Core), on-board memory storage (Random Access Memory – RAM), and programmable logic (RLA) as shown, is critical with respect to the success of the application. Also, sensor controllers and actuator controllers that drive the data from the external sensors to and from the PS or PL could be either part of the processing system or could be developed as part of programmable logic depending upon the application needs. The main aim of RLA is to provide an actuating signal to the respective robot actuators/motors in order to decide a future course of actions in the environment while depending upon the information it receives from the sensor sources and the processing system, in coordination with the elements of the reinforcement policy in place. The reinforcement policy in such a setup plays a key role in deciding the learning capabilities and rate of convergence. A few popular algorithms, such as Trust Region Policy Optimization (TRPO) [52], Proximal Policy Optimization (PPO) [53], Deep Deterministic Policy Gradient (DDPG) [54], Twin Delayed DDPG (TD3) [55], natural policy gradient-based algorithms [56,57], and asynchronous advantage actor critic (A3C) [58], are also being studied extensively at the

TABLE 12.2

Autonomous Robot Description for Different Modes of Operation

Power ON	Working Mode	Chemical Filling Mode	Docking Mode	Power OFF
This state indicates the start of the robot by activating all the necessary sensors.	This mode is responsible for reading the sensory data and processing it at both the processor and co-processor level.	When the chemical/ fertilizer in a container hosted on the body of the robot gets finished, this mode allows the robot to re-fill it at the designated dispensary point.	The docking mode is responsible for entering into a charging station when the power/ battery resource has reached a certain minimum level.	This state de-activates all the sensors.
This state also allows the robot to enter any of the working/ chemical filling/ docking modes depending upon the necessity (e.g. Scenario: If initially the chemical container is empty, it directly enters into chemical filling mode before entering to actual working mode).	The actual path planning and other operations related to plant disease detection and chemical spraying happens in this mode.	The activation of this mode halts the working mode temporarily till the chemical is filled in its respective container up to a particular level.	Docking mode is retained till the battery reaches the satisfactory level of charge with other working mode sensors temporarily suspended.	This state could be entered from any of the other states depending upon the necessity or emergency or as part of a safety measure or of completion of assigned tasks.

implementation and mathematical levels to decide on suitability to the application under consideration.

Work in the direction of extending the application to multi-agent or distributed RL algorithms is also being carried out, where multiple autonomous robots would share the knowledge while navigating the greenhouse environment. The main aim of this extension is to allow faster learning by knowledge sharing between the multiple-agents and significantly reduce the time for execution of tasks in comparison with that of a single agent setup. Initial work in the direction of fully cooperative behavior is carried out by studying algorithms and scenarios involving the robotic traits that needs to be in place while executing tasks corresponding to a given application. A brief

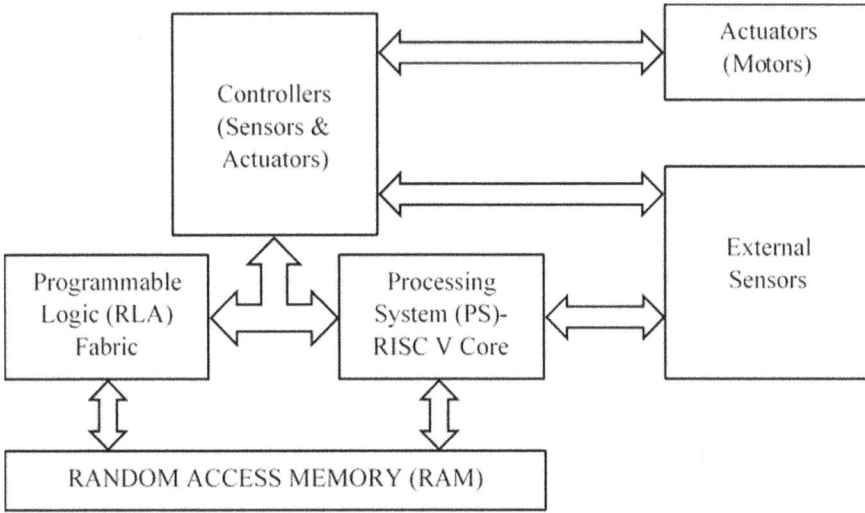

FIGURE 12.6   Block diagram describing the hardware design – an overview.

overview of such challenges and behaviors has already been provided in previous sections with reference to detailed sources available in the literature. Major challenges to be addressed as part of hardware design involve the parallelization of computational data to accommodate more agents and the reduction of delay in processing data and distributing data through various data channels.

A general agent–environment interface for MAS is shown in Figure 12.7, along with a representational diagram of how the multi-agent behavior could be achieved

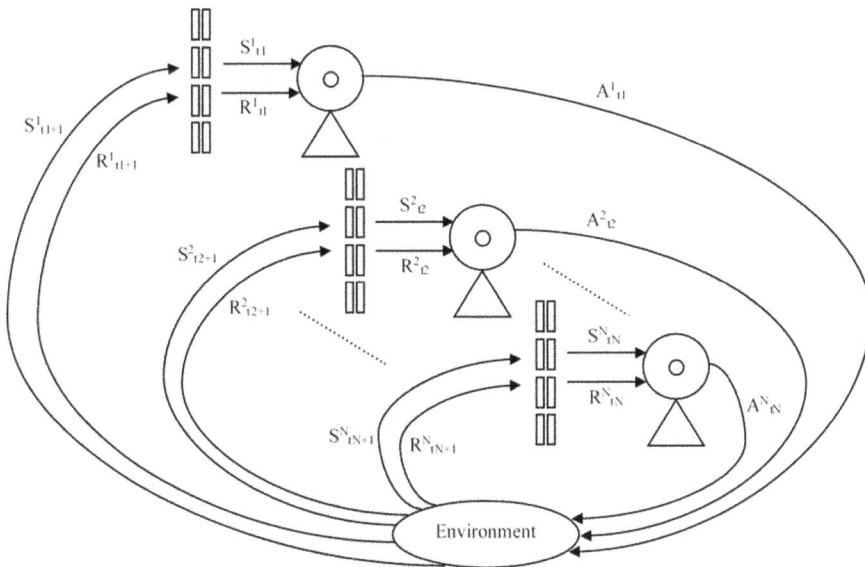

FIGURE 12.7   General agent–environment interface in multi-agent systems.

Green House Environment

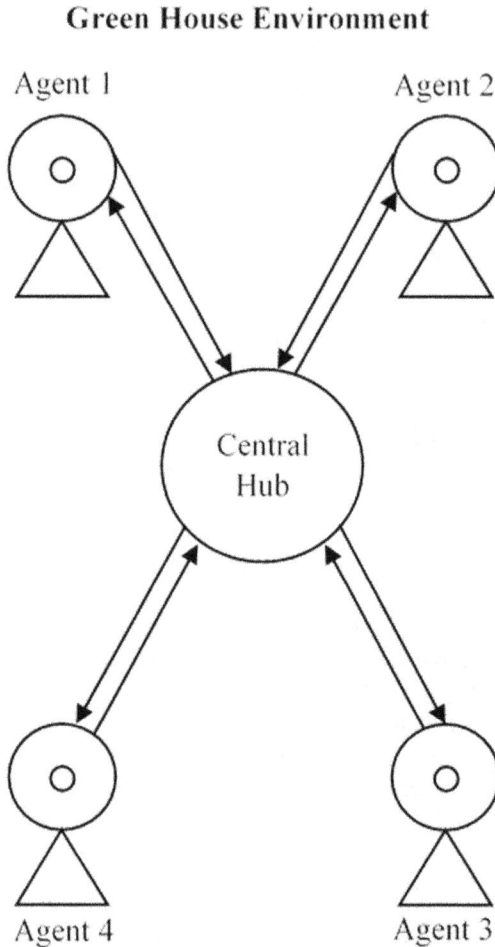

FIGURE 12.8 Representational diagram for multi-agent behavior in a greenhouse environment with a central hub.

in a greenhouse environment for agricultural applications with a central hub, shown in Figure 12.8. One can also relate this with actor-critic–based RL algorithms where the critic is centrally responsible for knowledge sharing between all the available agents. Though we have demonstrated the ability of RLA hardware as part of robotic application, the work could be easily extended towards different areas like financial services, medical facilities management, inventories, statistics, etc., with minimal modifications.

12.7 CONCLUSION

An attempt has been made to provide an overview with respect to RL algorithms, MDPs, specialized hardware architectures designed for the edge involving acceleration of ML algorithms in hardware along with a proposed hardware architecture for implementing SRL algorithms targeted for edge devices. A short survey of the available literature in the field of RL algorithms, hardware architectures, MAS, and agent behaviors in different application scenarios involving challenging social dilemmas was provided to equip an interested reader with necessary and important information involving several aspects of ML in conjunction with hardware, targeted towards the edge of the network. Applications of Q-learning in different fields were also discussed with the aim of its generalization towards the SRL algorithm. Information pertaining to both single-agent and multi-agent behavior in the environments of RL has also been provided with examples at the application level in the field of robotics targeted towards the agricultural and farming industry. The realization of RL agents as part of the FPGA fabric to accelerate RL policy for the embedded system with necessary sensory sources and controllers in place, has also been presented as future work, to provide knowledge about the implementation strategy of RL algorithms at the edge while realizing complex applications. The extension of applications to a multi-agent scenario was also presented to highlight the advantages that could be gained as part of knowledge sharing and distributed learning to speed up the tasks. As such, a representational situation of multi-agents in a greenhouse environment was also discussed. The field of ML in general and RL in particular is growing at a very high pace, generating a huge amount of research outputs and results every year. Though this could rapidly bring in changes at the levels of both algorithm and hardware realization, the central essence of learning and inference within ML algorithms is going to stay important in the future and bring in necessary impactful changes in society.

ACKNOWLEDGMENT

The authors would like to express sincere gratitude to IIITB as well as Machine Intelligence and Robotics Centre (Government of Karnataka) for supporting this work.

REFERENCES

[1] Richard S. Sutton and Andrew G. Barto, "Reinforcement Learning: An Introduction," A Bradford Book, Cambridge, MA, USA, 2018.

[2] Csaba Szepesvari, *"Algorithms for Reinforcement Learning – Synthesis lectures on Artificial Intelligence and Machine Learning,"* Morgan and Claypool Publishers, San Rafael, CA, 2010.

[3] Xiaofei Wang, Yiwen Han, Victor C. M. Leung, Dusit Niyato, Xueqiang Yan, and Xu Chen, "Edge AI - Convergence of Edge Computing and Artificial Intelligence," Springer, Singapore, 2020.

[4] NVIDIA GPU – Turing Architecture. Available online at (www.nvidia.com/en-us/geforce/turing/) - Accessed website on 29th November 2020.

[5] Coral Development Board – Edge TPU coprocessor. Available online at (https://coral.ai/products/dev-board/) - Accessed website on 29th November 2020.

[6] Intel Neural Compute Stick 2. Available online at (https://software.intel.com/content/www/us/en/develop/hardware/neural-compute-stick.html) – Accessed website on 29th November 2020.

[7]   Nvidia Jetson Nano Developer kit. Available online at (https://developer.nvidia.com/embedded/jetson-nano-developer-kit) - Accessed website on 29th November 2020.

[8]   Qualcomm Snapdragon 8 Series mobile platforms. Available online at (https://www.qualcomm.com/products/snapdragon-8-series-mobile-platforms) - Accessed website on 29th November 2020.

[9]   Samsung Exynos Mobile Processor. Available online at (https://www.samsung.com/semiconductor/minisite/exynos/products/mobileprocessor/exynos-9-series-9820/) - Accessed website on 29th November 2020.

[10] Mediatek Helio P60. Available online at (https://www.mediatek.com/products/smartphones/mediatek-helio-p60) - Accessed website on 29th November 2020.

[11] HiSilicon Ascend Series. Available online at (https://www.hisilicon.com/en/products/Ascend) - Accessed website on 29th November 2020.

[12] Baidu – OpenEdge. Available online at (https://github.com/DNGiveU/openedge) - Accessed website on 29th November 2020.

[13] Microsoft – Azure IoT Edge. Available online at (https://azure.microsoft.com/en-us/services/iot-edge/) - Accessed website on 29th November 2020.

[14] Amazon – AWS Greengrass. Available online at (https://aws.amazon.com/greengrass/) - Accessed website on 29th November 2020.

[15] "PyTorch: Tensors and Dynamic Neural Networks in Python with Strong GPU Acceleration." Available online at (https://github.com/pytorch/) - Accessed website on 29th November 2020.

[16] CoreML – Integrate Machine Learning Models into Your App. Available online at (https://developer.apple.com/documentation/coreml) - Accessed website on 29th November 2020.

[17] Qualcomm Neural Processing SDK for AI. Available online at (https://developer.qualcomm.com/software/qualcomm-neural-processing-sdk) - Accessed website on 29th November 2020.

[18] "Multi-platform Embedded Deep Learning Framework." Available online at (https://github.com/PaddlePaddle/Mobile) - Accessed website on 29th November 2020.

[19] "MACE – Mobile AI Compute Engine – Deep Learning Inference Framework Optimized for Mobile Heterogeneous Computing." Available online at (https://github.com/XiaoMi/mace) - Accessed website on 29th November 2020.

[20] M. Abadi, P. Barham et al., "TensorFlow: A System for Large-Scale Machine Learning," in Proceedings of the 12th USENIX Conference on Operating Systems Design and Implementation (OSDI 2016), pp. 265–283, 2016.

[21] "Deploy Machine Learning Models on Mobile and IoT Devices." Available online at (https://www.tensorflow.org/lite) - Accessed website on 29th November 2020.

[22] C. J. C. H. Watkins and P Dayan, "Q-learning," in Machine Learning, vol. 8, pp. 279–292, 1992.

[23] Zhenzehn Liu and I. Elhanany, "Large-Scale Tabular-Form Hardware Architecture for Q-Learning with Delays," in Proceedings of the 2007 50th Midwest Symposium on Circuits and Systems, Montreal, Que., pp. 827–830, 2007.

[24] K. Hwang, C. Lo, and W. Liu, "A Modular Agent Architecture for an Autonomous Robot," in IEEE Transactions on Instrumentation and Measurement, vol. 58, no. 8, pp. 2797–2806, Aug. 2009.

[25] W. Shi, J. Cao, Q. Zhang, Y. Li, and L. Xu, "Edge Computing: Vision and Challenges," in IEEE Internet of Things Journal, vol. 3, no. 5, pp. 637–646, Oct. 2016.

[26] X. Wang, Y. Han, V. C. M. Leung, D. Niyato, X. Yan, and X. Chen, "Convergence of Edge Computing and Deep Learning: A Comprehensive Survey," in IEEE Communications Surveys & Tutorials, vol. 22, no. 2, pp. 869–904, Secondquarter 2020.

[27] Hyungmin Cho, Pyeongseok Oh, Jiyoung Park, Wookeun Jung, and Jaejin Lee, "FA3C: FPGA-Accelerated Deep Reinforcement Learning," in Proceedings of the

Twenty-Fourth International Conference on Architectural Support for Programming Languages and Operating Systems (ASPLOS'19). Association for Computing Machinery, New York, NY, USA, pp. 499–513, 2019.

[28] L. Busoniu, R. Babuska, and B. De Schutter, "Multi-agent Reinforcement Learning: An Overview," in D. Srinivasan and L. C. Jain (eds), *Innovations in Multi-Agent Systems and Applications – 1*, vol. 310. Springer, Berlin, Germany, pp. 183–221, 2010.

[29] M. Matta et al., "Q-RTS: A Real-Time Swarm Intelligence Based on Multi-agent Q-Learning," in Electronics Letters, vol. 55, no. 10, pp. 589–591, 2019.

[30] Kaiqing Zhang, Zhuoran Yang, and Tamer Basar, "Multi-Agent Reinforcement Learning: A Selective Overview of Theories and Algorithms," in *AArxiv*, 2019. Available online at (https://arxiv.org/pdf/1911.10635.pdf).

[31] P. Hernandez-Leal, B. Kartal, and M. E. Taylor, "A Survey and Critique of Multiagent Deep Reinforcement Learning," in *Autonomous Agents and Multi-Agent Systems*, vol. 33, pp. 750–797, 2019.

[32] S. Shao et al., "Towards Hardware Accelerated Reinforcement Learning for Application-Specific Robotic Control," in Proceedings of the 2018 IEEE 29th International Conference on Application-specific Systems, Architectures and Processors (ASAP), Milan, pp. 1–8, 2018.

[33] Howard M. Schwartz, "Multi Agent Machine Learning – A Reinforcement Approach," John Wiley & Sons Inc., Hoboken, New Jersey, 2014.

[34] J. Liu, J. Wu, and L. Jain, "Multi-Agent Robotic Systems," CRC Press, Boca Raton, 2001.

[35] OpenAI – Gym Environments. Available online at (https://gym.openai.com/envs/# robotics) - Accessed website on 30th November 2020.

[36] Akira Notsu, K. Honda, H. Ichihashi, and Y. Komori, "Simple Reinforcement Learning for Small-Memory Agent," in Proceedings of the 2011 10th International Conference on Machine Learning and Applications and Workshops, Honolulu, HI, pp. 458–461, 2011.

[37] L. M. D. Da Silva, M. F. Torquato, and M. A. C. Fernandes, "Parallel Implementation of Reinforcement Learning Q-Learning Technique for FPGA," in IEEE Access, vol. 7, pp. 2782–2798, 2019.

[38] S. Spanò et al., "An Efficient Hardware Implementation of Reinforcement Learning: The Q-Learning Algorithm," in IEEE Access, vol. 7, pp. 186340–186351, 2019.

[39] "Vivado Design Suite Tutorial - Power Analysis and Optimization UG997 (v2019.2)" Available online at (https://www.xilinx.com/support/documentation/sw_ manuals/xilinx2019_2/ug997-vivado-power-analysis-optimization-tutorial.pdf)   - Accessed website on 30th November 2020.

[40] B. Jang, M. Kim, G. Harerimana, and J. W. Kim, "Q-Learning Algorithms: A Comprehensive Classification and Applications," in *IEEE Access*, vol. 7, pp. 133653–133667, 2019.

[41] J. Ho, D. W. Engels, and S. E. Sarma, "HiQ: A Hierarchical Q-Learning Algorithm to Solve the Reader Collision Problem," in Proceedings of the International Symposium on Applications and the Internet Workshops (SAINTW), pp. 4 and 91, Jan. 2006.

[42] L. Shoufeng, L. Ximin, and D. Shiqiang, "Q-Learning for Adaptive Trafc Signal Control Based on Delay Minimization Strategy," in Proceedings of the IEEE International Conference on Networking, Sensing and Control (ICNSC), p. 687691, Apr. 2008.

[43] V. Mnih, K. Kavukcuoglu, D. Silver, A. A. Rusu, J. Veness, M. G. Bellemare, A. Graves, M. Riedmiller, A. K. Fidjeland, G. Ostrovski, S. Petersen, C. Beattie, A. Sadik, I. Antonoglou, H. King, D. Kumaran, D. Wierstra, S. Legg, and D. Hassabis,

"Human-level Control Through Deep Reinforcement Learning," in *Nature*, vol. 518, no. 7540, p. 529, 2015.

[44] Kui-Hong Park, Yong-Jae Kim, and Jong-Hwan Kim, "Modular Q-Learning Based Multi-agent Cooperation for Robot Soccer," in *Robotics and Autonomous Systems*, vol. 35, no. 2, pp. 109–122, 2001.

[45] Tong Zhou, Bing-Rong Hong, Chao-Xia Shi, and Hong-Yu Zhou, "Cooperative Behavior Acquisition Based Modular Q Learning in Multi-Agent System," in Proceedings of the 2005 International Conference on Machine Learning and Cybernetics, Guangzhou, China, pp. 205–210, 2005.

[46] H. Iima and Y. Kuroe, "Swarm Reinforcement Learning Algorithms -Exchange of Information Among Multiple Agents," in Proceedings of the SICE Annual Conference 2007, Takamatsu, pp. 2779–2784, 2007.

[47] T. W. Sandholm and R. H. Crites, "On Multiagent Q-Learning in a Semi-Competitive Domain," in G. Weiß and S. Sen (eds), *Adaption and Learning in Multi-Agent Systems. IJCAI 1995. Lecture Notes in Computer Science (Lecture Notes in Artificial Intelligence)*, vol 1042. Springer, Berlin, Heidelberg, 1996.

[48] F. Zhang, J. Leitner, M. Milford, B. Upcroft, and P. Corke, "Towards Vision-Based Deep Reinforcement Learning for Robotic Motion Control," 2015. Available online at ⟨https://arxiv.org/abs/1511.03791⟩.

[49] Y. Zhu et al., "Target-Driven Visual Navigation in Indoor Scenes Using Deep Reinforcement Learning," in Proceedings of the 2017 IEEE International Conference on Robotics and Automation (ICRA), Singapore, pp. 3357–3364, 2017.

[50] Open AI - Spinning up in Deep RL. Available online at ⟨https://spinningup.openai.com/en/latest/⟩ - Accessed website on 1st December 2020.

[51] Xilinx Zynq UltraScale+ MPSoC Data Sheet: Overview, DS891 (v1.8)." Available online at ⟨https://www.xilinx.com/support/documentation/data_sheets/ds891-zynq-ultrascale-plus-overview.pdf⟩ - Accessed website on 1st December 2020.

[52] J. Schulman, S. Levine, P. Abbeel, M. Jordan, and P. Moritz, "Trust Region Policy Optimization," in Proceedings of the 32nd International Conference on Machine Learning in PMLR, vol. 37, pp. 1889–1897, 2015.

[53] John Schulman, Filip Wolski, Prafulla Dhariwal, Alec Radford, and Oleg Klimov, "Proximal Policy Optimization Algorithms," 2017. Available online at ⟨https://arxiv.org/pdf/1707.06347.pdf⟩.

[54] Timothy P. Lillicrap, Jonathan J. Hunt, Alexander Pritzel, Nicolas Heess, Tom Erez, Yuval Tassa, David Silver, and Daan Wierstra, "Continuous Control with Deep Reinforcement Learning," 2019.

[55] Scott Fujimoto, Herke van Hoof, and David Meger, "Addressing Function Approximation Error in Actor-Critic Methods," 2018.

[56] Sham Kakade, "A Natural Policy Gradient," 2001. Available online at ⟨https://papers.nips.cc/paper/2001/file/4b86abe48d358ecf194c56c69108433e-Paper.pdf⟩.

[57] Richard S. Sutton, David McAllester, Satinder Singh, and Yishay Mansour, "Policy Gradient Methods for Reinforcement Learning with Function Approximation," in Proceedings of the 12th International Conference on Neural Information Processing Systems (NIPS'99). MIT Press, Cambridge, MA, USA, pp. 1057–1063, 1999.

[58] Volodymyr Mnih, Adrià Puigdomènech Badia, Mehdi Mirza, Alex Graves, Timothy P. Lillicrap, Tim Harley, David Silver, and Koray Kavukcuoglu, "Asynchronous Methods for Deep Reinforcement Learning," 2016. Available online at ⟨https://arxiv.org/pdf/1602.01783.pdf⟩.

13 Deep Learning Techniques for Side-Channel Analysis

Varsha Satheesh Kumar[1],
S. Dillibabu Shanmugam[2], and
Dr. N. Sarat Chandra Babu[3]
[1]MS Student, CMU, USA
[2]Scientist SETS, India
[3]Executive Director, SETS, India

CONTENTS

DOI: 10.1201/9781003201038-13

13.1 INTRODUCTION

Side-channel attacks (SCAs) are becoming a cause for concern with the large-scale deployment and use of resource-constrained devices. SCAs are implementation attacks that gain information from the physical implementation of a computer system, rather than weaknesses in the implemented algorithm itself. These attacks were first pioneered by Paul Kocher in the 1990s, when it was discovered that the circuit consumes different amounts of power depending on the input data that are fed. These types of SCAs, where the variation in power consumption is exploited, are known as power analysis attacks [1]. During the implementation of a crypto-graphic algorithm, the plaintext (input) is combined with the key in different combinations to produce the ciphertext (output). An attacker exploits the data that leaks through side channels to reduce the hypothetical key search space and identify the dependency between the device and the secret key.

Hardware and software implementations of many block ciphers have been practically compromised by continuous side-channel analysis, and their security has been a long-standing issue for the embedded systems industry.

To protect devices from this attack, researchers have come up with efficient countermeasure techniques, namely, threshold implementation (TI) and private-t circuits. In 2006, Nikova et al. [2], proposed a scheme, TI based on secret-sharing, and this is provably secure against first-order differential power analysis attacks. Several TI techniques have been proposed over the years, with variations in secret sharing and implementation techniques. Another countermeasure was private circuits, proposed by Ishai et al. [3]. A circuit is t-probing secure, meaning any set of "t" intermediate variables is independent of the secret, which means circuit security is not dependent on the type of side-channel leakage but on the amount or the rate at which information leaks from the device. The main goal of this technique was to mask and wrap the building blocks with random values that remain secure against attacks even if the attacker is able to observe t bits during one computational clock cycle.

However, it does not stop exploring the attacks on the countermeasures, speci-fically after the advent of profiled (template) attacks [4,5]. Profiled SCAs play a significant role in the security assessment of cryptographic implementations. These are the most potent type of attacks as the adversary can characterize the side-channel leakage of the device before the attack. In this scenario, the adversary gains access to the inputs and keys of the profiling device and uses them to characterize the physical leakage. The profiling SCA happens in two steps. First is a profiling phase, where the adversary characterizes the leakage distribution function for all possible secret key values with the help of the traces acquired from the profiling device. Second is an attack phase, where he performs a key recovery attack on the target device. However, in a useful adversary model, many traces are required for the attack. Measurement handling and analysis expertise required for this attack are very high. Then, machine-learning–based attacks [6] widen up the possibility and exposure to handle the attack.

Recent works have highlighted the advantages of employing deep learning (DL) architectures [7] such as multi-layer perceptrons (MLPs) and convolutional neural networks (CNN) as an alternative to the existing profiling SCA attacks. Despite provably secure countermeasures ensuring protection against SCAs, the related experimental results have shown that these techniques are very efficient in conducting security evaluations of embedded systems. DL helps to establish the dependency between the device leakage behavior and the secret key. The trained DL model allows the attacker to use minimum leakage information (e.g. power traces in power analysis) to retrieve the key. Recent works have explored the SCA based on different DL techniques, including MLP network and CNN [8]. Again, countermeasure techniques are preferred over the first and last few rounds of ciphers, due to constrained resources. Therefore, vulnerabilities in the middle rounds are exploited using specific techniques [9]. These works indicate that SCA combined with DL algorithms can outperform the template attacks due to the following reasons: being able to extract the sensitive leakage part with bounded and misaligned traces in an automated fashion, thereby increasing accuracy and reducing time and manual intervention for the analysis.

Moreover, NIST initiated a program towards criteria for standardization of threshold schemes for crypto primitives, which would lead to a common evaluation process to validate crypto primitives.

Our Contribution. In this chapter, our main objective is to establish the effect of different protected and unprotected implementation profiles on the security of the cipher. We have taken the present inspired lightweight block cipher, GIFT, for our experiments and analysis. For analysis, a 64-bit version of GIFT was taken. We have discussed the performance of CNN on the GIFT algorithm with different implementation styles, particularly an Unrolled manner, and with the TI countermeasure. We noticed that the selection of the point of attack for our CNN model affects experimental results, i.e. the classification accuracy of the trained DL model. The point of interest (PoI) after the first-round output of the GIFT cipher requires significantly fewer traces than the PoI on the fifth round. It is evident that the key bits of the GIFT cipher get diffused by the fourth round (since there are 128 key bits in total and 28 bits of the key get diffused every round); hence, it is considerably more difficult to tap the information from the fifth round than it is from the first round. We have implemented the GIFT cipher in a (Un)Rolled manner with the TI countermeasure and noticed that the number of traces to retrieve the key bits is significantly more than the Rolled Implementation of the cipher

13.2 PRELIMINARIES

13.2.1 FRAMEWORK FOR IMPLEMENTATION VULNERABILITY ANALYSIS

Though cryptographic primitives are theoretically secure, their implementation might be vulnerable due to reasons such as lack of awareness about the adversity and knowledge about validation methodologies. The adopted countermeasure should be adequately vetted to avoid a chain of problems. The adopted evaluation method must be a standard one; otherwise, a proper framework has to be developed and augmented with existing standards for ease of invalidation. Validating security for countermeasure techniques needs a comprehensive testing process, which is lacking in existing standards such as FIPS-140-2 (conformance style testing) and Common Criteria

| Unprotected implementation of crypto primitives | Conformance style test: TVLA | Secret information/ Key recovery: CPA | Efficient Countermeasure: Threshold Implementation | Evaluate the protected implementation : TVLA / State-of-art-techniques (deep learning using CNN) |

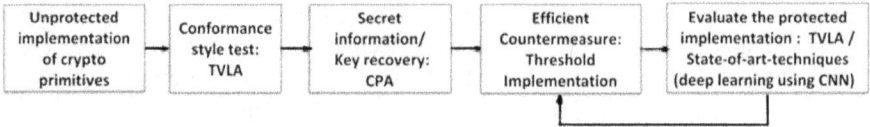

FIGURE 13.1   Implementation vulnerability analysis framework.

(evaluation style testing). The focus here is on the development of a framework for secure implementation of crypto-primitives against implementation attacks. The framework should be fast, effective, and reliable for analysis. To achieve this effectively, the objective is divided into multiple stages, as shown in Figure 13.1.

Naive implementation. Naive implementation involves identification of the algorithm building blocks, then realizing the blocks according to the application requirement. In general, serial, round, and unrolled implementations are preferred for constrained, conventional, and accelerator devices, respectively, along with optimized implementation techniques [10].

Conformance-style test. Generally, naive implementations are vulnerable against SCAs, in particular against power attacks. Conformance-style testing helps in identifying quickly whether the implementation leaks potential information or not. For instance, the test vector leakage assessment (TVLA) [11] explores the difference of means to standard error variation using two sets of traces. Though χ^2 test's [12] success rate [13] is effective, TVLA is simpler for analysis.

Exploitation using power analysis. The following points are essential to reveal the secret key of naïve implementation. First, identify a PoI in the algorithm. Second, capture power consumption during execution of the algorithm. Third, the power consumption model should reflect the actual power consumption of a circuit. Last, an appropriate statistical distinguisher should be used to correlate between modeled and captured power consumption of the circuit to reveal the secret key. Among many statistical methods [14], CPA is reliable and a widely adopted one.

Countermeasure. TI uses the secret sharing principle. Countermeasures break the correlation between modeled and captured power consumption. Masking and hiding are two preliminary approaches to protect the implementation. One such provable masking technique is TI based on secret sharing techniques. Other methods such as private-t circuits [3] and orthogonal direct sum masking come with several trade-offs. Therefore, TI is preferred for an efficient countermeasure realization.

Evaluation of protected implementation. A big challenge is how to evaluate the protected implementation efficiently. A straightforward way is to perform TVLA once more and compare with step 2's TVLA results to understand the attack complexity. Another way is to perform (non)-profiling techniques using advanced SCAs, like the DL [15] approach. Recent works have explored the SCA based on different DL techniques, particularly, CNNs [16,17] outperform other attacks. That's why the feedback mechanism is essential to improve countermeasure techniques towards higher order masking to increase attack complexity. Most of the steps evolved over the years. In this chapter, we focus on the evaluation of protected implementation, mainly using the DL architecture with their hyper-parameters for analysis. The rest of the section provides a brief introduction about profiled SCAs, including CNN, and the protected countermeasure techniques.

13.3 PROFILED SIDE-CHANNEL ATTACKS

Profiled SCAs are considered lethal in the context of SCAs since the attacker, in this case, has access to a clone device from which he can build leakage models and characterize the leakage associated with the device.

Let us assume the side-channel trace has D sample points. The PoI in the algorithm defined as variable $Z = f(P, K)$, where f denotes a cryptographic operation, P denotes a plaintext, and K denotes a part of the key, Z has all the possibilities based on inputs and operations, $Z \{S_1, \ldots S_{|Z|}\}$.

When targeting a device using a profiled SCA, two phases are preferred: a training phase and an attacking phase. During the training phase, adversaries have access to control the input and the secret key of the cryptographic algorithm to derive knowledge about sensitive leakages depending on Z. The PoIs are characterized by an adversarial model, $F: R^D \rightarrow R^{|Z|}$, that estimates the probability P_r $[T|Z = z]$ from a profiled set $T = \{(t_0, z_0),\ldots, (t_{Np-1}, z_{Np-1})\}$ of size N_p. By knowing the input and sensitive leakage variable, a rank vector is computed based on $F(t_i)$, i $\in [0, N-1]$, for each trace in attack traces set(N). Then, log-likelihood ranks are computed to make predictions for each key hypothesis. The candidate with the significant highest value gives the secret key. However, this mechanism needs much attention towards trace alignment, dimensionality reduction, and a bounded number of acquisitions for analysis. Class of DL architectures is reliable enough to overcome these problems as follows.

13.3.1 DEEP LEARNING ARCHITECTURE FOR ANALYSIS

In the neural network, the profiled SCA is a classification problem. The network constructs Function $F: R_n \rightarrow R_{|Z|}$ and predicts output(a) for the given inputs(t). For effective prediction, the network has to be trained with pairs (t_p^i, y_p^i) where t_p^i is the ith profiled input(trace) and y_p^i is the label (sensitive intermediate value) associated with the ith input. In general, the neural network constructs a Function(F) with three layers, namely input, multiple hidden, and output layers to estimate prediction vector(y), $y_p^i = F(t_p^i)$. The error in the classification is optimized using backward propagation [18–20] by varying loss function.

13.3.2 CONVOLUTIONAL NEURAL NETWORKS

A CNN [21] decomposes the problem into two parts: a feature extraction part and a classification part. The following formula can characterize a CNN:

$$|z| = s \cdot [\lambda]^{n^1} \cdot \left[\delta \cdot [\sigma \cdot \gamma]^{n^2} \right]^{n^3}$$

Where a CNN is composed of n^3 stacked convolutional blocks that correspond to n^2 convolutional layers (γ), an activation function(σ), one pooling (δ), n^1 fully connected (FCon) layers (λ), softmax layer(s). The deep neural network is made up of six convolutional layers and two fully connected layers. The convolutional layers extract features from the power traces. The activation function used is the Leaky Rectified Linear Unit (ReLU) function. It fixes the problem of dying because it does not have zero-slope parts. The convolutional and pooling layer function reduces the spatial set of the feature set, thereby reducing the number of parameters and the computation involved in the network. The average pooling function is preferred in DeepSCA. Two fully connected layers have 2048 units each, and the softmax layer is used to map the learned features to the label space.

13.4 PROTECTED COUNTERMEASURE TECHNIQUES

The following countermeasures are used in this chapter: (Un)rolled implementation and TI.

13.4.1 UNROLLED IMPLEMENTATION

The GIFT cipher implemented in a (Un)rolled manner. No register is used in this kind of implementation. Therefore, the leakage model has arrived between two intermediate [22,23] encryption states. All the encryption rounds of the cipher are executed in one clock cycle. The information has to be tapped from wires, and this requires a more robust hypothesis because of more resonant diffusion of the key bits.

13.4.2 THRESHOLD IMPLEMENTATION

The TI countermeasure works on the sharing principle. The number of shares is based on the algebraic degree of the S-Box, but the number of shares should at least be greater than or equal to one more than the degree. The GIFT cipher S-Box has a degree of 3, so the number of shares should be 4 or greater. This increases the circuit complexity and the associated area overhead. The S-Box is decomposed into smaller functions with a lower degree, as shown in Jati and Satheesh and Shanmugam [24,25]. The solutions need to satisfy the TI properties (correctness, non-completeness, uniformity) for secure, shared implementation. TI also was found to be vulnerable against a certain type of attack [26,27].

- Correctness: The sum of the output shares gives the desired output.
- Non-completeness: Every function is independent of at least one share of each of the input variables.
- Uniformity: The input, the output, and the distribution of its shared output values have to be uniform. In other words, each possible shared output has to be equally likely.

We adopted the GIFT cipher as a case study and evaluated profiled SCA using CNN as follows.

13.5   CASE STUDY OF GIFT CIPHER

13.5.1   GIFT Algorithm Description

GIFT is a substitution-permutation network (SPN) based cipher. GIFT-64 has 28 rounds of operation with a block size of 64 bits with 128-bit keys. Initialization: The cipher state, S, is first initialized from the 64-bit plaintext represented as 16 nibbles of 4-bit represented as W15,...W1, W0. The 128-bit key is divided into 16-bit words K7, K6,..., K0 and is used to initialize the key register K.

Round Function: Each round of the cipher comprises a substitution layer (S-layer) followed by a permutation layer (P-Layer) and an XOR with the round-key and predefined constants (AddRoundKey).

S-layer (S): Apply the same S-Box to each of the 4-bit nibbles of the state S (Table 13.1)

P-Layer (P): This operation permutes the bits of the cipher state S from position i to P(i) (Table 13.2)

AddRoundKey: A 32-bit Round Key (RK) and a 7-bit round constant (Rcon) is XORed to a part of the cipher state S in this operation.

13.5.2   Implementation Profiles

The GIFT cipher is implemented using Verilog. Three different implementation profiles of the algorithm are adopted for implementation to compare the effect of the applied countermeasures on the security of the cipher. The resultant power traces

TABLE 13.1
GIFT S-Box

x	0	1	2	3	4	5	6	7	8	9	a	b	c	d	e	f
S(x)	1	a	4	c	6	f	3	9	2	d	b	7	5	0	8	E

TABLE 13.2
GIFT P-Layer

i	0	1	2	3	4	5	6	7	8	9	10	11	12	13	14	15
P(i)	0	17	34	51	48	1	18	35	32	49	2	19	16	33	50	3
i	16	17	18	19	20	21	22	23	24	25	26	27	28	29	30	31
P(i)	4	21	38	55	52	5	22	39	36	53	6	23	20	37	54	7
i	32	33	34	35	36	37	38	39	40	41	42	43	44	45	46	47
P(i)	8	25	42	59	56	9	26	43	40	57	10	27	24	41	58	11
i	48	49	50	51	52	53	54	55	56	57	58	59	60	61	62	63
P(i)	12	29	46	63	60	13	30	47	44	61	14	31	28	45	62	15

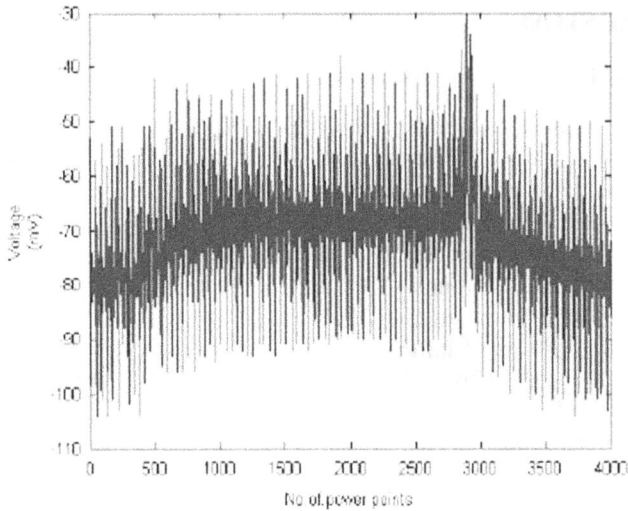

FIGURE 13.2 Rolled power consumption.

are divided into two sets: the traces that are going to be used for training of the CNN model and the traces that are going to be used for evaluating the functionality and classification accuracy of the trained DL model. GIFT is implemented in three distinct profiles as follows:

Profile 1: Naive Implementation (Unprotected)
Profile 2: (Un)Rolled Implementation
Profile 3: (Un)Rolled Implementation with TI Countermeasure

13.5.3 ROUND (NAIVE) IMPLEMENTATION

GIFT is implemented in the round-based fashion where the intermediate values are stored in the registers. Twenty-eight power consumption spikes can be seen in Figure 13.2, indicating the number of rounds of operation of the cipher.

13.5.4 (UN)ROLLED IMPLEMENTATION

GIFT is implemented in the (Un)Rolled manner, where all rounds of operation happen in a single clock cycle. There are no registers in between the rounds. Therefore, power consumption shows a steep peak in Figure 13.3.

13.5.5 PARTIALLY (UN)ROLLED IMPLEMENTATION WITH THRESHOLD IMPLEMENTATION COUNTERMEASURE

GIFT is implemented in a partially (Un)Rolled manner with the TI countermeasure applied for the first and last four rounds of operation, as given in Figure 13.3. Here, S-Box is decomposed into two functions, namely F and G [24,25]. The 128-bit key

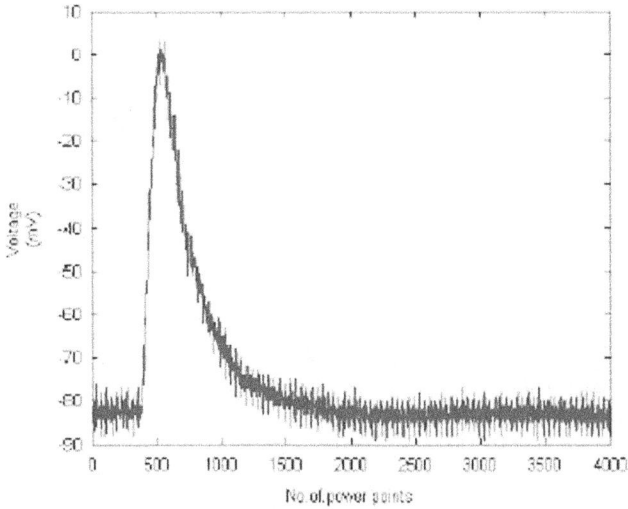

FIGURE 13.3 UnRolled power consumption.

of the cipher gets diffused by the fourth round, and a new key schedule begins from the fifth round onwards. Entropy increases as the rounds of operation increase.

13.5.6 EXPERIMENT SETUP

The target field programmable gate array (FPGA) used was the x2vP7 FPGA, and the power traces were observed on the MSO7104 oscilloscope. The Paramshavak supercomputer was configured and used to train the deep network models, and the model training script was written in Python using the Keras and Tensor Flow libraries.

13.6 DESCRIPTION OF PSCA ON GIFT USING DEEPSCA

The CNN models are trained using power traces from one board, and the testing traces are obtained from a different board to take into account inter-device variance. The CNN model takes into account the problem of trace misalignment. Once the model has been trained, the rank function can be evaluated to estimate the number of traces required to retrieve the key byte (where the rank converges to zero).

In the profiling phase, the dataset comprises the input traces and labels. The label of every trace is the intermediate PoI, which is used for analysis. The leakage model determines the number of outputs.

For each trace in the attack phase, the probability that the trace belongs to a particular class label is estimated. The maximum likelihood principle is used to calculate the set of traces that belong to a specific key. Every trace is associated with a probability.

The PoI and the labels for the three profiles of the cipher are as follows: For naive implementation, we decided to take the first-round output as the PoI to perform the attack.

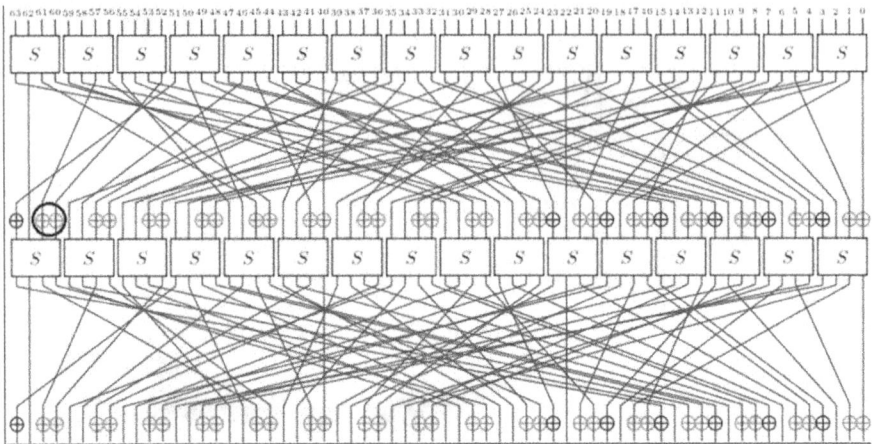

FIGURE 13.4 Attack on naïve implementation: point of interest.

The advantage of SCA is a divide-and-conquer approach. The round function provides confusion and diffusion by S-Box and P-layer operations, respectively. First-round output 64 bits divided into chunks of 4 bits. The 4-bit MSB value is considered as a label for each plaintext. Here, LSB 2-bit has key influence, such as K31 and K15, as highlighted in Figure 13.4. For the second profile, the implementation is divided into two sub-profiles depending on the PoI of attack: (Un) Rolled Implementation attacking the first-round (Un)Rolled Implementation attacking the fifth round. The third profile is similar to the second profile, where the first -and fifth-round outputs are the PoI for analysis.

The model is trained using 50,000 traces. The model is tested with the testing dataset consisting of 10,000 traces. Every trace is associated with a probability. These probabilities are outputs of the Softmax layer. Using the maximum likelihood principle, key enumeration using output probabilities takes place. The number of traces required to recover the key bits is estimated using the rank plot (when the rank function converges to zero).

13.6.1 VULNERABILITY ANALYSIS

We have used the GIFT-64 version of the cipher where there is 64-bit plaintext and a 128-bit key. According to the algorithm, every round utilizes 32 bits of the key. Hence, at the end of four rounds, all 128 bits of the key would have been diffused. Therefore, the attack complexity is increased with entropy or randomness as round iteration increases, which decreases the correlation between the trained model and testing traces due to key diffusion accruing over scheduling. Keeping this in mind, we have compared the results between the profiles having labels in the first and fifth rounds, respectively. Because of key diffusion, the number of traces required to retrieve the key bit when the fifth round is attacked is greater than the result obtained in the first round. This proves that the complexity of the attack increases with an increase in entropy.

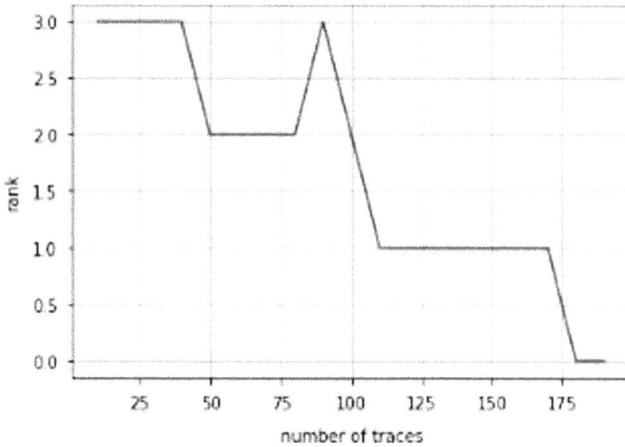

FIGURE 13.5 Rolled or naive implementation.

The CNN architecture and the number of training and testing traces remain the same for all three profiles. The following results were obtained for the three profiles of the cipher, respectively.

The number of traces required to retrieve two key bits of the GIFT cipher when implemented in a round-based manner was found to be around 175 traces, as shown in Figure 13.5.

Profile 2: Unrolled Implementation: Two sets of results were obtained for this profile, corresponding to attacking the first and fifth round, respectively.

Profile 2.1: Unrolled Implementation attacking the first round

The number of traces required to retrieve two key bits of the GIFT cipher when the first round of operation is attacked and when it is implemented in an Unrolled fashion comes to be around 1750 traces, as shown in Figure 13.6. This is significantly more than the number of traces required in the case of Round-based Implementation.

As expected, Unrolled Implementation is more challenging to break than naive implementation since there are no registers to store the intermediate state value outputs. Thus, Unrolled Implementation can also be viewed as a form of a countermeasure to increase the security and resistance of the cipher against side-channel vulnerabilities. Subsequently, in the conventional SCA scenario, the number of traces required to retrieve the key bits in an Unrolled Implementation is in 100,000 traces. This number is also substantially reduced in this case.

Profile 2.2: Unrolled Implementation attacking the fifth round

The number of traces required to retrieve two key bits of the GIFT cipher when the fifth round of operation is attacked and when it is implemented in an Unrolled

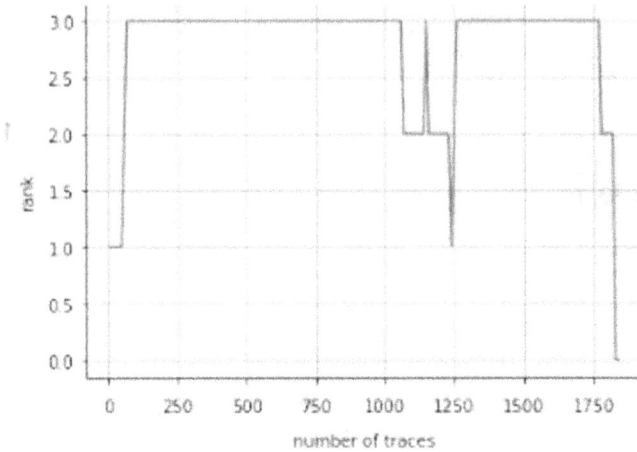

FIGURE 13.6   Unrolled implementation attacking the first round.

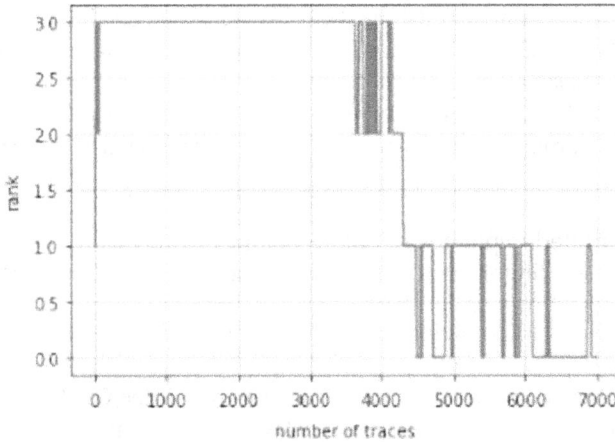

FIGURE 13.7   Unrolled implementation attacking the fifth round.

fashion comes to be around 4500 traces, as shown in Figure 13.7. This is the case since according to the algorithm, all the key bits get diffused by the end of the fourth round, and theoretically, it should become harder to retrieve key bits by attacking the fifth round.

Our result proves this concept wherein the number of traces required is substantially higher because of the increase in entropy, thereby increasing the security of the cipher.

Profile 2.3: Unrolled Implementation with TI countermeasure

For this profile, we combined the TI countermeasure with the Unrolled Implementation profile, and from our experiments, we found the implementation is secure up to 10,000 traces, proving that when we combine TI with the Unrolled Implementation of GIFT, it should be harder to break the algorithm.

13.7 CONCLUSION AND FUTURE WORK

DSCA on the protected profiles has become a cause of concern in the hardware security research domain. In this chapter, we perform profiled SCAs on three different profiles of the GIFT cipher using the DeepSCA neural network. The variation in key dependence between the first and the middle rounds can be observed through the results. With the help of DL techniques, it has become easier to perform hardware attacks. Therefore, proper countermeasures have to be incorporated before field deployment. As part of future work, we would like to delve into algebraic SCA research.

ACKNOWLEDGMENTS

Authors wishing to acknowledge technical staff and financial support from SETS.

REFERENCES

[1] P. C. Kocher, J. Jaffe, and B. Jun, "Differential power analysis," in Advances in Cryptology - CRYPTO '99, 19th Annual International Cryptology Conference, Santa Barbara, California, USA, August 15–19, 1999, Proceedings, ser. Lecture Notes in Computer Science, M. J. Wiener, Ed., vol. 1666. Springer, 1999, pp. 388–397. [Online]. Available: 10.1007/3–540-48405-125

[2] S. Nikova, V. Rijmen, and M. Schlaffer, "Secure hardware implementation of non-linear functions in the presence of glitches," in Information Security and Cryptology - ICISC 2008, 11th International Conference, Seoul, Korea, December 3–5, 2008, Revised Selected Papers, ser. Lecture Notes in Computer Science, P. J. Lee and J. H. Cheon, Eds., vol. 5461. Springer, 2008, pp. 218–234. [Online]. Available: 10.1 007/978-3-642-00730-914

[3] Y. Ishai, A. Sahai, and D. A. Wagner, "Private circuits: Securing hardware against probing attacks," in Advances in Cryptology - CRYPTO 2003, 23rd Annual International Cryptology Conference, Santa Barbara, California, USA, August 17–21, 2003, Proceedings, ser. Lecture Notes in Computer Science, D. Boneh, Ed., vol. 2729. Springer, 2003, pp. 463–481. [Online]. Available: 10.1007/978-3-540–45146-427

[4] S. Chari, J. R. Rao, and P. Rohatgi, "Template attacks," in Cryptographic Hardware and Embedded Systems - CHES 2002, 4th International Workshop, Redwood Shores, CA, USA, August 13–15, 2002, Revised Papers, ser. Lecture Notes in Computer Science, B. S. K. Jr., C̦. K. Koç, and C. Paar, Eds., vol. 2523. Springer, 2002, pp. 13–28. [Online]. Available: 10.1007/3–540-36400-53

[5] T. Bartkewitz and K. Lemke-Rust, "Efficient template attacks based on probabilistic multi-class support vector machines," in Smart Card Research and Advanced Applications - 11th International Conference, CARDIS 2012, Graz, Austria, November 28–30, 2012, Revised Selected Papers, ser. Lecture Notes in Computer Science, S. Mangard, Ed., vol. 7771. Springer, 2012, pp. 263–276. [Online]. Available: 10.1007/978-3-642–37288-918

[6] G. Hospodar, B. Gierlichs, E. D. Mulder, I. Verbauwhede, and J. Vandewalle, "Machine learning in side-channel analysis: A first study," *Journal of Cryptographic Engineering*, vol. 1, no. 4, pp. 293–302, 2011. [Online]. Available: 10.1007/s13389-011-0023-x

[7] H. Maghrebi, T. Portigliatti, and E. Prouff, "Breaking cryptographic implementations using deep learning techniques," in Security, Privacy, and Applied Cryptography Engineering - 6th International Conference, SPACE 2016, Hyderabad, India, December 14–18, 2016, Proceedings, ser. Lecture Notes in Computer Science, C. Carlet, M. A. Hasan, and V. Saraswat, Eds., vol. 10076. Springer, 2016, pp. 3–26. [Online]. Available: 10.1007/978-3-319–49445-61

[8] E. Cagli, C. Dumas, and E. Prouff, "Convolutional neural networks with data augmentation against jitter-based countermeasures - profiling attacks without preprocessing," in Cryptographic Hardware and Embedded Systems - CHES 2017 - 19th International Conference, Taipei, Taiwan, September 25–28, 2017, Proceedings, ser. Lecture Notes in Computer Science, W. Fischer and N. Homma, Eds., vol. 10529. Springer, 2017, pp. 45–68. [Online]. Available: 10.1007/978-3-319–66787-43

[9] S. Bhasin, J. Breier, X. Hou, D. Jap, R. Poussier, and S. M. Sim, "Sitm: See-in-the-middle–side-channel assisted middle-round differential cryptanalysis on spn block ciphers," Cryptology ePrint Archive, Report 2020/210, 2020, https://eprint.iacr.org/2020/210.

[10] T. Wollinger, J. Guajardo, and C. Paar, "Security on FPGAs: State-of-the-art implementations and attacks," *ACM Transactions on Embedded Computing Systems (TECS)*, vol. 3, no. 3, pp. 534–574, 2004.

[11] G. C. Becker, J. Cooper, E. DeMulder, G. Goodwill, J. Jaffe, G. Kenworthy, T. Kouzminov, A. Leiserson, M. E. Marson, P. Rohatgi, and S. Saab, "Test vector leakage assessment (TVLA) methodology in practice," 2013, https://eprint.iacr.org/2017/711.pdf

[12] S. Faust, V. Grosso, S. M. D. Pozo, C. Paglialonga, and F. Standaert, "Composable masking schemes in the presence of physical defaults & the robust probing model," *IACR Transactions on Cryptographic Hardware and Embedded Systems*, vol. 2018, no. 3, pp. 89–120, 2018. [Online]. Available: 10.13154/tches.v2018.i3.89-120

[13] D. B. Roy, S. Bhasin, S. Guilley, A. Heuser, S. Patranabis, and D. Mukhopadhyay, "Leak me if you can: Does TVLA reveal success rate" Cryptology ePrint Archive, Report 2016/1152, 2016, https://eprint.iacr.org/2016/1152.

[14] F.-X. Standaert, B. Gierlichs, and I. Verbauwhede, "Partition vs comparison side-channel distinguishers: An empirical evaluation of statistical tests for univariate side-channel attacks against two unprotected CMOS devices," in *Information Security and Cryptology – ICISC 2008*, P. J. Lee and J. H. Cheon, Eds. Berlin, Heidelberg: Springer Berlin Heidelberg, 2009, pp. 253–267.

[15] E. Prouff, R. Strullu, R. Benadjila, E. Cagli, and C. Dumas, "Study of deep learning techniques for side-channel analysis and introduction to ASCAD database," *IACR Cryptology ePrint Archive*, vol. 2018, p. 53, 2018.[Online]. Available: http://eprint.iacr.org/2018/053

[16] W. Rawat and Z. Wang, "Deep convolutional neural networks for image classification: A comprehensive review," *Neural Computation*, vol. 29, no. 9, pp. 2352–2449, 2017. [Online]. Available: 10.1162/necoa00990

[17] C. Whitnall and E. Oswald, "Robust profiling for DPA-style attacks," in Cryptographic Hardware and Embedded Systems - CHES 2015 - 17th International Workshop, Saint-Malo, France, September 13–16, 2015, Proceedings, ser. Lecture Notes in Computer Science, T. Guneysu and H. Handschuh, Eds., vol. 9293. Springer, 2015, pp. ¨ 3–21. [Online]. Available: 10.1007/978-3-662–48324-41

[18] L. Bottou, F. E. Curtis, and J. Nocedal, "Optimization methods for large-scale machine learning," arXiv:1606.04838, 2016.

[19] I. Goodfellow, Y. Bengio, and A. Courville, *Deep Learning*. The MIT Press, 2016.

[20] H. E. Robbins, "A stochastic approximation method," *Annals of Mathematical Statistics*, vol. 22, pp. 400–407, 2007.

[21] K. O'Shea and R. Nash, "An introduction to convolutional neural networks," ArXiv151108458, 2015.

[22] D. Shanmugam and S. Annadurai, "Crypto primitives ipcore implementation susceptibility in the cyber-physical system," in IEEE International Symposium on Smart Electronic Systems, iSES 2018 (Formerly iNiS), Hyderabad, India, December 17–19, 2018, IEEE, 2018, pp. 255–260. [Online]. Available: 10.1109/iSES.2018.00062

[23] V. Yli-Mayry, N. Homma, and T. Aoki, "Improved power analysis on unrolled architecture and its application to PRINCE block cipher," in Lightweight Cryptography for Security and Privacy - 4th International Workshop, LightSec 2015, Bochum, Germany, September 10–11, 2015, Revised Selected Papers, ser. Lecture Notes in Computer Science, T. Guneysu, G. Leander, and A. Moradi, Eds., vol. 9542. Springer, 2015, pp. 148–163. [Online]. Available: 10.1007/978-3-319–29078-29

[24] A. Jati, "Threshold implementations of gift: A trade-off analysis," in IEEE Transactions on Information Forensics and Security, 2017, pp. 2110–2120. [Online]. Available: https://eprint.iacr.org/2017/1040/20191110:101111

[25] V. Satheesh and D. Shanmugam, "Secure realization of lightweight block cipher: A case study using GIFT," in Security, Privacy, and Applied Cryptography Engineering - 8th International Conference, SPACE 2018, Kanpur, India, December 15–19, 2018, Proceedings, ser. Lecture Notes in Computer Science, A. Chattopadhyay, C. Rebeiro, and Y. Yarom, Eds., vol. 11348. Springer, 2018, pp. 85–103. [Online]. Available: 10.1007/978-3-030–05072-66

[26] T. Moos, A. Moradi, and B. Richter, "Static power side-channel analysis of a threshold implementation prototype chip," in Design, Automation & Test in Europe Conference & Exhibition, DATE 2017, Lausanne, Switzerland, March 27–31, 2017, 2017, pp. 1324–1329. [Online]. Available: 10.23919/DATE.2017.7927198

[27] S. Vaudenay, "Side-channel attacks on threshold implementations using a glitch algebra," in Cryptology and Network Security - 15th International Conference, CANS2016, Milan, Italy, November 14–16, 2016, Proceedings, 2016, pp. 55–70. [Online]. Available: 10.1007/978-3-319–48965-04

14 Machine Learning in Hardware Security of IoT Nodes

T Lavanya[1] and K Rajalakshmi[2]
[1]Research scholar, Department of ECE, PSG College of Technology, Coimbatore, Tamil Nadu
[2]Associate Professor, Department of ECE, PSG College of Technology, Coimbatore, Tamil Nadu

CONTENTS

DOI: 10.1201/9781003201038-14

14.1 INTRODUCTION

In semiconductor technology, there has been enormous progress, which has brought about a great amount of participation in the design stage and development stages of integrated circuits (ICs) [1]. The design complication of the circuits increasing continuously was that they required a specialized team worldwide to work on the complexity of the design to increase the manufacturability and efficiency of the ICs. However, the difficulty lies in the security of the circuits from an adversary at any stage of the manufacturing process inserting a malicious circuit [2]. In the design-fabrication process of ICs, various stages could be exploited by the adversary, such as attacks to perform an unwanted function in a circuit, called a hardware Trojan (HT), which impact the trustworthiness and security of the device. HT circuits are defined as an extra circuit present in a main circuit that consists of payload logic and trigger logic, as shown in Figure 14.1. The payload logic gets activated by a trigger signal that is sent by trigger logic from an adversary to perform an unwanted function in the main part of the circuit [3]. The attack may cause damage to hardware by means of change in the functionality of the device, obstruction caused during the execution, betrayal of information stored in the hardware, etc. These threats cause severe concern in various critical applications such as medical equipment, mobile communications, reactors, defense-based strategies like aerospace companies, devices connected to Internet of Things (IoTs), etc. [4]. HTs are evolving constantly beyond chips to layers of the design, circuit components, and even devices that lead to security issues in the entire hardware ecosystem. Thus, there is a need for security to protect hardware. Rather than software protection, the physical device is protected from hardware vulnerability by using hardware security. This hardware security adds a supplementary layer of security for an important hardware system [5].

In recent years, globalization in semiconductor technology has led to an increase in device manufacturing. This led to drastic achievements in wireless communication: embedded system application, microelectronics technology, sensor technology, and IoTs. These are widely used by governments, academic institutes, and industries worldwide [6]. For example, in smart healthcare systems, patients can register for an

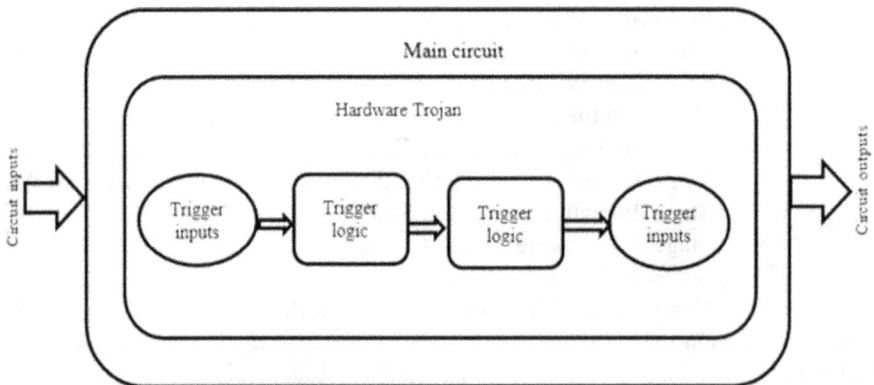

FIGURE 14.1 Hardware Trojan circuit [3].

appointment with a doctor for consultation in any area using IoT technology. In industrial systems, production is managed in various dimensions using IoT. IoT applications provide communication between humans and nature [7]. Accordingly, there are various methodologies to protect IoT nodes from adversaries. One such methodology is the machine learning approach. The machine learning approach is a recent trend in hardware security to protect against an adversary attack. Machine learning algorithms are considered to be the major protection method from any attack [8]. There are new models proposed for HT detection; even then, attacks from adversaries find a better method and insert an HT into the circuit. Thus, machine learning is more effective and used to detect the unknown HT present in the circuit by training the classifier to detect the new HT. Machine learning finds out the unwanted function in the circuit by determining variation in the power consumption of the device, run-time variation, etc. Accordingly, it is able to distinguish the HT-affected circuit from unaffected circuits. In this chapter, we discuss threats and countermeasures of hardware attacks, machine-learning–based approaches, and their application.

14.2 CLASSIFICATION OF HARDWARE ATTACKS

The multiple process of IC manufacturing and its complicated supply chain can be susceptible to adversary attack at any stage of the IC process, as shown in Figure 14.2. Figure 14.2 [9] shows that the IC manufacturing process has the possibility of three major threats:

- Design – This includes tools, IPs, standard cells, and models. Here, the models and cells will be used by the designer as per the design requirements

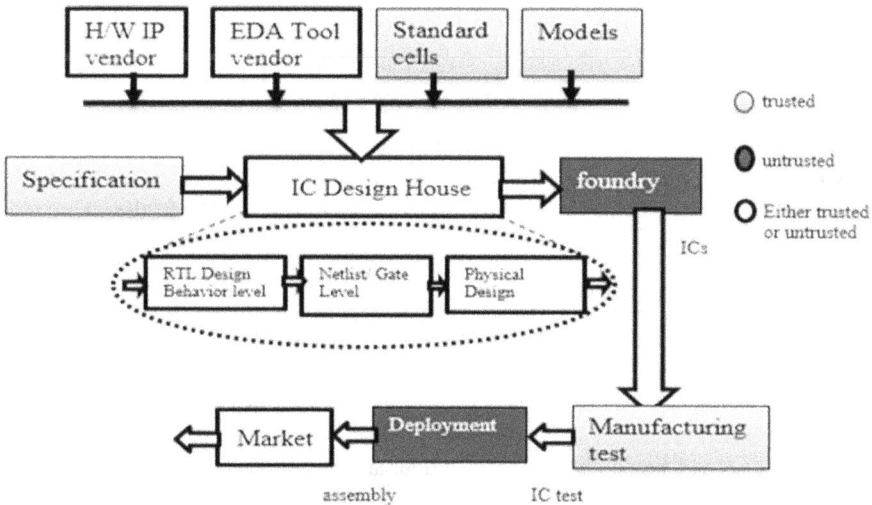

FIGURE 14.2 Attack by an adversary at the IC supply chain [3].
Note: H/W = hardware/software; EDA = electronic design and automation.

but there is the possibility of HT attack in the IPs supplied by the IP vendor if tools are provided by an untrusted vendor.

- Fabrication – This stage includes the process of masking, lithography, and packaging. The fabrication process may also be untrusted, and there is the possibility of the addition of an extra HT circuit in the main circuit.
- Manufacturing test – This stage includes verification and testing of the ICs. It is secured if the verification is done under a trusted unit. After testing, the ICs are sent to market, where the devices are distributed by a trusted distributor or possibly by an untrusted distributor, thus increasing vulnerability to HT attack.

The threats are unavoidable in these processes as the processes are expensive and consume more time because of an increase in the manufacturing units of ICs and their global requirements. Hence, it is challenging to determine the defensive approach to lessen the security risk caused by HT attacks [10]. These cause great economic loss and harm to society.

The attacks may be during fabrication or before fabrication; detection of these attacks is highly difficult because of various reasons [11]:

- First, the complexity of IP blocks in the circuit has made detection of minute modification in the IoT chips difficult.
- Second, physical inspection methods and reverse engineering methods consume a lot of time, and they are costly to operate in the circuit as the ICs are sized to nm scale. Thus, detection will be difficult. Reverse engineering is destructive in nature, and it is not an effective method as it fails to determine if the remaining ICs in a circuit are HT free.
- Third, the HT is activated only for the specified signal sent by the adversary as per the function requirement; hence, it is difficult to determine the affected signal.
- Fourth, faults present in the netlist of the design are determined by the design for test, but the test is capable of detecting only the stuck-at-faults. Thus, it fails to ensure the design is HT free.

Accordingly, there is a need to classify a netlist as an HT-affected or HT-free net [12]. However, because of the complexity of the design, there is an increase in the features of the design. Thus, by simple analysis, it is hard to detect the HT in the circuit.

14.2.1 HARDWARE TROJAN TAXONOMY

As new HT attacks have emerged, there has been an evolution in the taxonomy of HTs. Some taxonomies deal with the logical types of the HT, i.e. the HT may be combinational and sequential [13], but this type of taxonomy fails as the HT's impacts is not considered based on trigger conditions and payload. Other taxonomies deal with the classification of HT based on activation, action, and physical presence [14]. This type of taxonomy also fails as it contains only a limited number of HTs, and there is no obvious correlation. There is another taxonomy that classifies HTs into two categories: bug-based HTs and parasite-based HTs [15]. Even this taxonomy is said to be not very effective as trigger conditions are not known. The HTs are classified depending on the

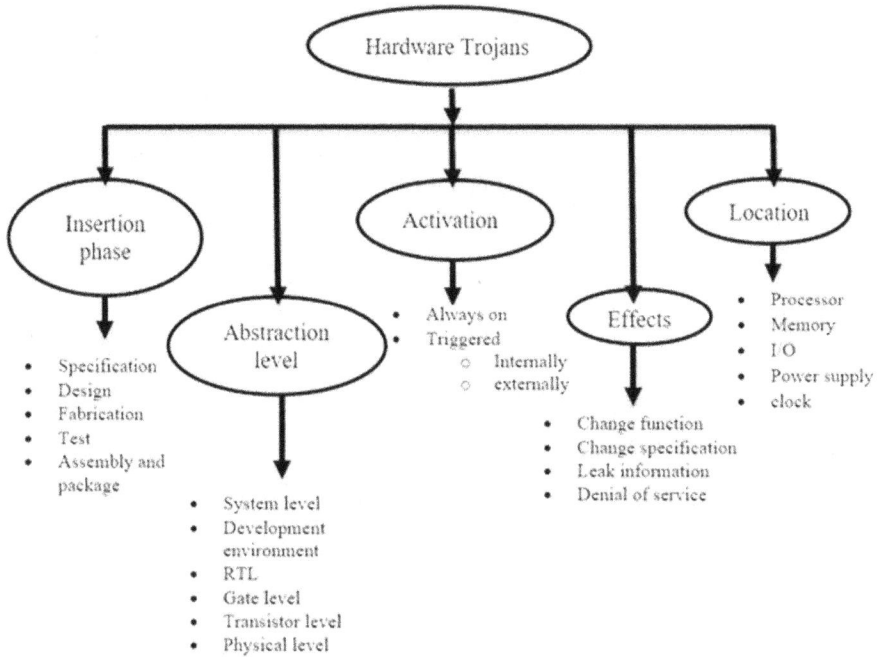

FIGURE 14.3 Hardware Trojan taxonomy [A detailed description of Trojan Taxonomy-trust-hub.org].
Note: RTL = register transfer level; I/O = input/output.

payload and trigger mechanism, thus differentiating them as analogue HTs and digital HTs [3]. Digital HTs include both combinational and sequential HTs. These types of HTs are integrated, extended, and further classified into an attribute-based HT taxonomy [16]. Attributes like design location, design phase, abstraction level, type of logic, layout, and design function have improved the classification of HTs [17]. The HT taxonomy is shown in Figure 14.3. This figure shows an organized taxonomy built on the relationship amongst the HT location injected in a target chip in addition to those targets that are overwhelmed through the trigger input.

14.2.1.1 Insertion Phase

HTs are classified based on the phase of insertion. As the HT insertion may take place during any development stage of the chip, it is necessary to know the stages of insertion.

- Specification phase: The system is designed with a specific parameter like power, delay, function of a design, area, etc. During this phase, there may be a possibility of the addition of an HT, which changes the parameter values. This indicates there is an existence of HTs.
- Design phase: In this phase, the designer creates a circuit using the available IP blocks and other standard cells that are supplied by the third-party seller. This phase mainly includes a mapping of constraints like logical function, timing constraints, and physical parameters in a specified technology.

- Fabrication phase: During this phase, the adversary can use his own masks over wafers. This may cause a serious effect. The adversary can change the chemical composition during fabrication. This leads to consumption of more power supply and leads to faster aging of the chip.
- Assembly phase: In this phase, the verified chips and the supplementary components are mounted on a printed circuit board (PCB). At this time, the adversary may add two or more components on the PCB, along with the HT-free ICs. These components later cause a change in the device function or the stored information may get leaked.
- Testing phase: In this phase, the circuit is tested to determine the circuit functions; however, even in this situation the HTs are not discovered as the adversary kept the nets secret. Due to these secret nets, the adversary can activate the HT when required to collect stored data.

14.2.1.2   Level of Description

The abstraction level of circuits is important as the functionality of the design depends on these levels. Thus, there may be a chance of HTs present in any of the levels.

- System level: The modules used in the target hardware design, such as interconnections, hardware blocks, and communication protocols, may be triggered by the HTs present in any of the modules. Thus, at system level there is a possibility of HT attack.
- Development atmosphere: This comprises verification, simulation, synthesis, and tools validation. The HT can be inserted using CAD tools and scripts.
- Register transfer level: At this level, the HT insertion can be done easily by the attacker. The adversary has control of the functional modules that are described in terms of signals, registers, and Boolean function.
- Gate level: At this level, the logic gates are interconnected to form a design. The HT affects the size and location of the design; thus, the attacker can alter the functionality of the design. These HTs can be sequential or combinational; thus, they are called "functional Trojans".
- Transistor level: To build logic gates, transistors are used. Thus, the attacker can remove the transistor or insert a transistor to alter the functionality of the circuit. The size of a transistor is also changed to modify the parameters of the circuit. This type of HT attack will produce a huge delay in the critical path of the circuit.
- Layout design: The circuit dimension and locations are described at this level. This is also called the physical level of the circuit. The adversary inserts the HT by changing the length of the wire, spacing among circuit fundamentals, and alignment of layers.

14.2.1.3   Activation Mechanism

The HTs inserted by an adversary have two activation mechanisms. First, they can be dormant until the circuit is triggered. Second, they can be always in the "ON" state, which is a serious issue for a hardware device. The most dangerous mechanism is one that is always in the "ON" state; this HT affects the device very badly and also leads to aging of the device. The dormant mechanism is activated by another circuit, which is either externally or internally triggered.

- Internally triggered mechanism: In this mechanism, when an event occurs in the circuit the HT gets activated in the targeted device. The events occurring in the circuit may be due to the physical condition of the circuit, i.e. the temperature, pressure, electromagnetic interference, altitude, etc. For example, when the temperature reaches a certain degree or if the circuit reaches a certain state, the HT is triggered. Another option is a time-based event, i.e. because of a counter present in the circuit, which results in a time bomb.
- Externally triggered mechanism: When an external input is given to the circuit, then the HT gets activated. This external input may be from the switches that are used in the circuit or it may be from the one block to the other; thus, during transition there may be the possibility of triggering the HT present in the circuit. To activate the HT externally, there should be some sensing module present in the circuit that receives the signal from any means to activate the HT circuit.

14.2.1.4   Effects of Hardware Trojans

The severity caused by HTs on a target device is characterized based on the effects. These effects can range from system failure or destruction of the target device.

- Change in the functionality: The presence of HTs in a device can cause an error in the circuit which leads to the change in the functionality of the hardware device. The change in functionality of the device can also be due to the modification of specification of the device by an advisory.
- Reduction in reliability: By changing the parameters of the device, the HT reduces the reliability of the device. The power and delay of the device are changed. The unwanted HT circuit present in the device degrades the performance of the device. This in turn causes aging of the device because of more power consumption.
- Leakage of information: The presence of HTs in a device can leak information stored in the device. The leakage of information is done through frequency transmission and external interfaces like JTAG and RS-232.
- Denial-of-service (DoS): The function or resource of the device is prevented by a denial-of-service Trojan. These DoS Trojans can cause physical damage to the device or may even disable and alter the configuration of the device. They may be permanent or temporary in nature.

14.2.1.5   Location

The HT can be present in any part of the chip. That is, it may be present in processors, I/O devices, memory, the clock, or the power supply. The HT can be in multiple components or in a single component. These HTs act independently where they are present.

- In the processing units, if an HT is present then it changes the order of execution of instructions.
- In the memory units, if an HT is present then there is a leakage of information that is stored in the memory. These HTs may even block reading or writing of data in the memory.
- In the I/O units, if an HT is present then this may lead to a disturbed communication between the processor and the peripheral.
- If an HT is present in the power supply, then this causes a variation in the voltage and the current supplied to the chip.
- If an HT is present in the clock circuit, this may cause an unwanted interruption in the chip.

14.2.2 TYPES OF HARDWARE TROJANS

HTs are categorized into three types, and they are given as follows, shown in Figure 14.4 [18], and summarized in Table 14.1:

- IP-level HT: Each individual IP core of the chip is implanted by an HT, and these are triggered by any of the internal conditions and by rare nets or signals. When the HT gets activated, it affects the IP core where it is present, as shown in Figure 14.4(a).

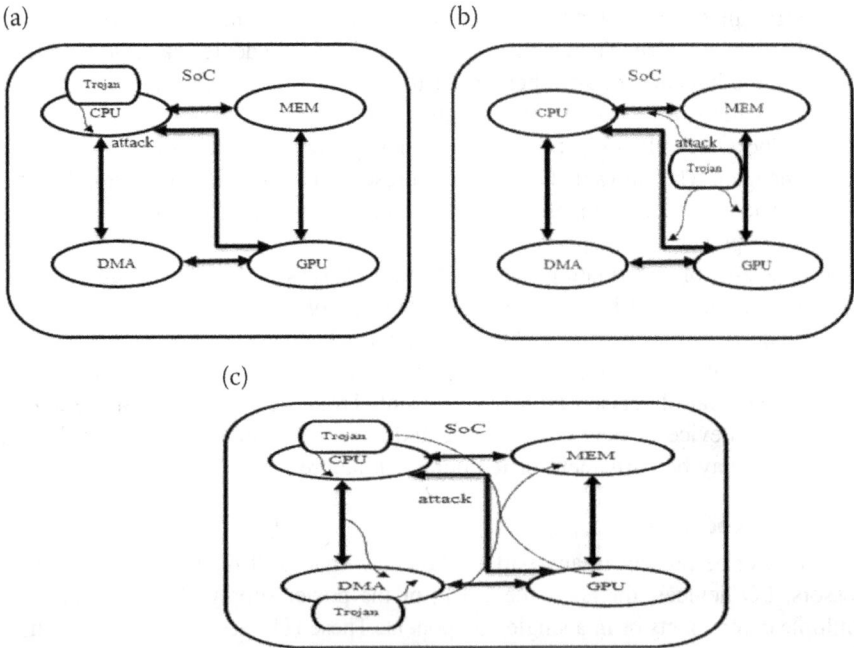

FIGURE 14.4 Trojan threats: (a) IP-level threat, (b) bus-level threat, (c) chip-level threat [31].

TABLE 14.1
Summary of Types of Hardware Trojans

Sl.no	Property	Type IP-level HTs	Bus-level HTs	SoC-level HTs
1	Location	Individual IP core	On-chip buses	One or more IPs
2	Target attack	HT-implanted IP core	Routers, linkers, interfaces	Other IPs or entire SoC
3	Trigger conditions	Internal rare signals	Internal rare signal or data flow through on chip buses	Internal rare signals or external signal or instructions
4	Attack pattern	Simple	Normal	Diverse
5	Impact of Trojan	Specific IP cores	Interconnecting buses or routers in SoCs	Impact on other IPs or overall SoC functions
6	Defense strategy	Static trust verification techniques	Identify patterns of on-chip buses	Identify suspicious activities
7	Effective stage	Design/test time	Run time	Run time
8	Deficiency	Error correction not addressed	On-chip bus features are exchanged continuously	Focused mainly on in-field protection

- Bus-level HT: In the network of buses, there is the possibility of an HT connected to any of these buses. This may tamper the integrity and reliability of the chip, as shown in Figure 14.4(b).
- Chip-level HT: The presence of an HT in any part of the chip, like the IP core, can affect the other IPs and thus change their behavior, which may hamper the chip's functionality. Thus, its efficiency gets reduced, as shown in Figure 14.4(c).

Thus, the affected IP core or the untrusted third-party IP core gets activated by the rare conditional signals, which severely damages the chip. IP-level HTs are similar to system on chip (SoC)-level HTs, but they differ in some aspects as the IP-level HTs are implanted in individual IPs but SoC-level HTs can be implanted in many IPs present in the chip. IP-level HTs affect the particular IP where they are embedded, whereas SoC-level HTs affect the other IPs, which affects the general function of the SoC. The detection of IP-level HTs can be done by determining functional and structural features, but SoC-level HTs are determined by dynamic analysis of the whole SoC.

Bus-level HTs also differ in some aspects from SoC-level HTs as these HTs are implanted in the circuit components that connect to buses but not in IPs, whereas SoC-level HTs are implanted in more than one IP. Bus-level HTs cause damage on interconnecting cores, whereas SoC-level HTs impact the whole SoC. Bus-level HTs are determined by linker and router behaviors, whereas SoC-level HTs are determined by the behaviors of IP cores.

14.3 COUNTERMEASURES FOR THREATS OF HARDWARE TROJANS IN IOT NODES

So far, we have discussed HT threats. To overcome these threats, there have been numerous countermeasures proposed. To deal with each threat, several defensive techniques have been proposed, and these are used in combination to ensure trustworthiness in the IC market. Essentially, three categories of countermeasures against hardware attacks are considered, and they are given as follows:

    i. HT detection approach: This is the process to determine the presence of HTs in the nets of the circuit.

   ii. HT diagnosis approach: This is the process to determine the presence of HTs in terms of location, inputs, and types.

  iii. HT prevention approach: This is the process that upsurges the complications of HT addition into the chip design, increases the efficiency of HT detection, and prevents the chip from receiving HT insertion.

14.3.1 HARDWARE TROJAN DETECTION APPROACHES

Earlier, HT detection was based on the golden model to represent the design free from an HT. If the circuit was suspected to be HT affected, then it deviated its characteristics or features from that of the golden IC. Thus, it was confirmed that the tested IC was affected by an HT [19]. With the increase in complexity of ICs, the approach used was destructive reverse engineering. This approach is said to be complicated and time consuming.

- Pre-silicon HT detection – In this stage, HTs inserted from the adversary in IP cores and electronic design and automation (EDA) tools are detected using verification techniques. The HTs are dormant during the functional verification; thus, it is difficult to determine during verification as HTs are resistant to traditional verification. Thus, pre-silicon detection uses a set of approaches to detect HTs. First, it uses formal verification or assertion-based verification along with Boolean satisfiability to find the redundant logic present in the circuit. Second, it uses dynamic and static verification techniques; it formulates HT detection as unused circuit identification (UCI), which is considered to be suspicious circuits. The UCI algorithm detects the signal pairs, compares the signals, and identifies suspicious HTs in the circuit. Recently, to detect HTs in the pre-silicon stage a new approach was proposed, i.e. feature analysis-based IP trust verification. This method detects HT free nets and HT-affected nets without using a golden netlist.
- Post-silicon HT detection – The HTs present in the design stage and the manufacturing stage of IC design are identified using the post-silicon process. In this process, the logic test and side-channel analysis approach are used to determine the HTs. Side-channel analysis – This method is used widely for HT detection, which determines the circuit's power consumption, delay, critical path, and electromagnetic interference. The SoC designer compares these parameters and

analyzes the presence of HTs in the circuit. Logic test – the side-channel analysis approach becomes ineffective due to process variations. Thus, the logic test is more effective as it generates test patterns of the circuit to detect HTs. As the adversary can insert any number of HTs in the circuit, it is difficult to determine all HTs, and the test patterns for each HT are difficult to generate. Thus, to generate the test vector, the statistical approach has been developed.

14.3.2 HARDWARE TROJAN DIAGNOSIS

This method is used to identify HT types, locations, and trigger signals in the circuit and to take away the HT from the circuit. HT diagnosis is dependant on gate-level classification and circuit division. This method is applied to large circuits, which are further divided into small sub-circuits with fewer gates to detect easily and accurately. This also diagnose HTs by evaluating their leakage power. This process is repeated to obtain the accuracy of HT nets. One approach of HT diagnosis is based on segmentation of circuits and gate-level characterization, as shown in Figure 14.5. The figure shows the three phases of diagnosing HTs: segmentation, HT diagnosis and post processing and validation. In the first phase, model training is done by using a controllability and correlation ratio for accuracy; these segmented properties are utilized for training the model. Further, the model is divided into small sub-circuits. The sub-circuits consist of a small number of gates; thus, it is easy to distinguish and analyze the HTs by determining the circuit leakage power. The post-process results are predicted and validated by applying statistical methods.

14.3.3 HARDWARE TROJAN PREVENTION

To improve the efficiency of the above approaches, prevention is necessary by designing the ICs with self-protection awareness. The techniques for SoC designers to prevent HT attacks are done by various methods such as obfuscation, layout filler, dummy circuit insertion, and split manufacturing.

- Obfuscation – The obfuscation method is a transformation from one source circuit to a new circuit that has the same functionality as that of the source circuit but is less intelligible in some senses. The two major requirements for obfuscation are as follows: i) efficient operation of the obfuscated circuit is similar to that of the native circuit, and ii) the obfuscation will not leak information. Obfuscation techniques are categorized combinatorial logic obfuscation, where there is insertion of additional gates, and sequential logic obfuscation, where there is an insertion of extra states.
- Layout filler – This technique is put forward to simplify HT detection and decrease HT insertion in the left-out space of layout design. The SoC designer and IP vendors are prohibited from changing the route for HT circuits. This method prohibits adversaries in foundries from inserting an HT intentionally or unintentionally in the circuit. In the layout, to avoid the insertion of additional HTs, built-in self-authentication techniques are used. In this technique, spare space in the layout design is removed and filled with a standard

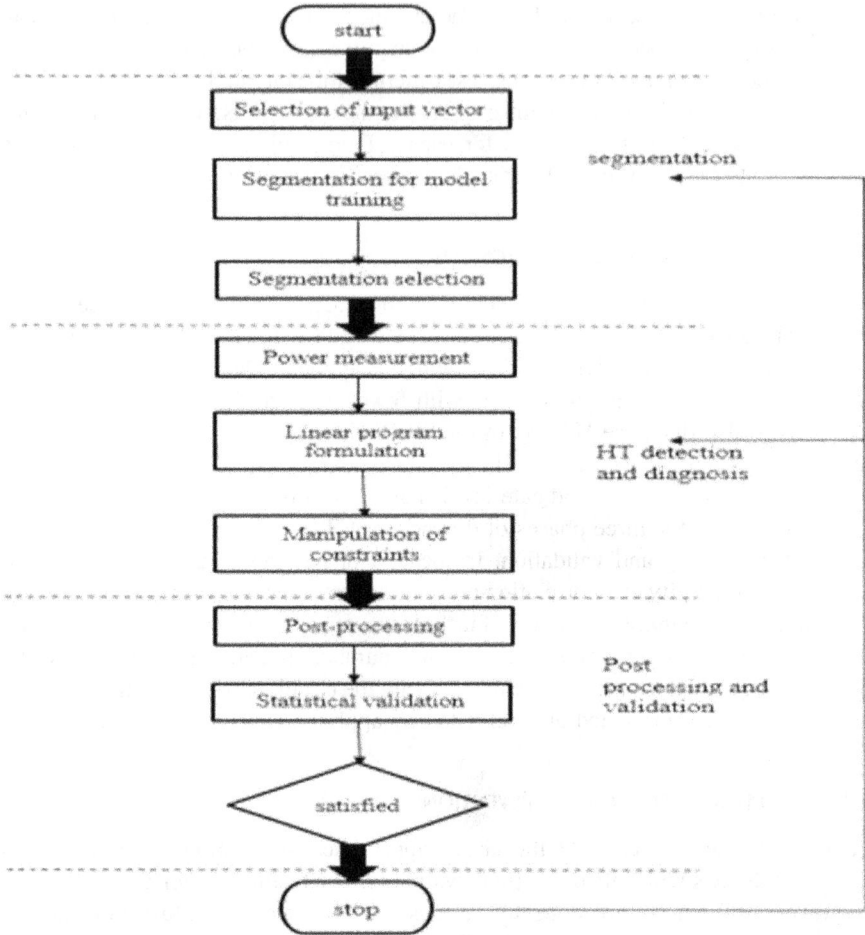

FIGURE 14.5   Flow chart of segmentation-based HT diagnosis and detection [11].

functional cell instead of nonfunctional cells. The standard cells that are inserted in the spare space are connected to form a circuit called "built-in self-authentication", which is an autonomous circuit.

- Split manufacturing – At the untrusted foundry, the transistors and lower metal layers are fabricated to improve the security of an IC. At the foundry, split manufacturing is used. That is, the attackers in the front-end on-line (FEOI) layers without back-end on-line (BEOI) cannot recognize the appropriate places in a circuit for inserting the HT. In routing, floor planning, and placement of the design, the insertion of an HT is avoided by split manufacturing.

14.4   MACHINE LEARNING MODELS

Machine learning is a class of algorithms that focuses on generating models like learning algorithms from a huge amount of previous data and applying these data for

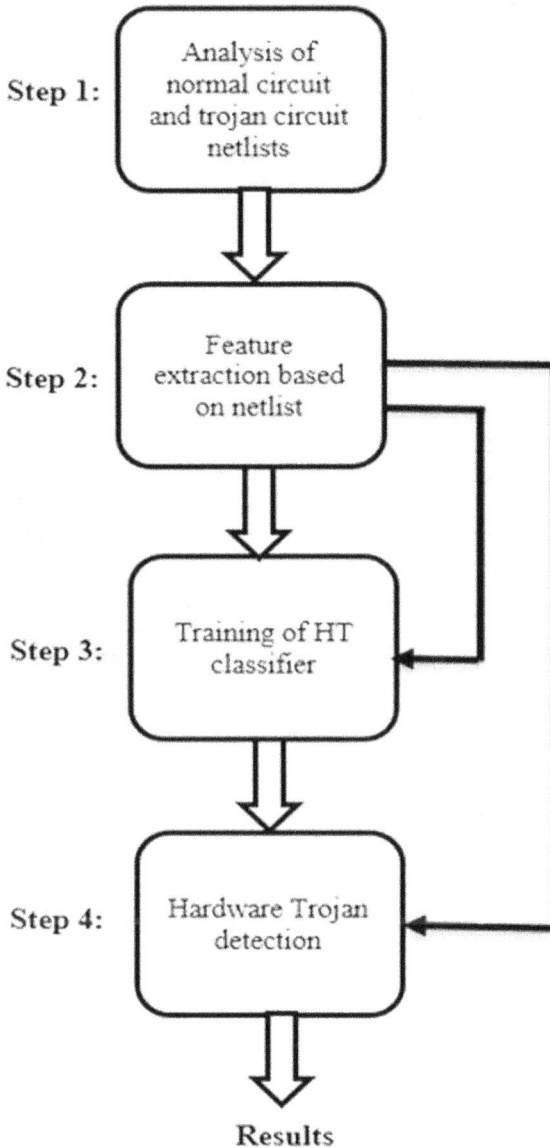

FIGURE 14.6 Flow of proposed methodologies.

training the model for prediction or classification [20]. The machine learning algorithm involves the general procedure shown in Figure 14.6, and it is given as follows:

i. Pre-processing phase: In this phase, first the related features of the design are selected, and later the data with these features are extracted from the selected exceptional data as they will be used to differentiate a changed value of the target outputs. To generate the samples for learning, scaling, feature selection, and dimensionality reductions are implemented.

ii. Learning phase: In this phase, to drive the models from the training dataset, the appropriate learning algorithms are selected and implemented. In order to obtain the final models, cross-validation, evaluation of results, and optimization, these operations are carried out to perform the models.

iii. Evaluation phase: In this phase, to evaluate their performance, the test datasets of the final models are tested.

iv. Prediction phase: In this phase, the target output values for the new inputs are predicted by using the final models.

According to the nature of the data types, machine learning processes the data; thus, the learning tasks are categorized into two types for HT defensive mechanisms, i.e. supervised machine learning and unsupervised machine learning. Supervised machine learning techniques perform model training by operating the labelled data, and the final model is selected to foresee the target classes over input data. Unsupervised machine learning techniques emphasize characterizing the data by learning from the labelled data for all classes of targets that are unavailable. Supervised and the unsupervised machine learning differ depending on the available data labels. An overview of machine learning algorithms, feature selections, optimization, and model enhancement techniques in defensive strategies against HTs are presented here.

14.4.1 Supervised Machine Learning

- **Artificial Neural Network (ANN):** ANN [21] is a computational algorithm inspired by biological neural networks similar to that of the human brain. It is a mathematical model and constructs numerous networks based on various connection patterns. ANNs are used in diverse applications, i.e. for classification and HT detection.
- **Support Vector Machine (SVM):** SVM [22] is a learning algorithm that is accomplished by carrying out classification, regression, and detection. It is a two-class classification model used to solve quadratic programming problems. SVM is widely used in the HT detection approach.
- **Bayesian classifiers:** Bayesian classifier [23] is a kind of classification algorithm that depends on misjudgment losses and statistics probability. Minimum error rate can be achieved by using this algorithm.
- **One-class classifier:** This technique has been proposed to regulate a new training set belonging to a specific class or not. Hence, it was proposed to obtain the novelty in the detection problem. It involves two classifiers: one is one-class ANN, and the other is one-class SVM.
- **Back Propagation Neural Networks (BPNN):** This is a type of artificial neural network, i.e. it is a type of feedforward ANN that consists of several hidden layers [24]. This algorithm has the ability to alter the weights and threshold while training the data to attain better generalization ability.

- **Extreme Learning Machine (ELM):** This is a kind of hidden single-layer feedforward ANN algorithm. It arbitrarily adjusts the weight of the input and attains the output weight accordingly [25]. The ELM has good generalization ability and high learning speed; hence, ELM is widely used in many applications.
- **Decision tree:** Decision tree [26] is a model based on tree-like structure. It consists of root nodes, leaf nodes, and internal nodes. A leaf node denotes the decision results, and the remaining nodes denote attribute tests. The learning process in the decision tree is the same as that of a human's method of choosing the choices. The commonly used decision tree models are regression tree and classification tree.

14.4.2 Unsupervised Machine Learning

- **Cluster algorithms:** Cluster algorithms [27] are proposed to form a different cluster by grouping the unlabeled data. This algorithm does not need knowledge of inputs. A cluster algorithm is used mainly in HT protection as it is unaffected if any issues are present.
- **K-means clustering:** This method [28] is a modified version of a prototype-dependant cluster algorithm that operates to identify the well-defined number (K) of clusters and divides a group of "n" data points into "K" number of clusters, i.e. inter-cluster resemblance is low and intra-cluster resemblance is high.
- **DBSCAN & OPTICS:** DBSCAN [29] & OPTICS [30] algorithms are density-depended cluster algorithms. They determine points based on data density, i.e. higher density points are estimated from the corresponding nodes, and connecting these high-density data points into one single block creates numerous clusters. Hence, this algorithm is more advantageous as it will produce various clusters of different shapes and sizes during data processing.

14.4.3 Dimensionality Reduction & Feature Selection

- **Genetic Algorithms:** Genetic algorithms [27] are said to be prevalent Heuristic algorithms. They provide the highest classification accuracy as they aim to determine a subset of variables from the available data; thus, a genetic algorithm is useful for feature selection of classification problems.
- **Principal Component Analysis (PCA):** This is a kind of reduced data dimensionality technique, and it is widely used for mapping where "n" dimensional features to a "k" dimensional space. Recently reassembled "k" dimensional features are called "principal components". The orthogonal principal components are the original variables, and they contain no repeated information. The k-dimension space is much lower than n.
- **Two-Dimensional PCA:** This is a modified version of the PCA method where a variance matrix is created by applying 2D matrices rather than a 1D vector. It overcomes the high computational complexity and time consumption of the PCA method.

14.4.4 Design Optimization

- **Adaptive Iterative Optimization Algorithm (AIOA):** This is a type of model improvement technique where an explicit classifier is iteratively enhanced by weighted input training data through a weight vector to decrease the errors.
- **Multi-objective Evolutionary Algorithm (MOEA):** MOEA is a type of improved optimization algorithm that initiates from arbitrarily-produced populations, and then the pareto-optimal solution is approached by implementing multiple-generation constant optimization processes along with adaptability.
- **Particle Swarm Optimization Algorithm:** This is an optimizer that puts in effort to determine the optimum results over the cooperation and information allocation between those individuals in an assembly. These algorithms are simple to implement as complex parameter adjustments are not required. The optimal solution quality is evaluated by determining the fitness.

The summary of machine learning algorithms is given in Table 14.2, which shows that supervised learning is effective compared to unsupervised learning because supervised learning is able to address cases with few features. Thus, it achieves better classification results. Exclusively, machine learning is suitable for HT detection as a definite output is analyzed for each input. Still, this process consumes more time, and multiple iterations are performed to attain optimum result.

14.5 PROPOSED METHODOLOGY

The machine learning algorithm with Bayesian classifiers is used to determine the presence of the HT circuit in the IoT chip. The following methodology is used to determine the HT, as shown in Figure 14.6.

14.5.1 Stage 1: Analysis of IoT Circuit Structure Features

The normal circuit and HT circuit structures and gate-level netlists are generated from the design. Based on these, net features are generated like features 1, features 2… features n.

14.5.2 Stage 2: Feature Extraction from Netlist

From the gate-level netlist generated from stage 1, the values for features of the recognized and unidentified netlist are obtained. The extracted feature is divided into training sets, and using the training set, the classifier is trained. By validation, the features of the trained classifiers are determined. The testing dataset will be an unknown netlist. The qualified classifiers classify the testing dataset and identify the nets based on the trained sets.

14.5.3 Stage 3: Hardware Trojan Classifier Training

The supervised machine learning algorithm is applied for HT detection in gate-level netlists. To detect the HT included in the netlist, the cross-validation method is

TABLE 14.2

Summary of Machine Learning Algorithms

Sl. No.	Machine Learning Algorithm	Advantages	Disadvantages
1	Supervised learning	• Cases with few features are performed quickly. • According to input, the output is clear. • Classification results are relatively better. • The learning models are not sensitive to noise.	• Overfitting and underfitting are easy. • It is unsuitable for multi-classification problems.
2	Unsupervised learning	• No golden ICs are required. • Large dataset is efficient and scalable. • Models are simple and easy to implement and independent of parameters.	• It has poor classification results. • The optimal solution is not accurate. • These models are noise sensitive. • It has the requirement of cluster values.
3	Selection of features	• Redundant features are removed. • It has better accuracy of HT detection. • Attribute space is removed.	• It consumes more time. • It loses HT features. • It manually determines threshold values.
4	Design Optimization	• It is used in the DFS strategy. Rare signals are decreased. • Learning model's performance is improved.	• It consumes more time. • It is effective for simple circuits. • To obtain optimal solutions, several iterations need to be performed.

used, where each one of the HT netlists is treated as a testing set and the remaining HT-included netlists are treated as a training set. To detect the HT-free netlist, one among that netlist is considered as a testing dataset, and the remaining netlists are treated as a training set.

Throughout the training procedure, classifier superiority is determined through the appropriate indicators. The classifier is retrained if the indicators and detection of HT nets are not satisfactory.

14.5.4 Stage 4: Detection of Hardware Trojan

Depending on the trained classifier given in stage 3 and the netlist given in stage 2, if the result does not contain the HT net, then the netlist is said to be the HT-free net, and if the result includes the HT, then it is said to be the HT-affected net.

14.5.5 COMPARISON OF HT DETECTION MODELS BASED ON ML

Machine-learning–based HT detection models are more advantageous than non-machine-learning–based HT models, but they still possess some disadvantages, as given in Table 14.3. Machine-learning–based detection models are applied in various models, such as reverse engineering, circuit feature analysis, and side channel analysis.

- In the reverse engineering technique [31], the minuscule imagings of individual layers are obtained. This is mainly applied to IC repackaging to

TABLE 14.3

Comparison of Machine Learning (ML) Approaches

Detection Technology of HT	Non-ML-based Strategy		ML-based Strategy	
	Advantages	Disadvantages	Advantage	Disadvantage
1. Reverse engineering	• Includes design and physical level • Irreversible physical level • Non-destructive design level • High accuracy	• Time consuming and costly • Destructive physical level • Effective only for simple circuits • Requires golden IC	• Golden IC not required • Simplified traditional reverse engineering • Avoids manual entering of gate-level netlist	• Performance of classifiers is based on parameters • Sensitive to noise • Only valid during test time
2. Circuit feature analysis	• A typical heuristic method • Detects suspicious modules • Easy and effective to implement • Extracts characteristics	• Requires golden design for reference • Fails to detect implicit HT • Suitable only for simple circuits	• Identifies HT net features automatically • Increases efficiency • Reduces feature vector size	• Requires golden IC for reference • Requires more executive time • Fails to detect implicit HT
3. Side channel analysis	• Applied widely and easy to implement • High detection accuracy • Effectively identifies large HTs	• Gets affected easily by noise • Has difficulty with small HT identification • Requires golden IC for reference	• Decreases effect of noise • Improves accuracy • Effectively extracts features	• Performance is based on parameters • Requires increased overhead time

recreate the novel design of the end products. The reverse engineering technique has the capability to determine minute changes to ICs and thus has a high accuracy of detection, but it is an irreversible method that consumes more time to apply to a complex IC. Even though it is time consuming, this method is applied on a design consisting of a smaller number of ICs. The consumption time of this traditional methodology is reduced by applying a favorable machine learning classifier that minimizes the number of steps to be carried out. Machine-learning–based reverse engineering helps to avoid manually entering the netlist, thus generating the netlist automatically by analyzing each layer of the IC design. The machine-learning–based non-destructive methodology also has some deficiencies, i.e. it is applicable to a simple design and is costly.

- In circuit feature analysis [32], the analysis is based on the circuit features extracted from the gate-level netlist. The extracted features are computed and further determine whether the net is suspicious or not. The HT is determined by analyzing the switching activity and net features. The suspicious nets get activated in rare conditions, which are supplied by the circuit inputs, which in turn act as trigger inputs and activate the HT in the circuit. The n extracted features from individual nets are used to segregate the infected HT nets from the standard nets. The machine-learning–based feature analysis makes use of the classifier to determine the infected nets from the standard nets. The classifier is trained with the training set, and the testing set from the unknown netlist is tested to identify the suspicious activity carried out. The true positive rate of this method is high, but the accuracy and the true negative rate is not achieved. The learning model has the combination of structural and functional features to generate a vigorous training set, thus improving the efficiency of HT detection.

- In side channel analysis [33,34], circuit parameters are measured, such as path delay, critical path, power and temperature, to differentiate the IC infected from the HT with that of the golden IC, which is free from HTs. The side channel analysis analyzes the variations in the parameters, and if there are any changes in the parameter values from that of the golden IC parameters, it determines the presence of the extra circuit in the design. The HT detection efficiency is based on the signal-to-noise ratio as side channel analysis depends on noise and process variation. To improve the deficiency of side channel analysis, machine learning is incorporated along with a traditional method like ANN with side channel, ELM with side channel, or BPNN with side channel to improve the signal-to-noise ratio in the circuit. However, they lack pre-processing of sampled features, and they are volatile in nature. This method analyzes various machine learning models and applies the most appropriate algorithm depending on detection cases and features of the circuit.

The machine learning algorithm is more effective, and higher precision rates are achieved. Machine learning techniques are not sensitive to noise, and thus the accuracy is achieved through special indicators, such as true positive rate, true negative rate, and accuracy.

14.6 CONCLUSION

The threat of HTs has raised concerns among designers. Thus, it has become a major focus in industrial and academic research. Here, we have explained the machine-learning–based HT approach by analyzing the various challenges and problems faced during research. Subsequently, the threats from adversaries have increased, and HT attacks are possible beyond the layers of the chip. Hence, a basic model is bestowed for an HT defensive approach of the hardware ecosystem. This chapter validates the application of machine learning techniques against HT attacks. In the future, the collaboration of two or more classifiers will be used to determine HT attacks in the system on a chip. Further, HT detection methodology will be used by the government, corporations, etc., to secure information from an adversary attack.

REFERENCES

[1] G. Sumathi, L. Srivani, D. T. Murthy, K. Madhusoodanan, and S. S. Murty, "A review on HT attacks in PLD and ASIC designs with potential defence solutions," *IETE Tech. Rev.*, vol. 35, no. 1, pp. 64–77, Jan. 2018, doi: 10.1080/02564602.201 6.1246385.
[2] O. Sinanoglu, "Do you trust your chip?" in Proc. Int. Conf. Design Technol. Integr. Syst. Nanosc. Era (DTIS), Apr. 2016, doi: 10.1109/dtis.2016.7483804.
[3] S. Bhunia, M. S. Hsiao, M. Banga, and S. Narasimhan, "Hardware Trojan attacks: Threat analysis and countermeasures," *Proc. IEEE*, vol. 102, no. 8, pp. 1229–1247, Aug. 2014, doi: 10.1109/jproc.2014.2334493.
[4] A. Antonopoulos, C. Kapatsori, and Y. Makris, "Trusted analog/mixedsignal/RF ICs: A survey and a perspective," *IEEE Des. Test*, vol. 34, no. 6, pp. 63–76, Dec. 2017, doi:10.1109/mdat.2017.2728366.
[5] M. Rostami, F. Koushanfar, and R. Karri, "A primer on hardware security: Models, methods, and metrics," *Proc. IEEE*, vol. 102, no. 8, pp. 1283–1295, Aug. 2014.
[6] R. Khan, S. U. Khan, R. Zaheer, and S. Khan, "Future Internet: The internet of things architecture, possible applications and key challenges," in Proc. 10th Int. Conf. Frontiers Inf. Technol., Dec. 2013, pp. 257–260.
[7] F. Conti et al., "An IoT endpoint system-on-chip for secure and energyefficient near-sensor analytics," *IEEE Trans. Circuits Syst. I, Reg. Papers*, vol. 64, no. 9, pp. 2481–2494, Sep. 2017.
[8] A. Kulkarni, Y. Pino, and T. Mohsenin, "Adaptive real-time Trojan detection framework through machine learning," in Proc. 2016 IEEE Int. Symp. Hardw. Oriented Secur. Trust. HOST 2016, 2016, pp. 120–123.
[9] H. Salmani, "COTD: Reference-free hardware trojan detection and recovery based on controllability and observability in gate-level netlist," *IEEE Trans. Inf. Forensics Security*, vol. 12, no. 2, pp. 338–350, Feb. 2017.
[10] H. Li, Q. Liu, and J. Zhang, "A survey of hardware Trojan threat and defense," *Integration*, vol. 55, pp. 426–437, Sep. 2016.
[11] H. Li, Q. Liu, J. Zhang, and Y. Lyu, "A survey of hardware Trojan detection, diagnosis and prevention," in Proc. 14th Int. Conf. Comput.-Aided Design Comput. Graph. (CAD/Graph.), Aug. 2015, pp. 173–180, doi: 10.1109/cadgraphics.2015.41.
[12] K. Xiao, D. Forte, Y. Jin, R. Karri, S. Bhunia, and M. Tehranipoor, "Hardware Trojans: Lessons learned after one decade of research," *ACM Trans. Des. Automat. Electron. Syst.*, vol. 22, no. 1, pp. 1–23, May 2016, doi: 10.1145/2906147.

[13] M. Banga and M. S. Hsiao, "A region based approach for the identification of hardware Trojans," in Proc. IEEE Int. Workshop Hardw.-Oriented Secur. Trust, Jun. 2008, pp. 40–47, doi: 10.1109/hst.2008.4559047.

[14] X. Wang, M. Tehranipoor, and J. Plusquellic, "Detecting malicious inclusions in secure hardware: Challenges and solutions," in Proc. IEEE Int. Workshop Hardw.-Oriented Secur. Trust, Jun. 2008, pp. 15–19, doi: 10.1109/hst.2008.4559039.

[15] J. Zhang, F. Yuan, L. Wei, Y. Liu, and Q. Xu, "VeriTrust: Verification for hardware trust," IEEE Trans. Comput.-Aided Design Integr. Circuits Syst., vol. 34, no. 7, pp. 1148–1161, Jul. 2015, doi: 10.1109/tcad.2015.2422836.

[16] S. Moein, S. Khan, T. A. Gulliver, F. Gebali, and M. W. El-Kharashi, "An attribute-based classification of hardware Trojans," in Proc. 10th Int. Conf. Comput. Eng. Syst. (ICCES), Dec. 2015, pp. 351–356, doi: 10.1109/icces.2015.7393074.

[17] S. Moein, T. A. Gulliver, F. Gebali, and A. Alkandari, "A new characterization of hardware Trojans," *IEEE Access*, vol. 4, pp. 2721–2731, 2016, doi: 10.1109/access.2016.2575039.

[18] A. Basak, S. Bhunia, T. Tkacik, and S. Ray, "Security assurance for system-on-chip designs with untrusted IPs," *IEEE Trans. Inf. Forensics Security*, vol. 12, no. 7, pp. 1515–1528, Jul. 2017, doi: 10.1109/tifs.2017.2658544

[19] S. Narasimhan and S. Bhunia, "Hardware Trojan detection," in M. Tehranipoor, C. Wang (eds.) *Introduction to Hardware Security and Trust*. New York, NY: Springer, 2012, pp. 51–57. https://doi.org/10.1007/978-1-4419-8080-9_15

[20] E. Alpaydin, "Introduction," in *Introduction to Machine Learning*. Cambridge, MA, USA: MIT Press, 2014, pp. 1–18.

[21] Y. Liu, G. Volanis, K. Huang, and Y. Makris, "Concurrent hardware Trojan detection in wireless cryptographic ICs," in Proc. IEEE Int. Test Conf. (ITC), Oct. 2015, pp. 1–8, doi: 10.1109/test.2015.7342386.

[22] K. Hasegawa, M. Yanagisawa, and N. Togawa, "A hardware-Trojan classification method using machine learning at gate-level netlists based on Trojan features," *IEICE Trans. Fundam.*, vol. E100-A, no. 7, pp. 1427–1438, Jul. 2017.

[23] X. Chen, L. Wang, Y. Wang, Y. Liu, and H. Yang, "A general framework for hardware Trojan detection in digital circuits by statistical learning algorithms," IEEE Trans. Comput.-Aided Design Integr. Circuits Syst., vol. 36, no. 10, pp. 1633–1646, Oct. 2017, doi: 10.1109/tcad.2016.2638442.

[24] J. Li, L. Ni, J. Chen, and E. Zhou, "A novel hardware Trojan detection based on BP neural network," in Proc. 2nd IEEE Int. Conf. Comput. Commun. (ICCC), Oct. 2016, pp. 2790–2794, doi: 10.1109/compcomm.2016.7925206.

[25] S. Wang, X. Dong, K. Sun, Q. Cui, D. Li, and C. He, "Hardware Trojan detection based on ELM neural network," in Proc. 1st IEEE Int. Conf. Comput. Commun. Internet (ICCCI), Oct. 2016, pp. 400–403, doi: 10.1109/cci.2016.7778952.

[26] F. K. Lodhi, S. R. Hasan, O. Hasan, and F. Awwadl, "Power profiling of micro-controller's instruction set for runtime hardware Trojans detection without golden circuit models," in Proc. Design, Automat. Test Europe Conf. Exhibit. (DATE), Mar. 2017, pp. 294–297, doi: 10.23919/date.2017.7927002.

[27] N. Karimian, F. Tehranipoor, M. T. Rahman, S. Kelly, and D. Forte, "Genetic algorithm for hardware Trojan detection with ring oscillator network (RON)," in Proc. IEEE Int. Symp. Technol. Homeland Secur. (HST), Apr. 2015, pp. 1–6, doi: 10.1109/ths.2015.7225334.

[28] C. X. Bao, Y. Xie, Y. Liu, and A. Srivastava, "Reverse engineeringbased hardware Trojan detection," *IEEE Trans. Comput.-Aided Design Integr. Circuits Syst.*, vol. 35, no. 1, pp. 49–57, Jan. 2016, doi: 10.1109/TCAD.2015.2488495.

[29] A. N. Nowroz, K. Hu, F. Koushanfar, and S. Reda, "Novel techniques for high-sensitivity hardware Trojan detection using thermal and power maps," *IEEE*

Trans. Comput.-Aided Design Integr. Circuits Syst., vol. 33, no. 12, pp. 1792–1805, Dec. 2014, doi: 10.1109/tcad.2014.2354293.

[30] B. Çakir and S. Malik, "Hardware Trojan detection for gate-level ICs using signal correlation-based clustering," in Proc. Design, Automat. Test Eur. Conf. Exhibit. (DATE), 2015, pp. 471–476.

[31] X. Wang, Y. Zheng, A. Basak, and S. Bhunia, "IIPS: Infrastructure IP for secure SoC design," *IEEE Trans. Comput.*, vol. 64, no. 8, pp. 2226–2238, Aug. 2015, doi: 10.1109/tc.2014.2360535.

[32] J. Smith, "Non-destructive state machine reverse engineering," in Proc. 6th Int. Symp. Resilient Control Syst. (ISRCS), Aug. 2013, pp. 120–124, doi: 10.1109/isrcs.2013.6623762.

[33] X. Chen, Q. Liu, S. Yao, J. Wang, Q. Xu, Y. Wang, Y. Liu, and H. Yang, "Hardware Trojan detection in third-party digital intellectual property cores by multilevel feature analysis," IEEE Trans. Comput.-Aided Design Integr. Circuits Syst., vol. 37, no. 7, pp. 1370–1383, Jul. 2018, doi: 10.1109/tcad.2017.2748021.

[34] D. Jap, W. He, and S. Bhasin, "Supervised and unsupervised machine learning for side-channel based Trojan detection," in Proc. IEEE 27th Int. Conf. Appl.-Specific Syst., Archit. Processors (ASAP), Jul. 2016, pp. 17–24, doi: 10.1109/asap.2016.7760768.

15 Integrated Photonics for Artificial Intelligence Applications

Ankur Saharia[1], Kamal Kishor Choure[2],
Nitesh Mudgal[2], Rahul Pandey[3], Dinesh Bhatia[2],
Manish Tiwari[1], and Ghanshyam Singh[2]
[1]Manipal University Jaipur, Jaipur, Rajasthan, India
[2]MNIT Jaipur, Jaipur, Rajasthan, India
[3]SKIT Jaipur, Jaipur, Rajasthan, India

CONTENTS

15.1 INTRODUCTION TO PHOTONIC NEUROMORPHIC COMPUTING

The human brain is a sophisticated and complex system that constitutes hundreds of billions of interconnected neurons and synapses, respectively. Neurons are the most important and basic element of the human brain, and they are connected through synapses. The information signal between these neurons has the form of a spike. Human brain architecture can provide unmatched high-speed parallel computing, super intelligence, self-learning, and an upgradation system with very low power consumption. The functionality of the human brain can be shown by spiking neural network (SNN) [1]. The signal between the neurons via synapses is in the form of a spike. Drawing inspiration from the architecture of the human brain and its unmatched functionality, the artificial neural network (ANN) was developed. The word "artificial neural" refers to the artificial neuron created with the same functionality as the brain neuron. The computation process performed with help from these brain-inspired artificial neurons is referred to as "neuromorphic computing". Today, ANNs are needed to process and compute large sets of data [2]. The era of ANNs started back in 1943 when Walter Pitts developed a mathematical model based on the nerve cell of the brain and its information processing system [3]. In recent years, the ANN has been extensively applied in artificial intelligence,

machine learning [4], and computing. The ANN is considered to mimic the architecture of the human brain. In von Neumann's model computer, there is physical
separation between computing and signal processing memory for information,
which limits computing efficiency while parallel processing a large set of signals.
On the other hand, to get high computing efficiency ANNs have massive accumulate computation (MAC) for parallel processing signals [5].

Researchers and the industry both are coming forward to develop a specific,
tailored electronic architecture that can be used to counter the shortcomings of the
present computer model for the ANN [6]. In the past few years, brain-inspired
neuromorphic computing has achieved tremendous popularity for its unmatched
computing efficiency. Researchers have developed several architectures and electronic devices that can perform similar functions as neurons and synapses for
neuromorphic computing [7–9], but there is a limitation on the computing speed of
electronic neuromorphic computing. More's law falters because nanotechnology
has reached atomic scale and high fabrication costs [10]. Photonics technology in the
form of integrated photonics provides the best alternative to fulfill the requirements of
present-day high-speed computing technologies like photonic neural network computing. Therefore, to overcome the limitation of electronic neuromorphic computing,
a parallel platform that has been introduced is photonic neuromorphic computing,
which uses photons rather than electrons for processing and computation. Photonic
neuromorphic computing has several advantages over electronics neuromorphic
computing; it provides wide bandwidth, has high processing and computational speed,
has a low power requirement, and has low latency [11,12]. Compared to electronic
neuromorphic computing, photonic neuromorphic computing is still in its early stage
of research and application. Silicon [13] and indium phosphide [14–16] are promising
materials for the integrated photonic platform for neuromorphic computing.
Figure 15.1 shows the simple architecture of neuromorphic computing reproduced
from Del Valle et al. [17]. The neuromorphic architecture consists of an input neuron
layer and output neuron layer connected via a synaptic connection, as shown in
Figure 15.1. The input signal is fed to an input neuron via synapses for processing and
computation. The main purpose of neurons is signal processing and computation, and
the purpose of synapses is weights and storage memory. The inputs are associated
with some weights, so the summation of input is done along with the weights. To get
the output close to the correct value, continuous training is performed. The mathematical representation is shown in Sui et al. [18] as

$$S = \phi\left(\sum_{i=1}^{n} Wi * Xi + bi\right)$$

Where S is output, Wi are weights, ϕ is threshold function, b is bias, and X is inputs.

15.2 CLASSIFICATION OF PHOTONIC NEURAL NETWORK

A very good taxonomy of photonic neural networks was presented in De Marinis et al.
[6]. The brief classification showed photonic neural networks as multilayer perceptrons,
convolutional neural networks, SNNs, and reservoir computing.

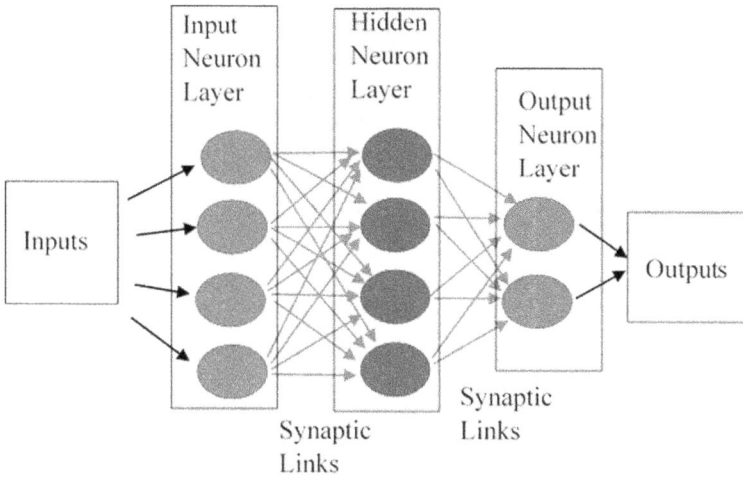

FIGURE 15.1 Neuromorphic architecture.

a. **Multilayer Perceptron**

The multilayer perceptron consists of a series of interconnected layers of neurons, as shown in Figure 15.2. The input is provided to an input layer, followed by the hidden layers, and finally the output is taken from the output layer. It can have one or more hidden layers. The connection between the neurons is done with the help of weights. The output is the weighted sum of input to the neurons, along with the modified activation function. The neurons of hidden layers are connected with each neuron of previous and successive layers, as shown. Due to the flow of information from one

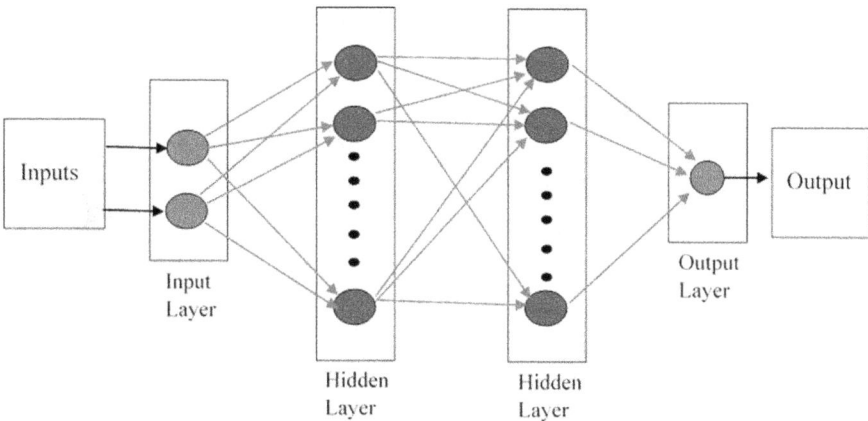

FIGURE 15.2 Multilayer perceptron.

layer to another successive layer, it is also called a feedforward neural network. The main purpose of the input layer is to pass the input to other connected layers instead of performing computation. To get the calculated correct output from the input, there should be the proper selection of weight and transfer function. Training data is required for supervised learning in multilayer perceptron [19].

b. Convolutional Neural Networks

Convolutional neural network is especially recommended for two-dimensional input data. It is a multiple-layer architecture in which each layer has multiple two-dimensional planes associated with multiple artificial neurons. Convolutional neural network is generally used for image categorization. Convolutional neural network architecture is a deep architecture that implies the hierarchical order of feature extraction. As shown in Figure 15.3, the convolutional neural network has an extraction feature part and a classification part that consists of five layers along with a two-dimensional plane named the convolutional layer, a pooling layer, followed by a fully connected layer, and an output layer. The main purpose of the input layer is to arrange the inputs in a proper format. The convolutional layer is a set of filters with learning properties, and it performs most of the computational work. The extracted features of the inputs are fed to the fully connected layers through which the output is classified [20].

c. Spiking Neural Networks

The advent of SNN started in 1907 when Louis Lapicque [21] proposed the integrate-and-fire neuron, the basic computational unit. The information in the SNN is represented through spikes. These spikes have proper timing and

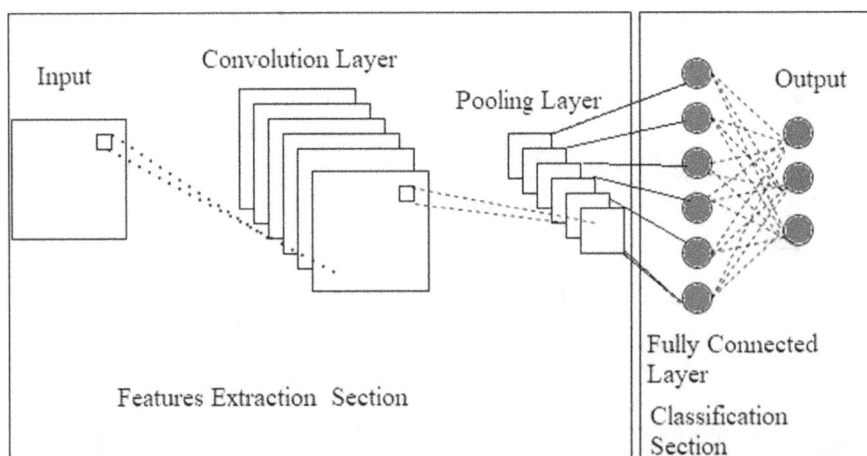

FIGURE 15.3   Convolutional neural network.

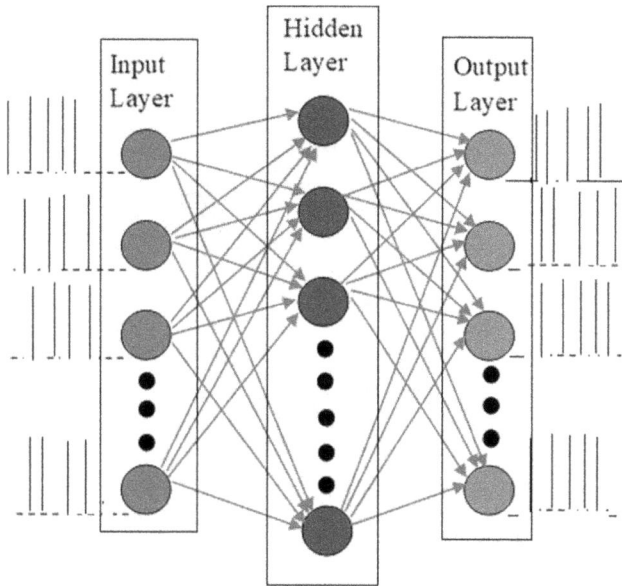

FIGURE 15.4   Spiking neural networks.

sequence for the transfer of information between artificial neurons. The architecture of SNNs consists of an input layer, a hidden layer, and an output layer, as shown in Figure 15.4. Every neuron of any layer connects to all the neurons of the next successive layer, and the output of neurons is the weighted sum of previous ones. The information transfer has a time difference. The uniqueness of the SNN is its time different delay property, which makes it more advantageous than other neural networks, but this time different delay makes the SNN more complex in terms of training and configuration [22].

d. **Reservoir Computing**

Reservoir computing is one of the most versatile computing types among the other types of neuromorphic computing. Reservoir computing is primarily a recurrent neural-network–based scheme. To improve the computational efficiency of reservoir computing, there needs to be a proper design for the recurrent neural network (RNN)-based reservoir. As shown in Figure 15.5, reservoir computing consists of an input layer through which the input data are transformed into high-dimensional space. The main purpose of the reservoir tank is to nonlinearly convert the provided input into a high-dimensional space that helps the learning algorithms. The internal points of the reservoir tank are also referred to as "reservoir states". The weights connected input layer and reservoir are represented as W_{in} while the weights between the reservoir and output layer are represented as W_{out}. In reservoir computing, the output weights are updated and the input and

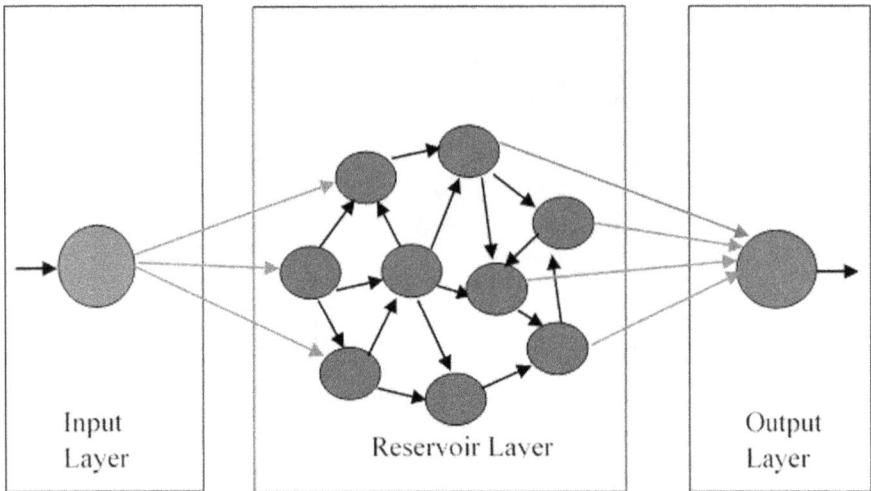

FIGURE 15.5   Reservoir computing.

reservoir are in fixed states. Reservoir computing can be realized with the help of components, devices, and substrates. In reservoir computing, the concept of time-delay reservoir computing is simple, has low power consumption, and uses less hardware [23,24].

15.3 PHOTONIC NEURON AND SYNAPSE

The basic fundamental unit of the brain is the neuron. The computation function of the brain is performed by the neurons. All the neurons are interconnected to each other for computation and processing of information. The synapse is the most important element, responsible for the transfer and storage of information between the neurons. Interconnection between two neurons is done by the synapse. This interconnection and memory are referred to as "weights". These weights keep on updating themselves and have learning capability based on the condition. Then, photonic synapses mimic biological synapses and perform a similar function in the optical domain.

The implementation of neurons and synapses through photonic components has been proposed, such as graphene excitable lasers [25], vertical-cavity surface-emitting lasers [26,27], and distributed feedback (DFB) lasers [28]. Researchers come up with photonic neurons based on vertical-cavity surface-emitting lasers with saturable absorber (VCSELs-SA) [29]. Chaoran et al. [30] proposed weight banks in micro ring resonators for photonic neural networks and microfiber-based proposed photonic synapses [31]. The on-chip photonic synapse was demonstrated in Zengguang et al. [32], and many more photonic components have been proposed for neuromorphic computing. Some of the recent proposed designs are mentioned in Table 15.1.

TABLE 15.1

Some of the Recent Advancements of Photonic Techniques for Neuromorphic Computing

Proposed Photonic Techniques	Neuromorphic Computing	Summary
Florian Denis-Le Coarer et al. [33] used nonlinear micro ring resonators as nodes for all optical reservoir computing on photonic chips. The photonic architecture for reservoir computing proposed is based on a silicon insulator on a micro ring resonator.	Reservoir computing	The proposed reservoir design is able to delay the XOR task at 20 Gb/s with bit error rates less than 10^{-3} and injection power less than 2.5 mW.
Changming Wu et al. [34] demonstrated on-waveguide metasurfaces made of phase-change materials $Ge_2Sb_2Te_5$ for multimode photonic convolutional neural computing.	Convolutional neural network	They built a photonic kernel based on arrays for convolutional neural networks for image processing and recognition tasks.
Laurent Larger et al. [35] implemented reservoir computing based on an electro-optic (EO) phase-delay dynamic built with telecom bandwidth devices for ultra-fast image processing.	Reservoir computing	They demonstrated the computation efficiency with the proposed model and experimentally got one million words per second for speech recognition with low error rate.
Indranil Chakraborty et al. [36] proposed a phase-change dynamic $Ge_2Sb_2Te_5$ [(GST) embedded on top of a micro ring resonator for fast neuromorphic computing.	Spiking neural network	They proposed the phase-change dynamics of GST based an all-photonic integrate-and-fire neuron, and this proposed neuron is compatible to integrate with synapses in all photonic spiking neural networks.
Matthew N. Ashner et al. [37] showed the uses of multimode micro ring resonators for standard reservoir computing and form with optical nonlinearity and optical feedback.	Reservoir computing	They demonstrated the equivalence of multimode micro ring resonators with the reservoir computer showing the equation of light through waveguide.
Charis Mesaritakis et al. [38] proposed all optical reservoir computing using an InGaAsp micro ring resonator for optical network application.	Reservoir computing	They presented two application-oriented benchmark tests. They used the nonlinear property i.e. the two-photon absorption and kerr effect for the micro ring resonator.
Mitsumasa Nakajima et al. [39] demonstrated scalable on chip photonic for reservoir computing.	Reservoir computing	They demonstrated parallel processing wavelength division multiplexing, which provides ultra-

(Continued)

TABLE 15.1 (Continued)

Some of the Recent Advancements of Photonic Techniques for Neuromorphic Computing

Proposed Photonic Techniques	Neuromorphic Computing	Summary
		high bandwidth for high-efficiency computation. The proposed device is able to achieve ten TMAC/s on a single photonic chip.
Jun Rong Ong et al. [40] presented a scalable photonic convolutional architecture based on the Fourier transform property of integrated star couplers.	Convolutional neural network	They compared their proposed architecture with the present design and found it less complex and fast.
Irene Estébanez et al. [41] demonstrated the numerical approach of bandwidth enhancement for high-speed time-delay reservoir computing.	Reservoir computing	They numerically demonstrated that the sampling time of computing can be reduced to 12 ns without compromising the performance of computing. They also presented the concept of optical injection for reservoir operating.

15.4 CONCLUSION

In this chapter, we presented the vital role of photonics for neuromorphic computing. The photonic components for neural networks were also discussed: photonic neuron and photonic synapse. Last, we presented a short survey on the recent developments of photonic techniques for neuromorphic computing. There still are many challenges that need to be addressed for efficient computing. There is a compatibility issue between photonic neurons and photonic synapses within neuromorphic computing that needs to be improved.

REFERENCES

[1] W. Maass, "Networks of spiking neurons: The third generation of neural network models," *Neur. Netw.*, vol. 10, no. 9, pp. 1659, 1997.

[2] D. Silver, A. Huang, C. J. Maddison, A. Guez, L. Sifre, G. van den Driessche, J. Schrittwieser, I. Antonoglou, V. Panneershelvam, M. Lanctot, S. Dieleman, D. Grewe, J. Nham, N. Kalchbrenner, I. Sutskever, T. Lillicrap, M. Leach, K. Kavukcuoglu, T. Graepel, and D. Hassabis, "Mastering the game of Go with deep neural networks and tree search," *Nature*, vol. 529, Jan. 2016.

[3] W. S. McCulloch and W. Pitts, "A logical calculus of the ideas immanent in nervous activity," *Bull. Math. Biophys.*, vol. 5, no. 4, pp. 115–133, Dec. 1943.

[4] S. Samarasinghe, "Neural networks for applied sciences and engineering: From fundamentals to complex pattern recognition." Boca Raton, USA: Auerbach Publications, 2016.

[5] B. W. Bai, H. W. Shu, X. J. Wang, et al., "Towards silicon photonic neural networks for artificial intelligence," *Sci. China Inf. Sci.*, vol. 63, no. 6, pp. 160403, 2020. 10.1 007/s11432-020-2872-3.

[6] L. De Marinis, M. Cococcioni, P. Castoldi, and N. Andriolli, "Photonic neural networks: A survey," in *IEEE Access*, vol. 7, pp. 175827–175841, 2019. 10.1109/ ACCESS.2019.2957245.

[7] C. D. Schuman, T. E. Potok, R. M. Patton, et al., "A survey of neuromorphic computing and neural networks in hardware," arXiv preprint arXiv: 1705. 06963, 2017.

[8] K. Roy, A. Jaiswal, and P. Panda, "Towards spike-based machine intelligence with neuromorphic computing," *Nature*, vol. 575, p. 607, 2019.

[9] J. D. Zhu, T. Zhang, Y. C. Yang, et al., "A comprehensive review on emerging artificial neuromorphic devices," *Appl. Phys. Rev.*, vol. 7, p. 011312, 2020.

[10] J. Shalf, "The future of computing beyond Moore's Law.Phil," *Trans. R. Soc. A.*, vol. 378, p. 20190061, 2020. 10.1098/rsta.2019.0061.

[11] P. R. Prucnal and B. J. Shastri, "*Neuromorphic photonics.*" Boca Raton: CRC Press, 2017.

[12] Q. Zhang, H. Yu, M. Barbiero, B. Wang, and M. Gu, "Artificial neural networks enabled by nanophotonics," *Light Sci. Appl.*, vol. 8, pp. 1–14, 2019.

[13] L. Chrostowski and M. Hochberg, "*Silicon photonics design: From devices to systems.*" Cambridge, UK: Cambridge Univ. Press, 2015.

[14] L. A. Coldren, S. C. Nicholes, L. Johansson, S. Ristic, R. S. Guzzon, E. J. Norberg, and U. Krishnamachari, "High performance InP-based photonic ICs: A tutorial," *J. Lightw. Technol.*, vol. 29, no. 4, pp. 554–570, 2011.

[15] M. Smit, X. Leijtens, H. Ambrosius, et al., "An introduction to InP-based generic integration technology," *Semicond. Sci. Technol.*, vol. 29, no. 8, 2014, Art. no. 083001.

[16] M. Smit, K. Williams, and J. van der Tol, "Past, present, and future of InP-based photonic integration," *APL Photon.*, vol. 4, no. 5, 2019, Art. no. 050901.

[17] Javier Del Valle, Juan Ramirez, Marcelo Rozenberg, and Ivan Schuller, "Challenges in materials and devices for resistive-switching-based neuromorphic computing," *J. Appl. Phys.*, vol. 124, p. 211101, 2018. 10.1063/1.5047800.

[18] X. Sui, Q. Wu, J. Liu, Q. Chen, and G. Gu, "A review of optical neural networks," in *IEEE Access*, vol. 8, pp. 70773–70783, 2020. 10.1109/ACCESS.2020.2987333.

[19] M. W. Gardner and S. R. Dorling, "Artificial neural networks (the multilayer perceptron)—a review of applications in the atmospheric sciences," *Atmos. Environ.*, vol. 32, no. 14–15, pp. 2627–2636, 1998. ISSN 1352-2310, 10.1016/ S1352-2310(97)00447-0.

[20] V. H. Phung and E. J. Rhee, "A deep learning approach for classification of cloud image patches on small datasets," *J. Inf. Commun. Converg. Eng.*, vol. 16, pp. 173–178, 2018. 10.6109/jicce.2018.16.3.173.

[21] L. F. Abbott, "Lapicque's introduction of the integrate-and-fire model neuron (1907)," *Brain Res. Bullet.*, vol. 50, pp. 303–304, 1999.

[22] T. Iakymchuk, A. Rosado, J. V. Frances, and M. Batallre, "Fast spiking neural network architecture for low-cost FPGA devices," in 7th International Workshop on Reconfigurable and Communication-Centric Systems-on-Chip (ReCoSoC), York, UK, pp. 1–6, 2012. 10.1109/ReCoSoC.2012.6322906.

[23] L. Appeltant, M. C. Soriano, G. Van der Sande, et al., "Information processing using a single dynamical node as complex system," *Nat. Commun.*, vol. 2, p. 468, 2011.

[24] F. Duport, A. Smerieri, A. Akrout, et al., "Fully analogue photonic reservoir computer," *Sci. Rep.*, vol. 6, p. 22381, 2016. 10.1038/srep22381.

[25] P. Y. Ma, B. J. Shastri, T. F. de Lima, A. N. Tait, M. A. Nahmias, and P. R. Prucnal, "All-optical digital-to-spike conversion using a graphene excitable laser," *Opt. Express*, vol. 25, p. 033504, 2017.

[26] T. Deng, J. Robertson, Z.-M. Wu, G.-Q. Xia, X.-D. Lin, X. Tang, Z.-J. Wang, and A. Hurtado, "Stable propagation of inhibited spiking dynamics in vertical cavity surface-emitting lasers for neuromorphic photonic networks," *IEEE Access*, vol. 6, pp. 67951–67958, 2018.

[27] J. Robertson, T. Deng, J. Javaloyes, and A. Hurtado, "Controlled inhibition of spiking dynamics in VCSELs for neuromorphic photonics: Theory and experiments," *Opt. Lett.*, vol. 42, pp. 1560–1563, 2017.

[28] H.-T. Peng, G. Angelatos, T. Ferreira de Lima, M. A. Nahmias, A. N. Tait, S. Abbaslou, B. J. Shastri, and P. R. Prucnal, "Temporal information processing with an integrated laser neuron," *IEEE J. Sel. Top. Quantum Electron.*, vol. 26, pp. 1–9, 2019.

[29] M. A. Nahmias, B. J. Shastri, A. N. Tait, and P. R. Prucnal, "A leaky integrate-and-fire laser neuron for ultrafast cognitive computing," *IEEE J. Sel. Top. Quantum Electron.*, vol. 19, no. 5, pp. 1–12, Sept.–Oct. 2013, Art no. 1800212. 10.1109/JSTQE.2013.2257700.

[30] Chaoran Huang, Simon Bilodeau, Thomas Ferreira de Lima, A. N. Tait, Philip Ma, Eric Blow, Aashu Jha, Hsuan-Tung Peng, Bhavin Shastri, and Paul Prucnal, "Demonstration of scalable microring weight bank control for large-scale photonic integrated circuits," *APL Photonics.*, vol. 5, p. 040803, 2020. 10.1063/1.5144121.

[31] Behrad Gholipour, Paul Bastock, Christopher Craig, Khouler Khan, Dan Hewak, and Cesare Soci, "Amorphous metal-sulphide microfibers enable photonic synapses for brain-like computing," *Adv. Opt. Mater.*, vol. 3, pp. 635–641, 2015. 10.1002/adom.201400472.

[32] Zengguang Cheng, Carlos Rios Ocampo, Wolfram Pernice, David Wright, and Harish Bhaskaran, "On-chip photonic synapse," Sci. Adv., vol. 3, p. e1700160, 2017. 10.1126/sciadv.1700160.

[33] Florian Coarer, Marc Sciamanna, Andrew Katumba, Matthias Freiberger, J. Dambre, Peter Bienstman, and Damien Rontani, "All-optical reservoir computing on a photonic chip using silicon-based ring resonators," IEEE J. Sel. Top. Quantum Electron., p. 1, 2018. 10.1109/JSTQE.2018.2836985.

[34] Changming Wu, Heshan Yu, Seokhyeong Lee, Ruoming Peng, I. Takeuchi, and M. Li, "Programmable phase-change metasurfaces on waveguides for multimode photonic convolutional neural network," *Nat. Commun.*, vol. 12, 2021. 10.1038/s41467-020-20365-z.

[35] Laurent Larger, Antonio Baylón-Fuentes, Romain Martinenghi, Vladimir Udaltsov, Yanne Chembo, and Maxime Jacquot, "High-speed photonic reservoir computing using a time-delay-based architecture: Million words per second classification," *Phys. Rev. X.*, p. 7, 2017. 10.1103/PhysRevX.7.011015.

[36] Indranil Chakraborty, Gobinda Saha, Abhronil Sengupta, and Kaushik Roy, "Toward fast neural computing using all-photonic phase change spiking neurons," *Sci. Rep.*, p. 8, 2018. 10.1038/s41598-018-31365-x.

[37] Matthew Ashner, Uttam Paudel, Marta Luengo-Kovac, Jacob Pilawa, Thomas Shaw, and George Valley, "Photonic reservoir computer with all-optical reservoir," *AI Opt. Data Sci. II*, p. 18, 2021. 10.1117/12.2577774.

[38] Charis Mesaritakis, Alexandros Kapsalis, and D. Syvridis, "All-optical reservoir computing system based on InGaAsP ring resonators for high-speed identification and optical routing in optical networks," in SPIE - The International Society for Optical Engineering, p. 9370, 2015. 10.1117/12.2078912.

[39] Mitsumasa Nakajima, Kenji Tanaka, and Toshikazu Hashimoto, "Scalable reservoir computing on coherent linear photonic processor," *Commun. Phys.*, p. 4, 2021. 10.1 038/s42005-021-00519-1.

[40] J. R. Ong, C. C. Ooi, T. Y. L. Ang, S. T. Lim, and C. E. Png, "Photonic convolutional neural networks using integrated diffractive optics," *IEEE J. Sel. Top. Quantum Electron.*, vol. 26, no. 5, pp. 1–8, Sept–Oct. 2020, Art no. 7702108. 10.11 09/JSTQE.2020.2982990.

[41] Irene Estébanez, Janek Schwind, Ingo Fischer, and Apostolos Argyris, "Accelerating photonic computing by bandwidth enhancement of a time-delay reservoir," *Nanophotonics*, vol. 9, no. 13, pp. 4163–4171, 2020. 10.1515/nanoph-2020-0184.

Index

Note: *Italicized* page numbers refer to figures, **bold** page numbers refer to tables

For Product Safety Concerns and Information please contact our EU
representative GPSR@taylorandfrancis.com
Taylor & Francis Verlag GmbH, Kaufingerstraße 24, 80331 München, Germany

www.ingramcontent.com/pod-product-compliance
Lightning Source LLC
Chambersburg PA
CBHW060327220326
41598CB00023B/2633

* 9 7 8 1 0 3 2 0 6 1 7 2 6 *